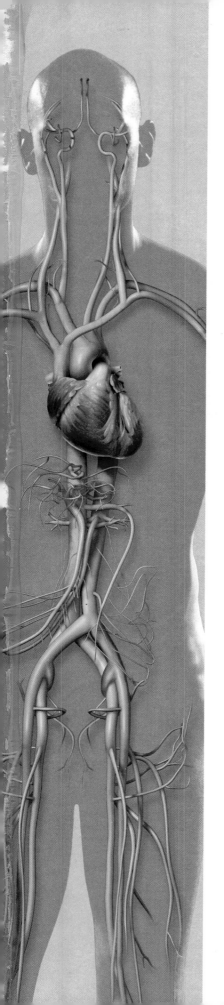
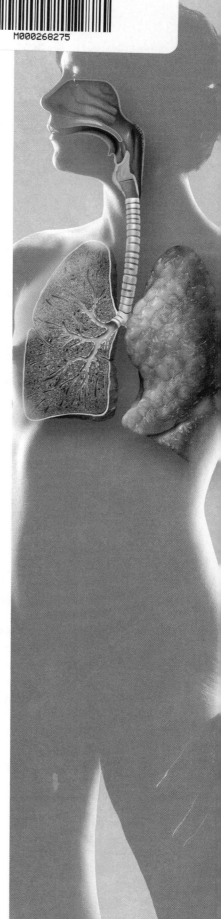

PRINCIPLES OF ANATOMY AND PHYSIOLOGY

Tenth Edition

Volume 1

Organization of the Human Body

Gerard J. Tortora

Bergen Community College

Sandra Reynolds Grabowski

Purdue University

John Wiley & Sons, Inc.

Senior Editor	Bonnie Roesch
Marketing Manager	Clay Stone
Developmental Editor	Ellen Ford
Production Director	Pamela Kennedy
Associate Production Manager	Kelly Tavares
Text and Cover Designer	Karin Gerdes Kincheloe
Art Coordinator	Claudia Durrell
Photo Editor	Hilary Newman
Chapter Opener Illustrations	Keith Kasnot

Cover photos: Photo of woman: ©PhotoDisc, Inc.
Photo of man: ©Ray Massey/Stone
Illustration and photo credits follow the Glossary.

This book was typeset by Progressive Information Technologies. It was printed and bound by Von Hoffmann Press, Inc. The cover was also printed by Von Hoffman Press, Inc.

The paper in this book was manufactured by a mill whose forest management programs include sustained yield harvesting of its timberlands. Sustained yield harvesting principles ensure that the number of trees cut each year does not exceed the amount of new growth.

This book is printed on acid-free paper. ∞

USA ISBN: 0-471-22931-8

Printed in the United States of America.
10 9 8 7 6 5 4 3 2 1

About The Cover

In choosing the cover illustration for the tenth edition of *Principles of Anatomy and Physiology,* we wanted to reflect the underlying principle in the textbook as well as our approach to presentation of the content. In a word, we wanted to suggest the idea of balance. Successfully practicing yoga requires finding the balance between the physical, mental, and spiritual aspects of the human body. Achieving this balance produces flexibility and strength, as demonstrated by the figures on the cover.

Similarly, balance has always been at the heart of the study of anatomy and physiology. Homeostasis, the underlying principle discussed throughout this textbook, is the state of equilibrium of the internal environment of the human body. During your study of anatomy and physiology you will learn about the structures and functions needed to regulate and maintain this balanced state and its importance in promoting optimal well being.

Balance also applies to the way we present the content in the tenth edition of *Principles of Anatomy & Physiology*. We offer a balanced coverage of structure and function; between normal and abnormal anatomy and physiology; between concepts and applications; between narrative and illustrations; between print and media; and among supplemental materials developed to meet the variety of teaching and learning styles. We hope that this balanced approach will provide you with the flexibility and strength to succeed in this course and in your future careers.

About the Authors

Gerard J. Tortora is Professor of Biology and former Coordinator at Bergen Community College in Paramus, New Jersey, where he teaches human anatomy and physiology as well as microbiology. He received his bachelor's degree in biology from Fairleigh Dickinson University and his master's degree in science education from Montclair State College. He is a member of many professional organizations, such as the Human Anatomy and Physiology Society (HAPS), the American Society of Microbiology (ASM), American Association for the Advancement of Science (AAAS), National Education Association (NEA), and the Metropolitan Association of College and University Biologists (MACUB).

Above all, Jerry is devoted to his students and their aspirations. In recognition of this commitment, Jerry was the recipient of MACUB's 1992 President's Memorial Award. In 1996, he received a National Institute for Staff and Organizational Development (NISOD) excellence award from the University of Texas and was selected to represent Bergen Community College in a campaign to increase awareness of the contributions of community colleges to higher education.

Jerry is the author of several best-selling science textbooks and laboratory manuals, a calling that often requires an additional 40 hours per week beyond his teaching responsibilities. Nevertheless, he still makes time for four or five weekly aerobic workouts that include biking and running. He also enjoys attending college basketball and professional hockey games and performances at the Metropolitan Opera House.

To my children, Lynne, Gerard, Kenneth, Anthony, and Andrew, who make it all worthwhile. — *G.J.T.*

Sandra Reynolds Grabowski is an instructor in the Department of Biological Sciences at Purdue University in West Lafayette, Indiana. For 25 years, she has taught human anatomy and physiology to students in a wide range of academic programs. In 1992 students selected her as one of the top ten teachers in the School of Science at Purdue.

Sandy received her BS in biology and her PhD in neurophysiology from Purdue. She is a founding member of the Human Anatomy and Physiology Society (HAPS) and has served HAPS as President and as editor of *HAPS News*. In addition, she is a member of the American Anatomy Association, the Association for Women in Science (AWIS), the National Science Teachers Association (NSTA), and the Society for College Science Teachers (SCST).

To students around the world, whose questions and comments continue to inspire the fine-tuning of this textbook. — *S.R.G.*

Preface

An anatomy and physiology course can be the gateway to a gratifying career in a host of health-related professions. As active teachers of the course, we recognize both the rewards and challenges in providing a strong foundation for understanding the complexities of the human body to an increasingly diverse population of students. Building on the unprecedented success of previous editions, the tenth edition of *Principles of Anatomy and Physiology* continues to offer a balanced presentation of content under the umbrella of our primary and unifying theme of homeostasis, supported by relevant discussions of disruptions to homeostasis. In addition, years of student feedback have convinced us that readers learn anatomy and physiology more readily when they remain mindful of the relationship between structure and function. As a writing team—an anatomist and a physiologist—our very different specializations offer practical advantages in fine-tuning the balance between anatomy and physiology.

Most importantly, our students continue to remind us of their needs for—and of the power of—simplicity, directness, and clarity. To meet these needs each chapter has been written and revised to include:

- clear, compelling, and up-to-date discussions of anatomy and physiology
- expertly executed and generously sized art
- classroom-tested pedagogy
- outstanding student study support.

As we revised the content for this edition we kept our focus on these important criteria for success in the anatomy and physiology classroom and have refined or added new elements to enhance the teaching and learning process.

Homeostasis: A Unifying Theme

The dynamic physiological constancy known as homeostasis is the prime theme in *Principles of Anatomy and Physiology*. We immediately introduce this unifying concept in Chapter 1 and describe how various feedback mechanisms work to maintain physiological processes within the narrow range that is compatible with life. Homeostatic mechanisms are discussed throughout the book, and homeostatic processes are clarified and reinforced through our well-received series of homeostasis feedback illustrations.

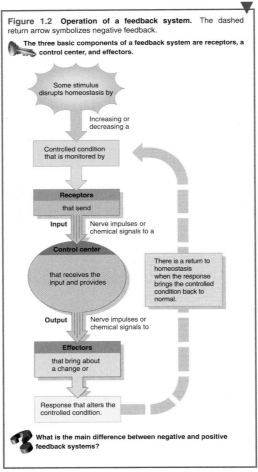

Figure 1.2 Operation of a feedback system. The dashed return arrow symbolizes negative feedback.

The three basic components of a feedback system are receptors, a control center, and effectors.

Some stimulus disrupts homeostasis by

Increasing or decreasing a

Controlled condition that is monitored by

Receptors
that send

Input — Nerve impulses or chemical signals to a

Control center
that receives the input and provides

There is a return to homeostasis when the response brings the controlled condition back to normal.

Output — Nerve impulses or chemical signals to

Effectors
that bring about a change or

Response that alters the controlled condition.

What is the main difference between negative and positive feedback systems?

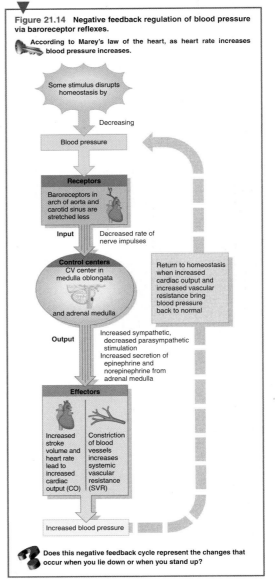

Figure 21.14 Negative feedback regulation of blood pressure via baroreceptor reflexes.

According to Marey's law of the heart, as heart rate increases blood pressure increases.

Some stimulus disrupts homeostasis by

Decreasing

Blood pressure

Receptors
Baroreceptors in arch of aorta and carotid sinus are stretched less

Input — Decreased rate of nerve impulses

Control centers
CV center in medulla oblongata

and adrenal medulla

Return to homeostasis when increased cardiac output and increased vascular resistance bring blood pressure back to normal

Output — Increased sympathetic, decreased parasympathetic stimulation
Increased secretion of epinephrine and norepinephrine from adrenal medulla

Effectors

Increased stroke volume and heart rate lead to increased cardiac output (CO)

Constriction of blood vessels increases systemic vascular resistance (SVR)

Increased blood pressure

Does this negative feedback cycle represent the changes that occur when you lie down or when you stand up?

circulatory routes, principal branches of the aorta, ascending aorta, branches of the brachiocephalic trunk in the neck, cerebral arterial circle, branches of the abdominal aorta, arteries of the pelvis and lower limbs, pulmonary circulation, fetal circulation, and development of blood vessels and blood cells.

Chapter 22 / The Lymphatic and Immune System and Resistance to Disease

New section on stress and immunity. New illustrations of lymphatic tissue, routes for lymph drainage, lymph node structure, development of the lymphatic system, phagocytosis, and stages of inflammation. New Clinical Applications on Ruptured Spleen, Edema and Lymph Flow, and Microbial Evasion of Phagocytosis. Autoimmune diseases, severe combined immunodeficiency disease, SLE, and lymphomas added to the disorders section. Descriptions of thymus and lymph nodes revised.

Chapter 23 / The Respiratory System

New illustrations of the structures of the respiratory system, location of peripheral chemoreceptors that help regulate respiration, and development of the bronchial tubes and lungs. New Clinical Applications on Laryngitis, Cancer of the Larynx, and Nebulization. New discussion and explanation of ventilation-perfusion coupling, nitrogen narcosis, and decompression sickness. Addition of ARDS to disorders section. Updated percentages of carbon dioxide dissolved in blood plasma and carried by bicarbonate ions and hemoglobin.

Chapter 24 / The Digestive System

New illustrations of organs of the gastrointestinal tract, peritoneal folds, salivary glands, histology of the stomach, secretion of HCl by stomach cells, histology of the small intestine, anatomy of the large intestine, and histology of the large intestine. New Clinical Applications on Root Canal Therapy, Jaundice, Occult Blood, and Absorption of Alcohol.

Chapter 25 / Metabolism

New section on energy homeostasis. New Clinical Application on Emotional Eating and new Medical Terminology list. Updated sections on regulation of food intake and obesity.

Chapter 26 / The Urinary System

New illustrations of a sagittal section through the kidney, blood supply of the kidneys, anatomy of the ureter, urinary bladder, and urethra, and development of the urinary system. More concise sections on reabsorption and secretion. New Clinical Applications on Nephroptosis and Loss of Plasma Proteins. Addition of Urinary Bladder Cancer to the disorders section.

Chapter 27 / Fluid, Electrolyte, and Acid–Base Homeostasis

New Summary table of factors that maintain body water balance. Discussion and related figure describing how kidneys contribute to acid–base balance by secreting hydrogen ions moved from Chapter 26 to this chapter.

Chapter 28 / The Reproductive Systems

New illustrations to compare mitosis and meiosis and to show development of internal reproductive organs. New table reviewing oogenesis and development of ovarian follicles; new explanation of mittelschmerz and the role of leptin in puberty. New Clinical Applications on Uterine Prolapse, Episiotomy, and Female Athlete Triad. Addition of genital warts and premenstrual dysphoric disorder to the disorders section.

Chapter 29 / Development and Inheritance

New illustrations of fertilization, cleavage, formation of primary germ layers, gastrulation, development of the notochordal process, neurulation, development of chorionic villi, embryo folding, and development of pharyngeal arches. Beautiful new photographs of embryonic and fetal development. New Clinical Applications on Anencephaly and Stem Cell Research. Addition of Trinucleotide Repeat Diseases to the disorders section. Updated and revised sections on embryonic and fetal development, noninvasive prenatal tests, and inheritance.

Enhancement to the Illustration Program

New Design A textbook with beautiful illustrations or photographs on most pages requires a carefully crafted and functional design. The new design for the tenth edition has been transformed to assist students in making the most of the text's many features and outstanding art. A larger trim size provides even more space for the already large and highly praised illustrations and accommodates the larger font size for easier reading. Each page is carefully laid out to place related text, figures, and tables near one another, minimizing the need for page turning while reading a topic. New to this edition is the red print used to indicate the first mention of a figure or table. Not only is the reader alerted to refer to the figure or table, but the color print also serves as a place locator for easy return to the narrative.

Distinctive icons incorporated throughout the chapters signal special features and make them easy to find during review. These include the **key** with Key Concepts Statements; the **ques-**

tion mark with the applicable questions that enhance every figure; the **stethoscope** indicating a clinical application within the chapter narrative; the **fetus icon** announcing the developmental anatomy section; the **running shoe** highlighting content relevant to exercise and the icons that indicate the study outline and distinctive types of **chapter-ending questions.**

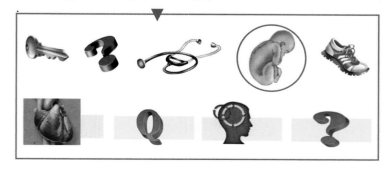

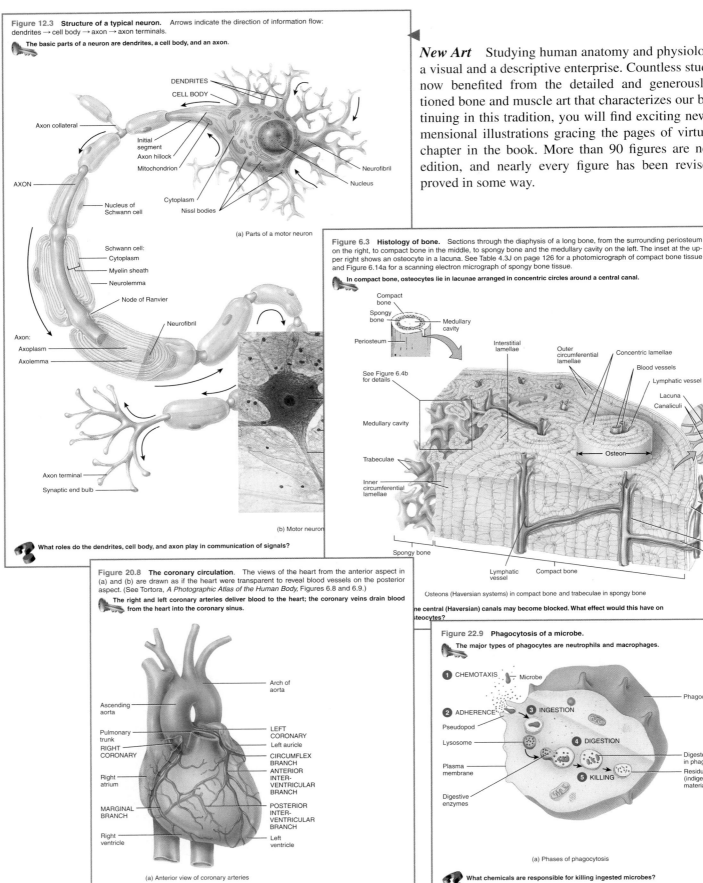

Figure 12.3 Structure of a typical neuron. Arrows indicate the direction of information flow: dendrites → cell body → axon → axon terminals.

The basic parts of a neuron are dendrites, a cell body, and an axon.

DENDRITES

CELL BODY

Axon collateral

Initial segment
Axon hillock
Mitochondrion

AXON

Nucleus of Schwann cell

Cytoplasm
Nissl bodies

Neurofibril
Nucleus

(a) Parts of a motor neuron

Schwann cell:
Cytoplasm
Myelin sheath
Neurolemma

Node of Ranvier

Neurofibril

Axon:
Axoplasm
Axolemma

Axon terminal
Synaptic end bulb

(b) Motor neuron

What roles do the dendrites, cell body, and axon play in communication of signals?

New Art Studying human anatomy and physiology is both a visual and a descriptive enterprise. Countless students have now benefited from the detailed and generously proportioned bone and muscle art that characterizes our book. Continuing in this tradition, you will find exciting new three-dimensional illustrations gracing the pages of virtually every chapter in the book. More than 90 figures are new to this edition, and nearly every figure has been revised or improved in some way.

Figure 6.3 Histology of bone. Sections through the diaphysis of a long bone, from the surrounding periosteum on the right, to compact bone in the middle, to spongy bone and the medullary cavity on the left. The inset at the upper right shows an osteocyte in a lacuna. See Table 4.3J on page 126 for a photomicrograph of compact bone tissue and Figure 6.14a for a scanning electron micrograph of spongy bone tissue.

In compact bone, osteocytes lie in lacunae arranged in concentric circles around a central canal.

Compact bone
Spongy bone
Medullary cavity
Periosteum

Interstitial lamellae
Outer circumferential lamellae
Concentric lamellae
Blood vessels
Lymphatic vessel
Lacuna
Canaliculi
Osteocyte

See Figure 6.4b for details

Medullary cavity

Trabeculae
Inner circumferential lamellae
Osteon

Periosteal vein
Periosteal artery
Periosteum:
Outer fibrous layer
Inner osteogenic layer
Central canal
Perforating canal

Spongy bone
Lymphatic vessel
Compact bone

Osteons (Haversian systems) in compact bone and trabeculae in spongy bone

The central (Haversian) canals may become blocked. What effect would this have on osteocytes?

Figure 20.8 The coronary circulation. The views of the heart from the anterior aspect in (a) and (b) are drawn as if the heart were transparent to reveal blood vessels on the posterior aspect. (See Tortora, *A Photographic Atlas of the Human Body*, Figures 6.8 and 6.9.)

The right and left coronary arteries deliver blood to the heart; the coronary veins drain blood from the heart into the coronary sinus.

Arch of aorta

Ascending aorta

Pulmonary trunk
RIGHT CORONARY

Right atrium

MARGINAL BRANCH

Right ventricle

LEFT CORONARY
Left auricle
CIRCUMFLEX BRANCH
ANTERIOR INTER-VENTRICULAR BRANCH

POSTERIOR INTER-VENTRICULAR BRANCH

Left ventricle

(a) Anterior view of coronary arteries

Figure 22.9 Phagocytosis of a microbe.

The major types of phagocytes are neutrophils and macrophages.

1 CHEMOTAXIS — Microbe

2 ADHERENCE
Pseudopod

3 INGESTION

Lysosome

4 DIGESTION

Plasma membrane

5 KILLING

Digestive enzymes

Phagocyte

Digested microbe in phagolysosome

Residual body (indigestible material)

(a) Phases of phagocytosis

What chemicals are responsible for killing ingested microbes?

New Histology-Based Art As part of our goal for continuous improvement, many of the anatomical illustrations based on histological preparations have been revised and redrawn for this edition.

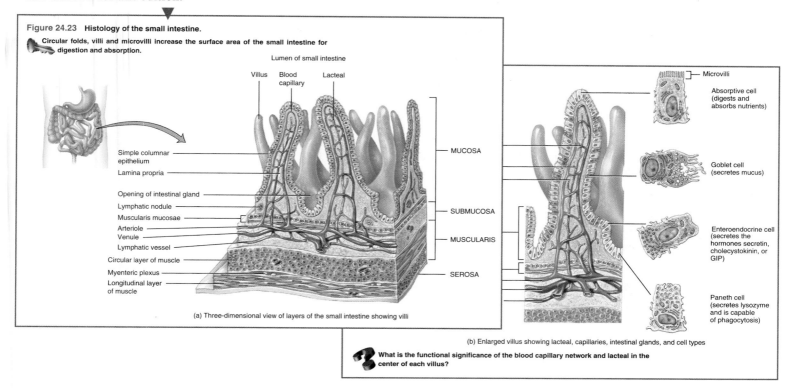

Figure 24.23 **Histology of the small intestine.**

Circular folds, villi and microvilli increase the surface area of the small intestine for digestion and absorption.

(a) Three-dimensional view of layers of the small intestine showing villi

(b) Enlarged villus showing lacteal, capillaries, intestinal glands, and cell types

What is the functional significance of the blood capillary network and lacteal in the center of each villus?

New Photomicrographs Dr. Michael Ross of the University of Florida has again provided us with more beautiful, customized photomicrographs of various tissues of the body. We have always considered Dr. Ross' photos among the best available and their inclusion in this edition greatly enhances the illustration program.

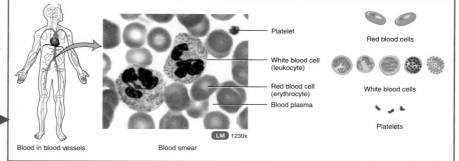

Cadaver Photos As before, we provide an assortment of large, clear cadaver photos at strategic points in many chapters. What is more, many anatomy illustrations are keyed to the large cadaver photos available in a companion text, *A Photographic Atlas of the Human Body*, by Gerard J. Tortora.

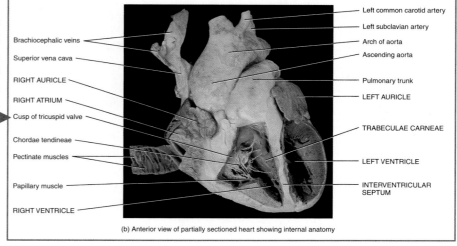

(b) Anterior view of partially sectioned heart showing internal anatomy

Helpful Orientation Diagrams Students sometimes need help figuring out the plane of view of anatomy illustrations — descriptions alone do not always suffice. An orientation diagram that depicts and explains the perspective of the view represented in the figure accompanies every major anatomy illustration. There are three types of diagrams: (1) planes used to indicate where certain sections are made when a part of the body is cut; (2) diagrams containing a directional arrow and the word "View" to indicate the direction from which the body part is viewed, and (3) diagrams with arrows leading from or to them that direct attention to enlarged and detailed parts of illustrations.

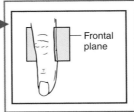
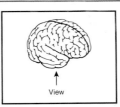
Frontal plane

View

Key Concept Statements This art-related feature summarizes an idea that is discussed in the text and demonstrated in a figure. Each Key Concept Statement is positioned adjacent to its figure and is denoted by a distinctive key icon.

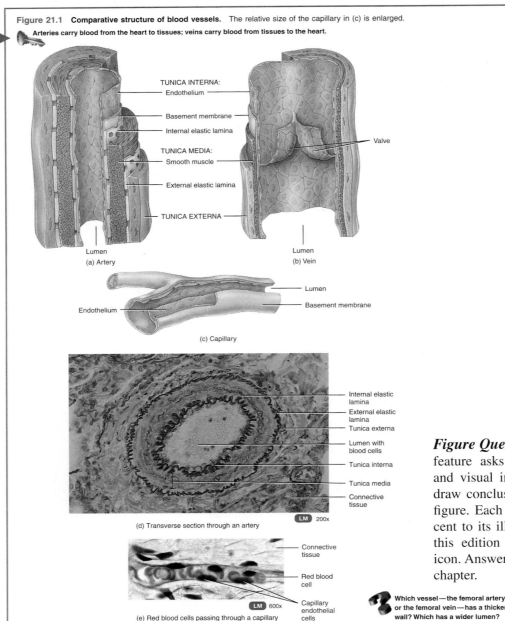

Figure 21.1 **Comparative structure of blood vessels.** The relative size of the capillary in (c) is enlarged.

Arteries carry blood from the heart to tissues; veins carry blood from tissues to the heart.

TUNICA INTERNA:
Endothelium
Basement membrane
Internal elastic lamina
TUNICA MEDIA:
Smooth muscle
External elastic lamina
TUNICA EXTERNA
Valve
Lumen
(a) Artery
Lumen
(b) Vein

Lumen
Basement membrane
Endothelium
(c) Capillary

Internal elastic lamina
External elastic lamina
Tunica externa
Lumen with blood cells
Tunica interna
Tunica media
Connective tissue
LM 200x
(d) Transverse section through an artery

Connective tissue
Red blood cell
Capillary endothelial cells
LM 600x
(e) Red blood cells passing through a capillary

Figure Questions This highly applauded feature asks readers to synthesize verbal and visual information, think critically, or draw conclusions about what they see in a figure. Each Figure Question appears adjacent to its illustration and is highlighted in this edition by the purple question mark icon. Answers are located at the end of each chapter.

Which vessel—the femoral artery or the femoral vein—has a thicker wall? Which has a wider lumen?

Correlation of Sequential Processes Correlation of sequential processes in text and art is achieved through the use of numbered lists in the narrative that correspond to numbered segments in the accompanying figure. This approach is used extensively throughout the book to lend clarity to the flow of complex processes.

Figure 20.10 **The conduction system of the heart.** Autorhythmic fibers in the SA node, located in the right atrial wall (a), act as the heart's pacemaker, initiating cardiac action potentials (b) that cause contraction of the heart's chambers.

The conduction system ensures that the chambers of the heart contract in a coordinated manner.

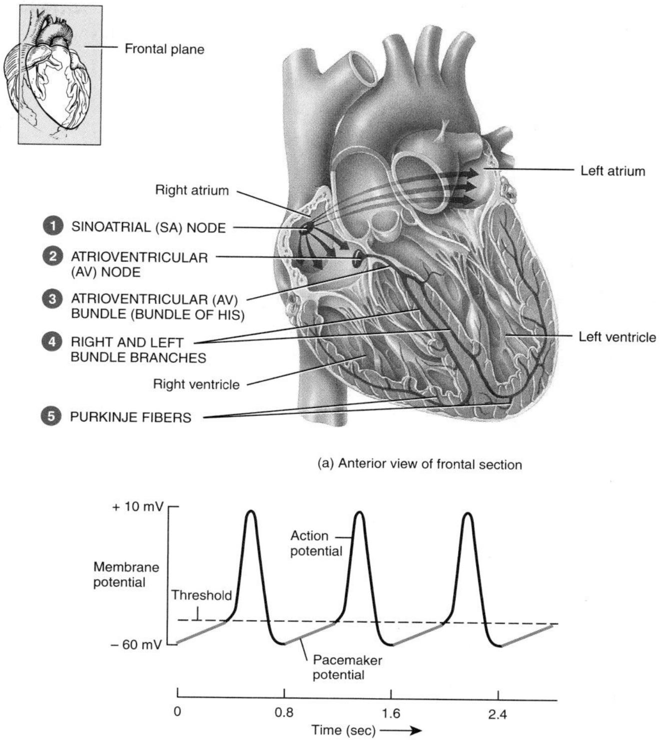

Frontal plane

Right atrium

Left atrium

1 SINOATRIAL (SA) NODE

2 ATRIOVENTRICULAR (AV) NODE

3 ATRIOVENTRICULAR (AV) BUNDLE (BUNDLE OF HIS)

4 RIGHT AND LEFT BUNDLE BRANCHES

Right ventricle

5 PURKINJE FIBERS

Left ventricle

(a) Anterior view of frontal section

+ 10 mV

Membrane potential

Threshold

− 60 mV

Action potential

Pacemaker potential

0   0.8   1.6   2.4
Time (sec)

(b) Pacemaker potentials and action potentials in autorhythmic fibers of SA node

Which component of the conduction system provides the only electrical connection between the atria and the ventricles?

Functions Overview This feature succinctly lists the functions of an anatomical structure or body system depicted within a figure. The juxtaposition of text and art further reinforces the connection between structure and function.

OVERVIEW OF KIDNEY FUNCTIONS

OBJECTIVE

- **List the functions of the kidneys.**

The kidneys do the major work of the urinary system. The other parts of the system are mainly passageways and storage areas. Functions of the kidneys include:

- ***Excreting wastes and foreign s*** ... urine, the kidneys help excrete ... have no useful function in the bod... in urine result from metabolic reac... include ammonia and urea from the deamination of amino acids; bilirubin from the catabolism of hemoglobin; creatinine from the breakdown of creatine phosphate in muscle fibers; and uric acid from the catabolism of nucleic acids.

Figure 26.1 **Organs of the urinary system in a female.** (See Tortora, *A Photographic Atlas of the Human Body,* Figure 13.2.)

Urine formed by the kidneys passes first into the ureters, then to the urinary bladder for storage, and finally through the urethra for elimination from the body.

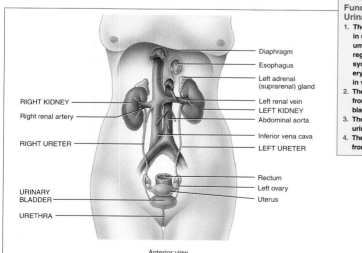

Diaphragm
Esophagus
Left adrenal (suprarenal) gland
Left renal vein
LEFT KIDNEY
Abdominal aorta
Inferior vena cava
LEFT URETER
Rectum
Left ovary
Uterus

RIGHT KIDNEY
Right renal artery
RIGHT URETER
URINARY BLADDER
URETHRA

Anterior view

Functions of the Urinary System

1. The kidneys excrete wastes in urine, regulate blood volume and composition, help regulate blood pressure, synthesize glucose, release erythropoietin, and participate in vitamin D synthesis.
2. The ureters transport urine from the kidneys to the urinary bladder.
3. The urinary bladder stores urine.
4. The urethra discharges urine from the body.

Which organs constitute the urinary system?

Hallmark Features

The tenth edition of *Principles of Anatomy and Physiology* builds on the legacy of thoughtfully designed and class-tested pedagogical features that provide a complete learning system for students as they navigate their way through the text and course. All have been revised to reflect the enhancements to the text.

Chapter-opening Pages Each chapter now begins with a chapter-opening page that includes a beautiful new piece of art related to the body system under consideration. A perspective on the chapter's content from either a recent student of the course or a practitioner in a health related field introduces current users to the relevance of the material or particularly interesting portions of the chapter. Suggestions for energizing their study with dynamic multimedia activities on the Foundations CD included with the text and specially developed internet activities on the companion website are highlighted.

Helpful Exhibits Students of anatomy and physiology need extra help learning the many structures that constitute certain body systems—most notably skeletal muscles, articulations, blood vessels, and nerves. As in previous editions, the chapters that present these topics are organized around **Exhibits,** each of which consists of an overview, a tabular summary of the relevant anatomy, and an associated suite of illustrations or photographs. Each Exhibit is prefaced by an Objective and closes with a Checkpoint activity. We trust you will agree that our spaciously designed Exhibits are ideal study vehicles for learning anatomically complex body systems.

Student Objectives and Checkpoints **Objectives** are found at the beginning of major sections throughout each chapter. Complementing this format, **Checkpoint** questions appear at strategic intervals within chapters to give students the chance to validate their understanding as they read.

Tools for Mastering Vocabulary Students—even the best ones—generally find it difficult at first to read and pronounce anatomical and physiological terms. Moreover, as teachers we are sympathetic to the needs of the growing ranks of college students who speak English as a second language. For these reasons, we have endeavored to ensure that this book has a strong and helpful vocabulary component. The key terms in every chapter are emphasized by use of **boldface type.** We include **pronunciation guides** when major, or especially hard-to-pronounce, structures and functions are introduced in the discussions, Tables, or Exhibits. **Word roots** citing the Greek or Latin derivations of anatomical terms are offered as an additional aid. As a further service to readers, we provide a list of **Medical Terminology** at the conclusion of most chapters, and a comprehensive **Glossary** at the back of the book. Also included at the end of the book is a list of the basic building blocks of medical terminology—**Combining Forms, Word Roots, Prefixes, and Suffixes.**

Study Outline As always, readers will benefit from the popular end-of-chapter **Study Outline** that is page referenced to the chapter discussions.

End-of-chapter Questions The **Self-Quiz Questions** are written in a variety of styles that are calculated to appeal to readers' different testing preferences. **Critical Thinking Questions** challenge readers to apply concepts to real-life situations. The style of these questions ought to make students smile on occasion as well as think! Answers to the Self-Quiz and Critical Thinking Questions are located in Appendix E.

Complete Teaching And Learning Package

Continuing the tradition of providing a complete teaching and learning package, the tenth edition of *Principles of Anatomy and Physiology* is available with a host of carefully planned supplementary materials that will help you and your students attain the maximal benefit from our textbook. Please contact your Wiley sales representative for additional information about any of these resources.

Media

Dynamic CD-ROMs and an enhanced Companion Website have been developed to complement the tenth edition of *Principles of Anatomy and Physiology*. These are fully described on the inside front cover and include:

- **Student Companion CD-ROM,** provided free with the purchase of a new textbook

- **Dedicated Book Companion Website** free to students who purchase a new textbook

- **Interactions: Exploring the Functions of the Human Body.** The first CD in this dramatic new series—Foundations—is provided free to adopters of the text and to students purchasing a new text.

For Instructors

- **Full-color Overhead Transparencies** This set of full-color acetates includes over 650 figures, including histology micrographs, from the text — complete with figure numbers and captions. Transparencies have enlarged labels for effective use as overhead projections in the classroom. (0-471- 25425-8)

- **Instructor's Resource CD-ROM** For lecture presentation purposes, this cross-platform CD-ROM includes all of the line art from the text in labeled and unlabeled formats. In addition, a pre-designed set of Powerpoint Slides is included with text and images. The slides are easily edited to customize your presentations to your specific needs. (0-471- 25350-2)

- **Professor's Resource Manual** by Lee Famiano of Cuyahoga CC. This on-line manual provides instructors with tools to enhance their lectures. Key features include: Lecture Outlines, What's New and Different in this Chapter, Critical Thinking Questions, Teaching Tips, and a guide to other educational resources. The entire manual is found on the text's dedicated companion website and is fully downloadable. The electronic format allows you to customize lecture outlines or activities for delivery either in print or electronically to your students. (0-471-25355-3)
- **Printed Test Bank** by P. James Nielsen of Western Illinois University. A testbank of nearly 3,000 questions, many new to the tenth edition, is available. A variety of formats—multiple choice, short answer, matching and essay—are provided to accommodate different testing preferences. (0-471-25421-5)
- **Computerized Test Bank** An electronic version of the printed test bank is available on a cross-platform CD-ROM. (0-471-25146-1)
- **Interactions: Exploring the Functions of the Human Body**
- **WebCT** or **Blackboard** Course Management Systems with content prepared by Sharon Simpson are available.
- **Faculty Resource Network** – New from Wiley is a support structure to help instructors implement the dynamic new media that supports this text into their classrooms, laboratories, or online courses. Consult with your Wiley representative for details about this program.

For Students
- **Student Companion CD-ROM**
- **Interactions: Exploring the Functions of the Human Body**
- **Book Companion Website**
- **Learning Guide (0-471-43447-7)** by Kathleen Schmidt Prezbindowski, College of Mount St. Joseph. Designed specifically to fit the needs of students with different learning styles, this well-received guide helps students to more closely examine important concepts through a variety of activities and exercises. The 29 chapters in the *Learning Guide* parallel those of the textbook and include many activities, quizzes and tests for review and study.
- **Illustrated Notebook (0-471-25150-X)** A true companion to the text, this unique notebook is a tool for organized note taking in class and for review during study. Following the sequence in the textbook, each left-handed page displays an unlabeled black and white copy of every text figure. Students can fill in the labels during lecture or lab at the instructor's directions and take additional notes on the lined right-handed pages.
- **Anatomy and Physiology: A Companion Coloring Book (0-471-39515-3)** This helpful study aid features over 500 striking, original illustrations that, by actively coloring them, give students a clear understanding of key anatomical structures and physiological processes.

For the Laboratory
- **A Brief Atlas of the Human Skeleton** This brief photographic review of the human skeleton is provided free with every new copy of the text.
- **Laboratory Manual for Anatomy and Physiology** by Connie Allen and Valerie Harper, Edison Community College **(0-471-39464-5)** This new laboratory manual presents material covered in the 2-semester undergraduate anatomy & physiology laboratory course in a clear and concise way, while maintaining a student-friendly tone. The manual is very interactive and contains activities and experiments that enhance students' ability to both visualize anatomical structures and understand physiological topics.
- **Cat Dissection Manual** by Connie Allen and Valerie Harper, Edison Community College **(0-471-26457-1)** This manual includes photographs and illustrations of the cat along with guidelines for dissection. It is available independently as well as bundled with the main manual depending upon your an adoption needs.
- **Fetal Pig Dissection Manual** by Connie Allen and Valerie Harper, Edison Community College **(0-471-26458-X)** This manual includes photographs and illustrations of the fetal pig along with guidelines for dissection. It is available independently as well as bundled with the main manual depending upon your an adoption needs.
- **Photographic Atlas of the Human Body with Selected Cat, Sheep, and Cow Dissections** by Gerard Tortora **(0-471-37487-3)** This four-colored atlas is designed to support both study and laboratory experiences. Organized by body systems, the clearly labeled photographs provide a stunning visual reference to gross anatomy. Histological micrographs are also included. Many of the illustrations within *Principles of Anatomy and Physiology*, 10e are cross-referenced to this atlas.

Like each of our students, this book has a life of its own. The structure, content, and production values of *Principles of Anatomy and Physiology* are shaped as much by its relationship with educators and readers as by the vision that gave birth to the book ten editions ago. Today you, our readers, are the "heart" of this book. We invite you to continue the tradition of sending in your suggestions to us so that we can include them in the eleventh edition.

Gerard J. Tortora
Department of Science and Health, S229
Bergen Community College
400 Paramus Road
Paramus, NJ 07652

Sandra Reynolds Grabowski
Department of Biological Sciences
1392 Lilly Hall of Life Sciences
Purdue University
West Lafayette, IN 47907-1392
email: Sgrabows@bilbo.bio.purdue.edu

Acknowledgements

For the tenth edition of *Principles of Anatomy and Physiology*, we have enjoyed the opportunity of collaborating with a group of dedicated and talented professionals. Accordingly, we would like to recognize and thank the members of our book team, who often worked evenings and weekends, as well as days, to bring this book to you. At John Wiley & Sons, Inc., our Editor Bonnie Roesch again illuminates the path toward ever better books with her creative ideas and dedication. Bonnie was our valued editor during the seventh and eighth editions—welcome back and thanks! Karin Kincheloe is the Wiley designer whose vision for the tenth edition is a larger, more colorful, more user-friendly new style. Moreover, Karin laid out each page of the book to achieve the best possible placement of text, figures, and other elements. Both instructors and students will appreciate and benefit from the pedagogically effective and visually pleasing design elements that augment the content changes made to this edition. Danke sehr! Claudia Durrell, our Art Coordinator, has collaborated on this text since its seventh edition. Her artistic ability, organizational skills, attention to detail, and understanding of our illustration preferences greatly enhance the visual appeal and style of the figures. She remains a cornerstone of our projects— thank you, Claudia, for all your contributions. Kelly Tavares, Senior Production Editor, demonstrated her untiring expertise during each step of the production process. She coordinated all aspects of actually making and manufacturing the book. Kelly also was "on press" as the book was being printed to ensure the highest possible quality. Muito obrigada, Kelly for all the extra hours you spent to implement book-improving changes! Ellen Ford, Developmental Editor, shepherded the manuscript and electronic files during the revision process. She also contacted many reviewers for their comments and sought out former students to obtain their perspectives on learning anatomy and physiology for the new chapter-opening pages. Hillary Newman, Photo Editor, provided us with all of the photos we requested and did it with efficiency, accuracy, and professionalism. Mary O'Sullivan, Assistant Editor, coordinated the development of many of the supplements that support this text. We are most appreciative! Wiley Editorial Assistants Justin Bow and Kelli Coaxum helped with various aspects of the project and took care of many details. Thanks to all of you!

Jerri K. Lindsey, Tarrant County Junior College, Northeast and Caryl Tickner, Stark State College wrote the end-of-chapter Self-Quiz Questions. Joan Barber of Delaware Technical and Community College contributed the Critical Thinking Questions. Thanks to all three for writing questions that students will appreciate. And a special thank you to Kathleen Prezbindowski, who has authored the *Learning Guide* for many editions. The high quality of her study activities ensures student success.

Outstanding illustrations and photographs have always been a signature feature of *Principles of Anatomy and Physiology*. Respected scientific and medical illustrators on our team of exceptional artists include Mollie Borman, Leonard Dank, Sharon Ellis, Wendy Hiller Gee, Jean Jackson, Keith Kasnot, Lauren Keswick, Steve Oh, Lynn O'Kelley, Hilda Muinos, Tomo Narashima, Nadine Sokol, and Kevin Somerville. Mark Nielsen of the University of Utah provided many of the cadaver photos that appear in this edition. Artists at Imagineering created the amazing computer graphics images and provided all labeling of figures.

Reviewers

We are extremely grateful to our colleagues who reviewed the manuscript and offered insightful suggestions for improvement. The contributions of all these people, who generously provided their time and expertise to help us maintain the book's accuracy and clarity, are acknowledged in the list that follows.

Patricia Ahanotu, Georgia Perimeter College
Cynthia L. Beck, George Mason University
Clinton L. Benjamin, Lower Columbia College
Anna Berkovitz, Purdue University
Charles J. Biggers, University of Memphis
Mark Bloom, Tyler Junior College
Michele Boiani, University of Pennsylvania
Bruce M. Carlson, University of Michigan
Barbara Janson Cohen, Delaware County Community
    College
Matthew Jarvis Cohen, Delaware County Community
    College
Marcia Carol Coss, George Mason University
Victor P. Eroschenko, University of Idaho
Lorraine Findlay, Nassau Community College
Candice Francis, Palomar College
Christina Gan, Rogue Community College
Gregory Garman, Centralia College

Alan Gillen, Pensacola Christian College
Chaya Gopalan, St. Louis Community College
Janet Haynes, Long Island University
Clare Hayes, Metropolitan State College of Denver
James Junker, Campbell University
Gerald Karp, University of Florida
William Kleinelp, Middlesex County College
John Langdon, University of Indianapolis
John Lepri, University of North Carolina at Greensboro
Jerri K. Lindsey, Tarrant County College
Mary Katherine K. Lockwood, University of
    New Hampshire
Jennifer Lundmark, California State University, Sacramento
Paul Malven, Purdue University
G. K. Maravelas, Bristol Community College
Jane Marks, Paradise Valley Community College
Lee Meserve, Bowling Green State University
Javanika Mody, Anne Arundel Community College
Robert L. Moskowitz, Community College of Philadelphia
Shigihiro Nakajima, University of Illinois at Chicago
Jerry D. Norton, Georgia State University
Justicia Opoku, University of Maryland
Weston Opitz, Kansas Wesleyan University
Joann Otto, Purdue University
David Parker, Northern Virginia Community College
Karla Pouillon, Everett Community College
Linda Powell, Community College of Philadelphia
C. Lee Rocket, Bowling Green State University
Esmond J. Sanders, University of Alberta
Louisa Schmid, Tyler Junior College
Hans Schöler, University of Pennsylvania
Charles Sinclair, University of Indianapolis
Dianne Snyder, Augusta State University
Dennis Strete, McLennan Community College
Eric Sun, Macon State College
Antonietto Tan, Worcester State College
Jim Van Brunt, Rogue Community College
Jyoti Wagle, Houston Community College
Curt Walker, Dixie State College

DeLoris M. Wenzel, The University of Georgia
David Westmoreland, US Air Force Academy
Frederick E. Williams, University of Toledo

In addition, every chapter was read and reviewed by either an Anatomy & Physiology student or a health care professional now working in his or her chosen career. Their comments on the relevance and interest of the chapter topics is presented on each chapter opening page. We appreciate their time and effort, and especially their enthusiasm for the subject matter. We think that enthusiasm is catching! Thanks go out to the following:

Tamatha Adkins, RN
Molly Causby, Georgia Perimeter College
Margaret Chambers, George Mason University
RoNell Coco, Clark College
John Curra, Respiratory Technologist
Stephanie Hall Ford, Barry University
Helen Hart Ford, Radiological Technologist
Elizabeth Garrison, Modesto Junior College
Emily Gordon, Edison Community College
Kim Green, Edison Community College
Mike Grosse, Licensed Physical Therapy Assistant
Caroline Guerra, Broward Community College
Jill Haan, Cardiographic Technician
Dorie Hart, RN
Lisa P. Hubbard, Macon State College
Mark Johnson, Edison Community College
Wendy Lawrence, Troy State University
Candice Machado, Modesto Junior College
Susan Mahoney, Physical Therapist
Christine McGrellis, Mohawk Valley Community College
Jacqueline Opera, Medical Laboratory Technologist
Denise Pacheco, Modesto Junior College
Joan Petrokovsky, Licensed Massage Therapist
Toni Sheridan, Pharmacist
Barbara Simone, RN, FNP
Tiffany Smith, Stark State College of Technology
Sabrina von Brueckwitz, Edison Community College

To the Student

Your book has a variety of special features that will make your time studying anatomy and physiology a more rewarding experience. These have been developed based on feedback from students – like you – who have used previous editions of the text. A review of the preface will give you insight, both visually and in narrative, to all of the text's distinctive features.

Our experience in the classroom has taught us that student's appreciate a hint – both visually and verbally – at the beginning of each chapter about what to expect from its contents. Each chapter of your book begins with a stunning illustration depicting the system or main content being covered in the chapter. In addition, a short introduction to the chapter contents – usually written by a student who has recently completed the course, but occasionally by a practitioner in an allied health field – offers you an insight into some of the most intriguing or relevant aspects of the chapter. Links to activities on your **Foundations** CD, and special web-based activities called **Insights and Explorations** are suggested to make your study time worthwhile and interesting.

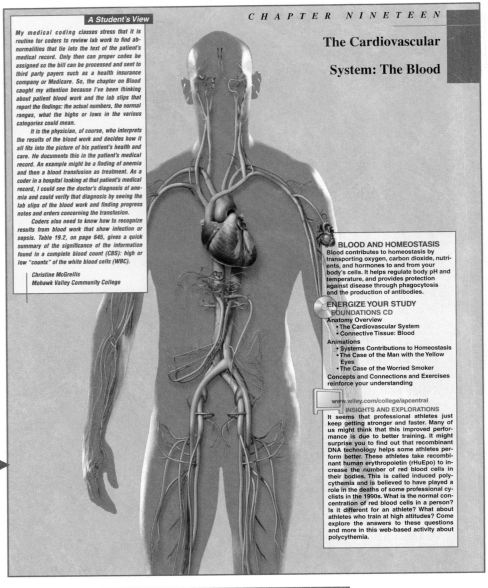

A Student's View

My medical coding classes stress that it is routine for coders to review lab work to find abnormalities that tie into the text of the patient's medical record. Only then can proper codes be assigned so the bill can be processed and sent to third party payers such as a health insurance company or Medicare. So, the chapter on Blood caught my attention because I've been thinking about patient blood work and the lab slips that report the findings: the actual numbers, the normal ranges, what the highs or lows in the various categories could mean.

It is the physician, of course, who interprets the results of the blood work and decides how it all fits into the picture of his patient's health and care. He documents this in the patient's medical record. An example might be a finding of anemia and then a blood transfusion as treatment. As a coder in a hospital looking at that patient's medical record, I could see the doctor's diagnosis of anemia and could verify that diagnosis by seeing the lab slips of the blood work and finding progress notes and orders concerning the transfusion.

Coders also need to know how to recognize results from blood work that show infection or sepsis. Table 19.2, on page 645, gives a quick summary of the significance of the information found in a complete blood count (CBS): high or low "counts" of the white blood cells (WBC).

Christine McGrellis
Mohawk Valley Community College

CHAPTER NINETEEN

The Cardiovascular System: The Blood

BLOOD AND HOMEOSTASIS
Blood contributes to homeostasis by transporting oxygen, carbon dioxide, nutrients, and hormones to and from your body's cells. It helps regulate body pH and temperature, and provides protection against disease through phagocytosis and the production of antibodies.

ENERGIZE YOUR STUDY
FOUNDATIONS CD
Anatomy Overview
• The Cardiovascular System
• Connective Tissue: Blood
Animations
• Systems Contributions to Homeostasis
• The Case of the Man with the Yellow Eyes
• The Case of the Worried Smoker
Concepts and Connections and Exercises reinforce your understanding

www.wiley.com/college/apcentral
INSIGHTS AND EXPLORATIONS
It seems that professional athletes just keep getting stronger and faster. Many of us might think that this improved performance is due to better training. It might surprise you to find out that recombinant DNA technology helps some athletes perform better. These athletes take recombinant human erythropoietin (rHuEpo) to increase the number of red blood cells in their bodies. This is called induced polycythemia and is believed to have played a role in the deaths of some professional cyclists in the 1990s. What is the normal concentration of red blood cells in a person? Is it different for an athlete? What about athletes who train at high altitudes? Come explore the answers to these questions and more in this web-based activity about polycythemia.

FUNCTIONS AND PROPERTIES OF BLOOD

OBJECTIVES

• **Describe the functions of blood.**
• **Describe the physical characteristics and principal components of blood.**

Most cells of a multicellular organism cannot move around to obtain oxygen and nutrients or eliminate carbon dioxide and other wastes. Instead, these needs are met by two fluids: blood and interstitial fluid. **Blood** is a connective tissue composed of a liquid matrix ca...

As you begin each narrative section of the chapter, be sure to take note of the **objectives** at the beginning of the section to help you focus on what is important as you read it.

CHECKPOINT
1. In what ways is blood plasma similar to and different from interstitial fluid?
2. How many kilograms or pounds of blood is present in your body?
3. How does the volume of blood plasma in your body compare to the volume of fluid in a two-liter bottle of Coke?
4. What are the major solutes in blood plasma? What does each do?
5. What is the significance of lower-than-normal or higher-than-normal hematocrit?

At the end of the section, take time to try and answer the **Checkpoint** questions placed there. If you can, then you are ready to move on to the next section. If you experience difficulty answering the questions, you may want to re-read the section before continuing.

Studying the figures (illustrations that include artwork and photographs) in this book is as important as reading the text. To get the most out of the visual parts of this book, use the tools we have added to the figures to help you understand the concepts being presented. Start by reading the **legend**, which explains what the figure is about. Next, study the **key concept statement**, which reveals a basic idea portrayed in the figure. Added to many figures you will also find an **orientation diagram** to help you understand the perspective from which you are viewing a particular piece of anatomical art. Finally, at the bottom of each figure you will find a **figure question**. If you try to answer these questions as you go along, they will serve as self-checks to help you understand the material. Often it will be possible to answer a question by examining the figure itself. Other questions will encourage you to integrate the knowledge you've gained by carefully reading the text associated with the figure. Still other questions may prompt you to think critically about the topic at hand or predict a consequence in advance of its description in the text.

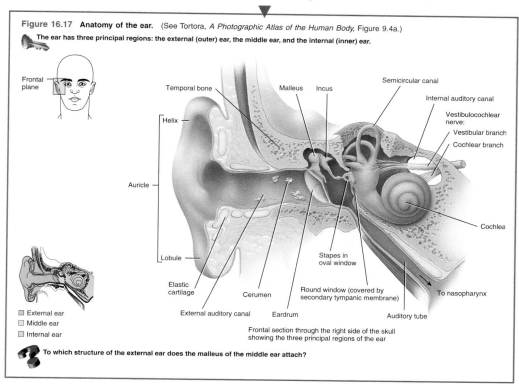

Figure 16.17 Anatomy of the ear. (See Tortora, *A Photographic Atlas of the Human Body*, Figure 9.4a.)

The ear has three principal regions: the external (outer) ear, the middle ear, and the internal (inner) ear.

Frontal plane

Temporal bone — Malleus — Incus — Semicircular canal — Internal auditory canal — Vestibulocochlear nerve: — Vestibular branch — Cochlear branch

Helix

Auricle

Cochlea

Lobule

Stapes in oval window

Elastic cartilage — Cerumen — Round window (covered by secondary tympanic membrane) — To nasopharynx

☐ External ear
☐ Middle ear
☐ Internal ear

External auditory canal — Eardrum — Auditory tube

Frontal section through the right side of the skull showing the three principal regions of the ear

To which structure of the external ear does the malleus of the middle ear attach?

At the end of each chapter are other resources that you will find useful. **The Study Outline** is a concise statement of important topics discussed in the chapter. Page numbers are listed next to key concepts so you can easily refer to the specific passages in the text for clarification or amplification.

STUDY OUTLINE

SPINAL CORD ANATOMY (p. 420)

1. The spinal cord is protected by the vertebral column, the meninges, cerebrospinal fluid, and denticulate ligaments.
2. The three meninges are coverings that run continuously around the spinal cord and brain. They are the dura mater, arachnoid mater, and pia mater.
3. The spinal cord begins as a continuation of the medulla oblongata and ends at about the second lumbar vertebra in an adult.
4. The spinal cord contains cervical and lumbar enlargements that serve as points of origin for nerves to the limbs.
5. The tapered inferior portion of the spinal cord is the conus medullaris, from which arise the filum terminale and cauda equina.
6. Spinal nerves connect to each segment of the spinal cord by two

roots. The posterior or dorsal root contains sensory axons, and the anterior or ventral root contains motor neuron axons.
7. The anterior median fissure and the posterior median sulcus partially divide the spinal cord into right and left sides.
8. The gray matter in the spinal cord is divided into horns, and the white matter into columns. In the center of the spinal cord is the central canal, which runs the length of the spinal cord.
9. Parts of the spinal cord observed in transverse section are the gray commissure; central canal; anterior, posterior, and lateral gray horns; and anterior, posterior, and lateral white columns, which contain ascending and descending tracts. Each part has specific functions.
10. The spinal cord conveys sensory and motor information by way of ascending and descending tracts, respectively.

The **Self-quiz Questions** are designed to help you evaluate your understanding of the chapter contents. **Critical Thinking Questions** are word problems that allow you to apply the concepts you have studied in the chapter to specific situations.

SELF-QUIZ QUESTIONS

Fill in the blanks in the following statements.

1. _____ are fast, predictable, automatic responses to changes in the environment.

2. Because they contain both sensory and motor axons, spinal nerves are considered to be _____ nerves.

Indicate whether the following statements are true or false.

3. Reflexes permit the body to make exceedingly rapid adjustments to homeostatic imbalances.

4. Autonomic reflexes involve responses of smooth muscle, cardiac muscle, and glands.

Choose the one best answer to the following questions.

(a) endoneurium, (b) fascicle, (c) perineurium, (d) epineurium, (e) neurolemma.

6. Which of the following is *not* a function of the spinal cord? (a) reflex center, (b) integration of EPSPs and IPSPs, (c) conduction pathway for sensory impulses, (d) conduction pathway for motor impulses, (e) interpretation of sensory stimuli.

7. Which of the following are true? (1) The anterior (ventral) gray horns contain cell bodies of neurons that cause skeletal muscle contraction. (2) The gray commissure connects the white matter of the right and left sides of the spinal cord. (3) Cell bodies of autonomic motor neurons are located in the lateral gray horns. (4) Sensory (ascending) tracts conduct motor impulses down the spinal _____. (5) Gray matter in the spinal cord consists of cell bodies of

CRITICAL THINKING QUESTIONS

1. Pearl, a lifeguard at the local beach, stepped on a discarded lit cigarette with her bare foot. Trace the reflex arcs set in motion by her accident. Name the reflex types.
   HINT *How will she remain standing when she picks up her foot?*

2. Why doesn't your spinal cord creep up toward your he_____ time you bend over? Why doesn't it get all twisted out o_____ when you exercise?
   HINT *How would you prevent a boat from floating awa_____ dock?*

3. Jose's severe headaches and other symptoms were suggestive of meningitis, so his physician ordered a spinal tap. List the structures that the needle will pierce from the most superficial to the deepest. Why would the physican order a test in the spinal region to check a problem in Jose's head?

You will also find the **Answers to Figure Questions** at the end of chapters.

ANSWERS TO FIGURE QUESTIONS

13.1 The superior boundary of the spinal dura mater is the foramen magnum of the occipital bone. The inferior boundary is the second sacral vertebra.

13.2 The cervical enlargement connects with sensory and motor nerves of the upper limbs.

13.3 A horn is an area of gray matter, and a column is a region of white matter in the spinal cord.

13.4 The anterior corticospinal tract is located on the anterior side of the spinal cord, originates in the cortex of the cerebrum, and

the motor impulses leave the spinal cord on the side opposite the entry of sensory impulses.

13.10 All spinal nerves are mixed (have sensory and motor components) because the posterior root containing sensory axons and the anterior root containing motor axons unite to form the spinal nerve.

13.11 The anterior rami serve the upper and lower limbs.

13.12 Severing the spinal cord at level C2 causes respiratory arrest because it prevents descending nerve impulses from reaching the

Learning the language of anatomy and physiology can be one the more challenging aspects of taking this course. Throughout the text we have included pronunciations, and sometimes, Word Roots, for many terms that may be new to you. These appear in parentheses immediately following the new words, and the pronunciations are repeated in the glossary at the back of the book. You companion CD includes the complete glossary and offers you the opportunity to hear the words pronounced. In addition, the Companion Website offers you a review of these terms by

chapter, pronounces them for you, and allows you the opportunity to create flash cards or quiz yourself on the many new terms.

Look at the words carefully and say them out loud several times. Learning to pronounce a new word will help you remember it and make it a useful part of your medical vocabulary. Take a few minutes to review the following pronunciation key, so it will be familiar to you when you encounter new words. The key is repeated at the beginning of the Glossary, as well.

Pronounciation Key

1. The most strongly accented syllable appears in capital letters, for example, bilateral (bī-LAT-er-al) and diagnosis (dī-ag-NŌ-sis).

2. If there is a secondary accent, it is noted by a prime ('), for example, constitution (kon'-sti-TOO-shun) and physiology (fiz'-ē-OL-ō-jē). Any additional secondary accents are also noted by a prime, for example, decarboxylation (dē'-kar-bok'-si-LĀ-shun).

3. Vowels marked by a line above the letter are pronounced with the long sound, as in the following common words:
   ā as in *māke*   ō as in *pōle*
   ē as in *bē*   ū as in *cute*
   ī as in *īvy*

4. Vowels not marked by a line above the letter are pronounced with the short sound, as in the following words:
   a as in *above* or *at*   o as in *not*
   e as in *bet*   u as in *bud*
   i as in *sip*

5. Other vowel sounds are indicated as follows:
   oy as in *oil*
   oo as in *root*

6. Consonant sounds are pronounced as in the following words:
   b as in *bat*   m as in *mother*
   ch as in *chair*   n as in *no*
   d as in *dog*   p as in *pick*
   f as in *father*   r as in *rib*
   g as in *get*   s as in *so*
   h as in *hat*   t as in *tea*
   j as in *jump*   v as in *very*
   k as in *can*   w as in *welcome*
   ks as in *tax*   z as in *zero*
   kw as in *quit*   zh as in *lesion*
   l as in *let*

Brief Table of Contents

Contents

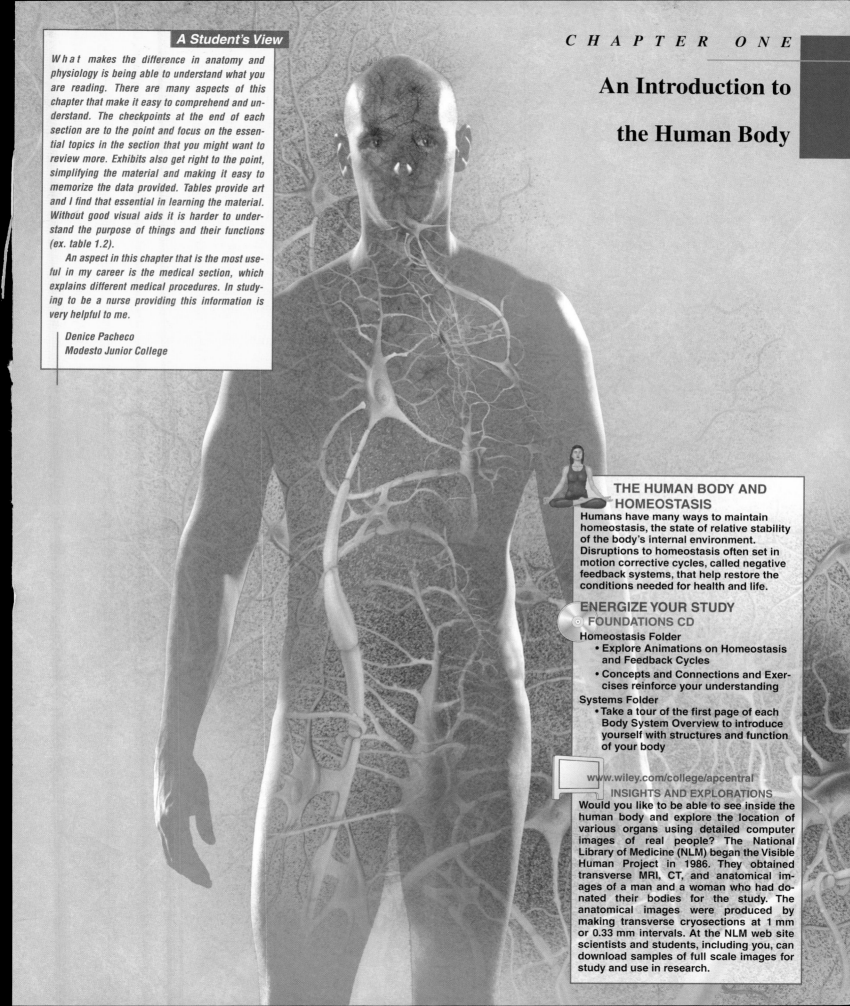

An Introduction to the Human Body

THE HUMAN BODY AND HOMEOSTASIS

Humans have many ways to maintain homeostasis, the state of relative stability of the body's internal environment. Disruptions to homeostasis often set in motion corrective cycles, called negative feedback systems, that help restore the conditions needed for health and life.

ENERGIZE YOUR STUDY
FOUNDATIONS CD

Homeostasis Folder
- **Explore Animations on Homeostasis and Feedback Cycles**
- **Concepts and Connections and Exercises reinforce your understanding**

Systems Folder
- **Take a tour of the first page of each Body System Overview to introduce yourself with structures and function of your body**

www.wiley.com/college/apcentral

INSIGHTS AND EXPLORATIONS

Would you like to be able to see inside the human body and explore the location of various organs using detailed computer images of real people? The National Library of Medicine (NLM) began the Visible Human Project in 1986. They obtained transverse MRI, CT, and anatomical images of a man and a woman who had donated their bodies for the study. The anatomical images were produced by making transverse cryosections at 1 mm or 0.33 mm intervals. At the NLM web site scientists and students, including you, can download samples of full scale images for study and use in research.

The fascinating journey that begins the exploration of the human body starts with an introduction to the disciplines of anatomy and physiology. Then, we will consider the organization of living things and explore the properties that all living things share. Next, you will discover how the body regulates its own internal environment, a ceaseless process called homeostasis that is a major theme in every chapter of this book. Finally, we introduce the basic vocabulary that will help you speak about the body in a way that is understood by scientists and health-care professionals alike.

ANATOMY AND PHYSIOLOGY DEFINED

▶ O B J E C T I V E

- **Define anatomy and physiology, and name several subdisciplines of these sciences.**

Two branches of science—anatomy and physiology—provide the foundation for understanding the body's parts and functions. **Anatomy** (a-NAT-ō-mē; *ana-* = up; *-tomy* = process of cutting) is the science of body structures and the relationships among structures. It was first studied by **dissection** (dis-SEK-shun; *dis-* = apart; *-section* = act of cutting), the careful cutting apart of body structures to study their relationships. Today, a variety of imaging techniques also contribute to the advancement of anatomical knowledge. We will describe and compare some common imaging techniques in Table 1.4. Whereas anatomy deals with structures of the body, **physiology** (fiz′-ē-OL-o-jē; *physio-* = nature; *-logy* = study of) is the science of body func-

tions—how the body parts work. Table 1.1 describes several subdisciplines of anatomy and physiology.

You will learn about the human body by studying its anatomy and physiology together. Always keep in mind that structure and function are intertwined. The structure of a part of the body allows performance of certain functions. For example, the bones of the skull join tightly to form a rigid case that protects the brain. The bones of the fingers, by contrast, are more loosely joined, which allows a variety of movements. The walls of the air sacs in the lungs are very thin, permitting rapid movement of inhaled oxygen into the blood. By contrast, the lining of the urinary bladder is much thicker. Yet its construction allows for considerable stretching as the urinary bladder fills with urine.

▶ C H E C K P O I N T

1. What body function might a respiratory therapist strive to improve?
2. Give an example of how the structure of a part of the body is related to its function.

Table 1.1	Selected Subdisciplines of Anatomy and Physiology		
Subdisciplines of Anatomy	**Study of**	**Subdisciplines of Physiology**	**Study of**
Embryology (em′-brē-OL-ō-jē; *embry-* = embryo; *-logy* = study of)	Structures that emerge from the time of the fertilized egg through the eighth week in utero.	**Neurophysiology** (NOOR-ō-fiz-ē-ol′-ō-jē; *neuro-* = nerve)	Functional properties of nerve cells.
Developmental biology	Structures that emerge from the time of the fertilized egg to the adult form.	**Endocrinology** (en′-dō-kri-NOL-ō-jē; *endo-* = within; *-crin* = secretion)	Hormones (chemical regulators in the blood) and how they control body functions.
Histology (hiss′-TOL-ō-jē; *hist-* = tissue)	Microscopic structure of tissues.	**Cardiovascular physiology** (kar-dē-ō-VAS-kū-lar; *cardi-* = heart; *-vascular* = blood vessels)	Functions of the heart and blood vessels.
Surface anatomy	Anatomical landmarks on the surface of the body through visualization and palpation.	**Immunology** (im′-ū-NOL-ō-jē; *immun-* = not susceptible)	How the body defends itself against disease-causing agents.
Gross anatomy	Structures that can be examined without using a microscope.	**Respiratory physiology** (RES-pir-a-to′-rē; *respira-* = to breathe)	Functions of the air passageways and lungs.
Systemic anatomy	Structure of specific systems of the body such as the nervous or respiratory systems.	**Renal physiology** (RĒ-nal; *ren-* = kidney)	Functions of the kidneys.
Regional anatomy	Specific regions of the body such as the head or chest.	**Exercise physiology**	Changes in cell and organ functions as a result of muscular activity.
Radiographic anatomy (rā′-dē-ō-GRAF-ik; *radio-* = ray; *-graphic* = to write)	Body structures that can be visualized with x rays.		
Pathological anatomy (path′-ō-LOJ-i-kal; *path-* = disease)	Structural changes (from gross to microscopic) associated with disease.	**Pathophysiology** (PATH-ō-fiz-ē-ol′-ō-jē)	Functional changes associated with disease and aging.

LEVELS OF BODY ORGANIZATION

▶ O B J E C T I V E S

- **Describe the levels of structural organization that make up the human body.**
- **List the eleven systems of the human body, the organs present in each, and their general functions.**

The levels of organization of a language—letters of the alphabet, words, sentences, paragraphs, and so on—provide a useful comparison to the levels of organization of the human body. Your exploration of the human body will extend from some of the smallest body structures and their functions to the largest structure—an entire person. From the smallest to the largest size of their components, six levels of organization are relevant to understanding anatomy and physiology: the chemical, cellular, tissue, organ, system, and organismal levels of organization (Figure 1.1).

1 The *chemical level* includes **atoms,** the smallest units of matter that participate in chemical reactions, and **molecules,** two or more atoms joined together. Certain atoms, such as carbon (C), hydrogen (H), oxygen (O), nitrogen (N),

Figure 1.1   Levels of structural organization in the human body.

The levels of structural organization are chemical, cellular, tissue, organ, system, and organismal.

1 CHEMICAL LEVEL

Atoms
(C, H, O, N, P)

Molecule
(DNA)

2 CELLULAR LEVEL

Smooth muscle cell

3 TISSUE LEVEL

Smooth muscle tissue

Serous membrane

4 ORGAN LEVEL

Smooth muscle tissue layers

Stomach

Epithelial tissue

5 SYSTEM LEVEL

Esophagus
Liver
Stomach
Pancreas
Gallbladder
Small intestine
Large intestine

Digestive system

6 ORGANISMAL LEVEL

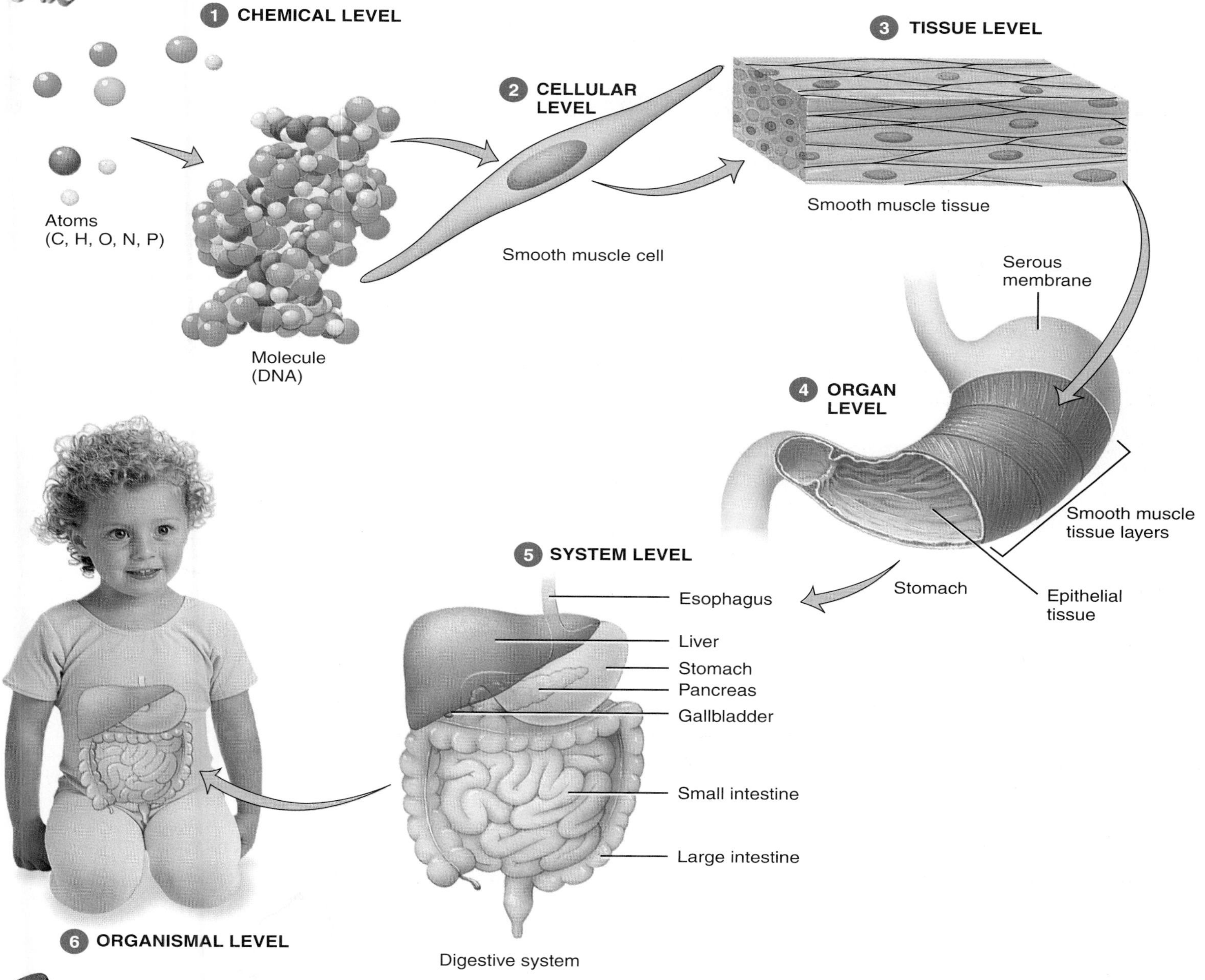

Which level of structural organization is composed of two or more different types of tissues that work together to perform a specific function?

phosphorus (P), calcium (Ca), and sulfur (S), are essential for maintaining life. Two familiar examples of molecules found in the body are deoxyribonucleic acid (DNA), the genetic material passed from one generation to the next, and glucose, commonly known as blood sugar. Chapters 2 and 25 focus on the chemical level of organization.

2 At the *cellular level,* molecules combine to form **cells,** the basic structural and functional units of an organism. Cells are the smallest living units in the human body. Among the many kinds of cells in your body are smooth muscle cells, nerve cells, and epithelial cells. The focus of Chapter 3 is the cellular level of organization.

3 The next level of structural organization is the *tissue level.* **Tissues** are groups of cells and the materials surrounding them that work together to perform a particular function. There are just four basic types of tissue in your body:

epithelial tissue, connective tissue, muscle tissue, and *nervous tissue.* Chapter 4 describes the tissue level of organization. Shown here is smooth muscle tissue, which consists of tightly packed smooth muscle cells.

4 At the *organ level,* different kinds of tissues are joined together. **Organs** are structures that are composed of two or more different types of tissues; they have specific functions and usually have recognizable shapes. Examples of organs are the stomach, heart, liver, lungs, and brain. Figure 1.1 shows how several tissues make up the stomach. The stomach's outer covering is a layer of epithelial tissue and connective tissue that reduces friction when the stomach moves and rubs against other organs. Underneath are the *smooth muscle tissue layers,* which contract to churn and mix food and then push it into the next digestive organ, the small intestine. The innermost lining is an *epithelial tissue layer*

Table 1.2   The Eleven Systems of the Human Body

Integumentary System

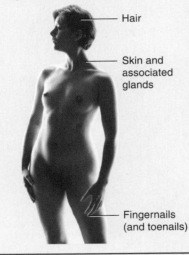

Hair

Skin and associated glands

Fingernails (and toenails)

Components: Skin, and structures derived from it, such as hair, nails, sweat glands, and oil glands.

Functions: Protects the body; helps regulate body temperature; eliminates some wastes; helps make vitamin D; and detects sensations such as touch, pain, warmth, and cold.

Muscular System

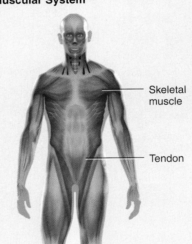

Skeletal muscle

Tendon

Components: Muscles composed of skeletal muscle tissue, so-named because it is usually attached to bones.

Functions: Produces body movements, such as walking; stabilizes body position (posture); generates heat.

Skeletal System

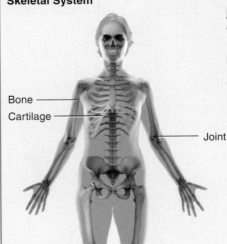

Bone

Cartilage

Joint

Components: Bones and joints of the body and their associated cartilages.

Functions: Supports and protects the body; aids body movements; houses cells that produce blood cells; stores minerals and lipids (fats).

Nervous System

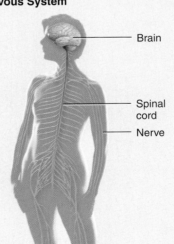

Brain

Spinal cord

Nerve

Components: Brain, spinal cord, nerves, and special sense organs, such as the eye and ear.

Functions: Generates action potentials (nerve impulses) to regulate body activities; detects changes in the body's internal and external environment, interprets the changes, and responds by causing muscular contractions or glandular secretions.

that produces fluid and chemicals responsible for digestion in the stomach.

5 The next level of structural organization in the body is the *system level.* A **system** consists of related organs that have a common function. An example is the digestive system, which breaks down and absorbs food. Its organs include the mouth, salivary glands, pharynx (throat), esophagus, stomach, small intestine, large intestine, liver, gallbladder, and pancreas. Sometimes an organ is part of more than one system. The pancreas, for example, is part of both the digestive system and the hormone-producing endocrine system.

6 The largest organizational level is the *organismal level.* An **organism** is any living individual. All the parts of the human body functioning together constitute the total organism — one living person.

In the chapters that follow, you will study the anatomy and physiology of the major body systems. Table 1.2 lists the components and introduces the functions of these systems. You will also discover that all body systems influence one another. As an example, consider how just two body systems — the integumentary and skeletal systems — cooperate. The integumentary system, which includes the skin, hair, and nails, protects all other body systems, including the skeletal system, which is composed of all the bones and joints of the body. The skin serves as a barrier between the outside environment and internal tissues and organs. It also participates in the production of vitamin D, which is needed for proper deposition of calcium and other minerals into bone. The skeletal system, in turn, provides support for the integumentary system. It also serves as a reservoir for calcium, storing calcium in times of plenty and releasing it for other tissues in times of need. In addition, red bone marrow, present within some bones, generates the white blood cells that

---

Endocrine System

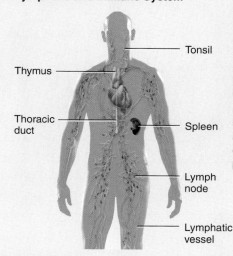

Pituitary gland
Pineal gland
Thyroid gland
Thymus
Adrenal gland
Pancreas
Ovary
Testis

Components: Hormone-producing glands (pineal gland, hypothalamus, pituitary gland, thymus, thyroid gland, parathyroid glands, adrenal glands, pancreas, ovaries, and testes) and hormone-producing cells in several other organs.

Functions: Regulates body activities by releasing hormones, which are chemical messengers transported in blood from an endocrine gland to a target organ.

Lymphatic and Immune System

Thymus
Thoracic duct
Tonsil
Spleen
Lymph node
Lymphatic vessel

Components: Lymphatic fluid and vessels; also includes spleen, thymus, lymph nodes, and tonsils.

Functions: Returns proteins and fluid to blood; carries lipids from gastrointestinal tract to blood; includes structures where lymphocytes that protect against disease-causing organisms mature and proliferate.

Cardiovascular System

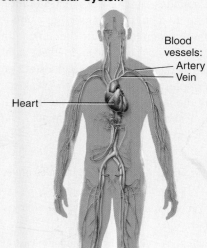

Blood vessels:
Artery
Vein
Heart

Components: Blood, heart, and blood vessels.

Functions: Heart pumps blood through blood vessels; blood carries oxygen and nutrients to cells and carbon dioxide and wastes away from cells and helps regulate acid-base balance, temperature, and water content of body fluids; blood components help defend against disease and mend damaged blood vessels.

Respiratory System

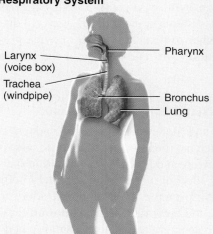

Larynx (voice box)
Trachea (windpipe)
Pharynx
Bronchus
Lung

Components: Lungs and air passageways such as the pharynx (throat), larynx (voice box), trachea (windpipe), and bronchial tubes leading into and out of them.

Functions: Transfers oxygen from inhaled air to blood and carbon dioxide from blood to exhaled air; helps regulate acid–base balance of body fluids; air flowing out of lungs through vocal cords produces sounds.

(continues)

Table 1.2 The Eleven Systems of the Human Body (*continued*)

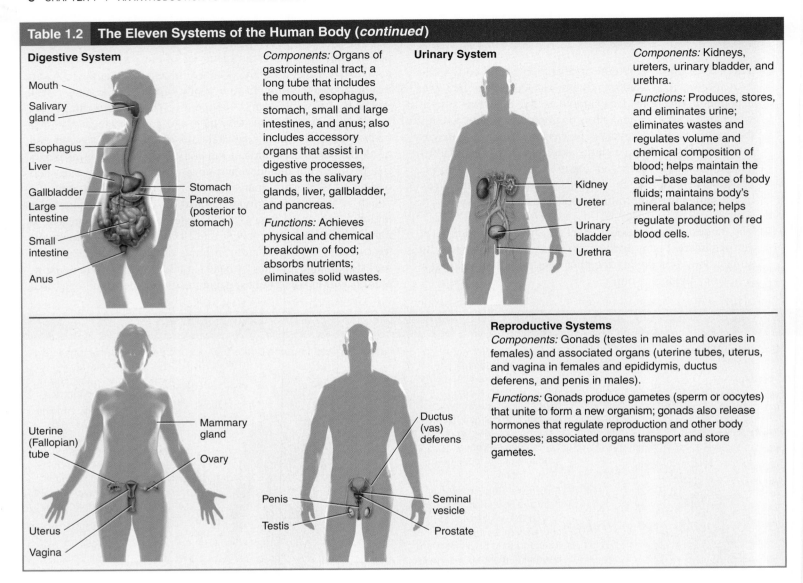

Digestive System

Mouth
Salivary gland
Esophagus
Liver
Gallbladder
Large intestine
Small intestine
Anus
Stomach
Pancreas (posterior to stomach)

Components: Organs of gastrointestinal tract, a long tube that includes the mouth, esophagus, stomach, small and large intestines, and anus; also includes accessory organs that assist in digestive processes, such as the salivary glands, liver, gallbladder, and pancreas.

Functions: Achieves physical and chemical breakdown of food; absorbs nutrients; eliminates solid wastes.

Urinary System

Kidney
Ureter
Urinary bladder
Urethra

Components: Kidneys, ureters, urinary bladder, and urethra.

Functions: Produces, stores, and eliminates urine; eliminates wastes and regulates volume and chemical composition of blood; helps maintain the acid–base balance of body fluids; maintains body's mineral balance; helps regulate production of red blood cells.

Uterine (Fallopian) tube
Mammary gland
Ovary
Uterus
Vagina

Ductus (vas) deferens
Penis
Seminal vesicle
Testis
Prostate

Reproductive Systems

Components: Gonads (testes in males and ovaries in females) and associated organs (uterine tubes, uterus, and vagina in females and epididymis, ductus deferens, and penis in males).

Functions: Gonads produce gametes (sperm or oocytes) that unite to form a new organism; gonads also release hormones that regulate reproduction and other body processes; associated organs transport and store gametes.

---

help the skin resist invasion by disease-causing organisms. As you study each of the body systems in more detail, you will discover how they work together to maintain health, provide protection from disease, and allow for reproduction of the human species.

Palpation, Auscultation, and Percussion

Health-care professionals and students of anatomy and physiology commonly use three noninvasive techniques to assess certain aspects of body structure and function. In **palpation** (pal-PĀY-shun; *palp-* = gently touching) the examiner feels body surfaces with the hands. An example is palpating an artery to find the pulse and measure the heart rate. In **auscultation** (auscul-TĀY-shun; *auscult-* = listening) the examiner listens to body sounds to evaluate the functioning of certain organs, often using a stethoscope to amplify the sounds. An example is auscultation of the lungs during breathing to check for crackling sounds associated with abnormal fluid accumulation in the lungs. In **percussion** (pur-KUSH-un; *percus-* = beat through) the examiner taps on the body surface with the fingertips and listens to the resulting echo. For example, percussion may reveal the abnormal presence of fluid in the lungs or air in the intestines. It may also provide information about the size, consistency, and position of an underlying structure. ■

▶ C H E C K P O I N T

3. Define the following terms: atom, molecule, cell, tissue, organ, system, and organism.

4. At what levels of organization would an exercise physiologist study the human body? *(Hint: Refer to Table 1.1.)*

5. Referring to Table 1.2, which body systems help eliminate wastes?

CHARACTERISTICS OF THE LIVING HUMAN ORGANISM

▶ O B J E C T I V E S

- **Define the important life processes of the human body.**
- **Define homeostasis and explain its relationship to interstitial fluid.**

Basic Life Processes

Organisms carry on certain processes that distinguish them from nonliving things. Following are the six most important life processes of the human body:

Metabolism (me-TAB-ō-lizm) is the sum of all the chemical processes that occur in the body. One phase of metabolism is **catabolism** (ka-TAB-ō-lizm; *catabol-* = throwing down; *-ism* = a condition), the breaking down of complex chemical substances into simpler ones. The other phase of metabolism is **anabolism** (a-NAB-ō-lizm; *anabol-* = a raising up), the building up of complex chemical substances from smaller, simpler ones. In metabolic processes, oxygen taken in by the respiratory system and nutrients broken down in the digestive system provide the chemical energy to power cellular activities. Metabolism also includes breaking down large, complex molecules into smaller, simpler ones and using the resulting building blocks to assemble the body's structural components. For example, digestive processes split proteins in food into amino acids, which are the building blocks of proteins. Amino acids then are used to build new proteins that make up body structures such as muscles and bones.

Responsiveness is the body's ability to detect and respond to changes in its internal or external environment. A decrease in body temperature would be an example of a change in the internal environment. Turning your head toward the sound of squealing brakes is an example of responsiveness to a change in the external environment. Different cells in the body respond to environmental changes in characteristic ways. Nerve cells respond by generating electrical signals known as nerve impulses (action potentials). Muscle cells respond by contracting, which generates force to move body parts.

Movement includes motion of the whole body, individual organs, single cells, and even tiny structures inside cells. For example, the coordinated action of leg muscles moves your whole body from one place to another when you walk or run. After you eat a meal that contains fats, your gallbladder contracts and squirts bile into the gastrointestinal tract. The bile aids in the digestion of fats. When a body tissue is damaged or infected, certain white blood cells move from the blood into the affected tissue to help clean up and repair the area. Even inside individual cells, various cellular parts move from one position to another to carry out their functions.

Growth is an increase in body size that results from an increase in the size of existing cells, the number of cells, or both. In addition, a tissue sometimes increases in size because the amount of material between cells increases. In a growing bone, for example, mineral deposits accumulate around the bone cells, causing the bone to enlarge in length and width.

Differentiation is the process a cell undergoes to develop from an unspecialized to a specialized state. As you will see later, each type of cell in the body has a specialized structure and function. Specialized cells differ in structure and function from the ancestor cells that gave rise to them. For example, red blood cells and several types of white blood cells differentiate from the same unspecialized ancestor cells in red bone marrow. Such ancestor cells, which can divide and give rise to progeny that undergo differentiation, are known as **stem cells.** Also through differentiation, a fertilized egg (ovum) develops into an embryo, and then into a fetus, an infant, a child, and finally an adult.

Reproduction refers either to the formation of new cells for tissue growth, repair, or replacement or to the production of a new individual. Through the fertilization of an ovum by a sperm cell, life continues from one generation to the next.

When the life processes cease to occur properly, the result is death of cells and tissues, which may lead to death of the organism. Clinically, loss of the heartbeat, absence of spontaneous breathing, and loss of brain functions indicate death in the human body.

 **Autopsy**

An **autopsy** (AW-top-sē = seeing with one's own eyes) is a postmortem (after death) examination of the body and dissection of its internal organs to confirm or determine the cause of death. An autopsy can uncover the existence of diseases not detected during life, determine the extent of injuries, and explain how those injuries may have contributed to a person's death. It also may provide more information about a disease, assist in the accumulation of statistical data, and educate healthcare students. Moreover, an autopsy can reveal conditions that may affect offspring or siblings (such as congenital heart defects). Sometimes, an autopsy is legally required, such as during a criminal investigation. It may also be useful in resolving disputes between beneficiaries and insurance companies about the cause of death. ■

Homeostasis

The French physiologist Claude Bernard (1813–1878) first proposed that the cells of many-celled organisms flourish because they live in the relative constancy of *"le milieu interieur"*—the internal environment—despite continual changes in the organisms' external environment. The American physiologist Walter B. Cannon (1871–1945) coined the term *homeostasis* to describe this dynamic constancy. **Homeostasis** (hō'mē-ō-STĀ-sis;

homeo- = sameness; *-stasis* = standing still) is the condition of equilibrium in the body's internal environment. It occurs due to the ceaseless interplay of all the body's regulatory processes. Homeostasis is a dynamic condition. In response to changing conditions, the body's equilibrium point can change over a narrow range that is compatible with maintaining life. For example, the level of glucose in blood normally stays between 70 and 110 milligrams of glucose per 100 milliliters of blood.* Each body structure, from the cellular level to the systemic level, contributes in some way to keeping the internal environment within normal limits.

Body Fluids

An important aspect of homeostasis is maintaining the volume and composition of **body fluids,** which are dilute, watery solutions found inside cells as well as surrounding them. The fluid within cells is **intracellular fluid** (*intra-* = inside), abbreviated **ICF.** The fluid outside body cells is **extracellular fluid** (*extra-* = outside), abbreviated **ECF.** Dissolved in the water of ICF and ECF are oxygen, nutrients, proteins, and a variety of *ions* (electrically charged chemical particles). All these substances are needed to maintain life. The ECF that fills the narrow spaces between cells of tissues is known as **interstitial fluid** (in´-ter-STISH-al; *inter-* = between). As you progress with your studies, you will learn that the ECF within blood vessels is termed **blood plasma,** that within lymphatic vessels is called **lymph,** that in and around the brain and spinal cord is called **cerebrospinal fluid,** that in joints is called **synovial fluid,** and that in the eyes is called **aqueous humor** and **vitreous body.**

As Bernard predicted, the proper functioning of body cells depends on precise regulation of the composition of their surrounding fluid. Because interstitial fluid surrounds all body cells, it is often called the body's *internal environment.* The composition of interstitial fluid changes as substances move back and forth between it and blood plasma. Such exchange of materials occurs across the thin walls of the smallest blood vessels in the body, the *blood capillaries.* This movement in both directions across capillary walls provides needed materials, such as glucose, oxygen, ions, and so on, to tissue cells. It also removes wastes, such as carbon dioxide, from interstitial fluid.

▶ C H E C K P O I N T
6. Which life process in the human body sustains all the others?
7. Describe the locations of intracellular fluid, extracellular fluid, interstitial fluid, and blood plasma.
8. Why is interstitial fluid called the internal environment of the body?

_____
*Appendix A describes metric measurements.

CONTROL OF HOMEOSTASIS

▶ O B J E C T I V E S

• **Describe the components of a feedback system.**
• **Contrast the operation of negative and positive feedback systems.**
• **Explain why homeostatic imbalances cause disorders.**

Homeostasis in the human body is continually being disturbed. Some disruptions come from the external environment (outside the body) in the form of physical insults such as intense heat or lack of oxygen. Other disruptions originate in the internal environment (within the body), such as a blood glucose level that is too low. Homeostatic imbalances may also occur due to psychological stresses in our social environment—the demands of work and school, for example. In most cases the disruption of homeostasis is mild and temporary, and the responses of body cells quickly restore balance in the internal environment. In some cases the disruption of homeostasis may be intense and prolonged, as in poisoning, overexposure to temperature extremes, or severe infection. Under these circumstances, regulation of homeostasis may fail.

Fortunately, the body has many regulating systems that usually bring the internal environment back into balance. Most often, the nervous system and the endocrine system, working together or independently, provide the needed corrective measures. The nervous system regulates homeostasis by sending messages in the form of *nerve impulses* to organs that can counteract deviations from the balanced state. The endocrine system includes many glands that secrete messenger molecules called *hormones* into the blood. Whereas nerve impulses typically cause rapid changes, hormones usually work more slowly. Both means of regulation, however, work toward the same end, namely maintaining homeostasis. As you will see, both systems carry out their mission mainly through negative feedback systems.

Feedback Systems

The body can regulate its internal environment through a multitude of feedback systems. A **feedback system** is a cycle of events in which the status of a body condition is continually monitored, evaluated, changed, remonitored, reevaluated, and so on. Each monitored variable, such as body temperature, blood pressure, or blood glucose level, is termed a *controlled condition.* Any disruption that changes a controlled condition is called a *stimulus.* Three basic components make up a feedback system—a receptor, a control center, and an effector (Figure 1.2).

• A **receptor** is a body structure that monitors changes in a controlled condition and sends input to a control center. Typically, the input is in the form of nerve impulses or chemical signals. For example, nerve endings in the skin that sense temperature are one of many different types of receptors in the body.

Figure 1.2 Operation of a feedback system. The dashed return arrow symbolizes negative feedback.

The three basic components of a feedback system are receptors, a control center, and effectors.

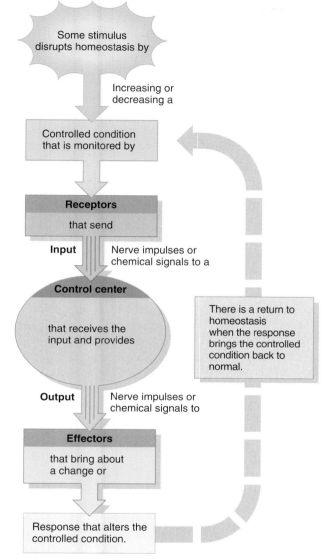

Some stimulus disrupts homeostasis by

Increasing or decreasing a

Controlled condition that is monitored by

Receptors

that send

Input Nerve impulses or chemical signals to a

Control center

that receives the input and provides

There is a return to homeostasis when the response brings the controlled condition back to normal.

Output Nerve impulses or chemical signals to

Effectors

that bring about a change or

Response that alters the controlled condition.

What is the main difference between negative and positive feedback systems?

- A **control center** in the body sets the range of values within which a controlled condition should be maintained, evaluates the input it receives from receptors, and generates output commands when they are needed. Output from the control center typically occurs as nerve impulses, hormones, or other chemical signals.

- An **effector** is a body structure that receives output from the control center and produces a *response* or effect that changes the controlled condition. Nearly every organ or tis-

sue in the body can behave as an effector. For example, when your body temperature drops sharply, your brain (control center) sends nerve impulses (output) to your skeletal muscles (effectors). The result is shivering, which generates heat and raises your body temperature.

A group of receptors and effectors communicating with their control center forms a feedback system that can regulate a controlled condition in the body's internal environment. In a feedback system, or *feedback loop,* the response of the system "feeds back" to change the controlled condition in some way. Feedback systems can produce either negative feedback or positive feedback. If the response reverses the original stimulus, as in the body temperature regulation example, the system is operating by *negative feedback.* If the response enhances or intensifies the original stimulus, the system is operating by *positive feedback.*

Negative Feedback Systems

A **negative feedback system** reverses a change in a controlled condition. First, a stimulus disrupts homeostasis by altering the controlled condition. The receptors that are part of the feedback system detect the change and send input to a control center. The control center evaluates the input and, if necessary, issues output commands to an effector. The effector produces a physiological response that is able to return the controlled condition to its normal state.

Consider one negative feedback system that helps regulate blood pressure. Blood pressure (BP) is the force exerted by blood as it presses against the walls of blood vessels. When the heart beats faster or harder, BP increases. If some internal or external stimulus causes blood pressure (controlled condition) to rise, the following sequence of events occurs (Figure 1.3). *Baroreceptors,* which are pressure-sensitive nerve cells (the receptors) that are located in the walls of certain blood vessels, detect the higher pressure. The baroreceptors send nerve impulses (input) to the brain (control center), which interprets the impulses and responds by sending nerve impulses (output) to the heart (the effector). Heart rate decreases, which causes BP to decrease (response). This sequence of events quickly returns the controlled condition—blood pressure—to normal, and homeostasis is restored. Notice that the activity of the effector causes BP to drop, a result that is opposite to the original stimulus (an increase in BP). Thus, this is a negative feedback system.

Positive Feedback Systems

A **positive feedback system** tends to strengthen or reinforce a change in one of the body's controlled conditions. A positive feedback system operates similarly to a negative feedback system, except for the way the response affects the controlled condition. A stimulus alters a controlled condition, which is monitored by receptors that send input to a control center. The control center provides commands to an effector, but this time

Figure 1.3 Homeostatic regulation of blood pressure by a negative feedback system. Note that the response is fed back into the system, and the system continues to lower blood pressure until there is a return to normal blood pressure (homeostasis).

If the response reverses the stimulus, a system is operating by negative feedback.

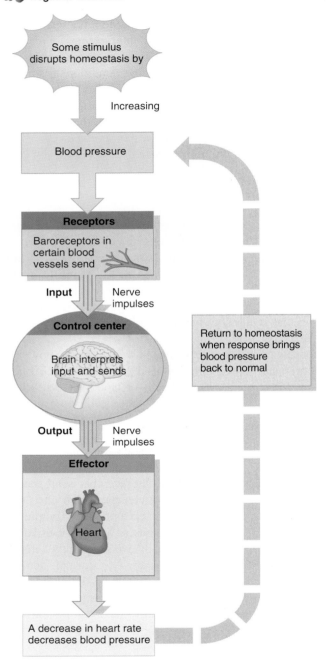

 What would happen to heart rate if some stimulus caused blood pressure to decrease? Would this occur by way of positive or negative feedback?

the effector produces a physiological response that *reinforces* the initial change in the controlled condition. The action of a positive feedback system continues until it is interrupted by some mechanism outside the system.

Normal childbirth provides a good example of a positive feedback system (Figure 1.4). The first contractions of labor (stimulus) push part of the baby into the cervix, the lowest part of the uterus, which opens into the vagina. Stretch-sensitive nerve cells (receptors) monitor the amount of stretching of the cervix (controlled condition). As stretching increases, they send more nerve impulses (input) to the brain (control center), which in turn releases the hormone oxytocin (output) into the blood. Oxytocin causes muscles in the wall of the uterus (effector) to contract even more forcefully. The contractions push the baby farther down the uterus, which stretches the cervix even more. The cycle of stretching, hormone release, and ever-stronger contractions is interrupted only by the birth of the baby. Then, stretching of the cervix ceases and oxytocin is no longer released. As you will see later, blood clotting is also an example of a positive feedback system.

These examples suggest some important differences between positive and negative feedback systems. Because a positive feedback system continually reinforces a change in a controlled condition, some event outside the system must shut it off. If the action of a positive feedback system is not stopped, it can "run away" because it cannot be controlled. It may even produce life-threatening conditions in the body. The action of a negative feedback system, by contrast, slows and then stops as the controlled condition returns to its normal state. Usually, positive feedback systems reinforce conditions that do not happen very often whereas negative feedback systems regulate conditions in the body that remain fairly stable over long periods.

Homeostatic Imbalances

As long as all the body's controlled conditions remain within certain narrow limits, body cells function efficiently, negative feedback loops maintain homeostasis, and the body stays healthy. Should one or more components of the body lose their ability to contribute to homeostasis, however, the normal equilibrium among body processes may be disturbed. If the homeostatic imbalance is moderate, a disorder or disease may occur; if it is severe, death may result.

A **disorder** is any derangement or abnormality of function. **Disease** is a more specific term for an illness characterized by a recognizable set of signs and symptoms. A *local disease* affects one part or a limited region of the body; a *systemic disease* affects either the entire body or several parts of it. Diseases alter body structures and functions in characteristic ways. A person with a disease may experience **symptoms,** which are subjective changes in body functions that are not apparent to an observer. Examples of symptoms are headache, nausea, and anxiety. Objective changes that a clinician can observe and measure, on the other hand, are called **signs.** Signs of disease can be either

Figure 1.4 Positive feedback control of labor contractions during birth of a baby.   The solid return arrow symbolizes positive feedback.

 If the response enhances or intensifies the stimulus, a system is operating by positive feedback.

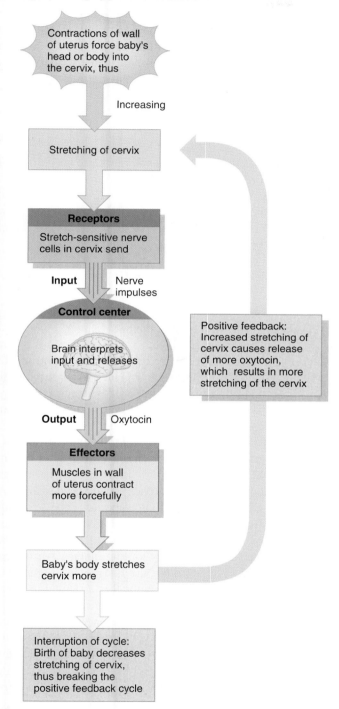

Why do positive feedback systems that are part of a normal physiological response include a termination mechanism?

anatomical, such as swelling or a rash, or physiological, such as fever, high blood pressure, or paralysis.

The science that deals with why, when, and where diseases occur and how they are transmitted among individuals in a community is known as **epidemiology** (ep′-i-dē-mē-OL-ō-jē; *epi-* = upon; *-demi* = people). **Pharmacology** (far′-ma-KOL-ō-jē; *pharmac-* = drug) is the science that deals with the effects and uses of drugs in the treatment of disease.

Diagnosis of Disease

Diagnosis (dī′ag-NŌ-sis; *dia-* = through; *-gnosis* = knowledge) is the science and skill of distinguishing one disorder or disease from another. The patient's signs and symptoms, his or her medical history, a physical exam, and laboratory tests provide the basis for making a diagnosis. Taking a *medical history* consists of collecting information about events that might be related to a patient's illness. These include the chief complaint, history of present illness, past medical problems, family medical problems, social history, and review of symptoms. A *physical examination* is an orderly evaluation of the body and its functions. This process includes inspection (looking at or into the body using various instruments), palpation, auscultation, percussion, measuring vital signs (temperature, pulse, respiratory rate, and blood pressure), and sometimes laboratory tests. ■

▶ C H E C K P O I N T

9. What types of disturbances can act as stimuli that initiate a feedback system?

10. How are negative and positive feedback systems similar? How are they different?

11. Contrast and give examples of symptoms and signs of a disease.

AGING AND HOMEOSTASIS

▶ O B J E C T I V E

• **Describe some of the effects of aging.**

As you will see later, aging is a normal process characterized by a progressive decline in the body's responses to restore homeostasis. Aging produces observable changes in structure and function and increases vulnerability to stress and disease. The changes associated with aging are apparent in all body systems. Examples include wrinkled skin, gray hair, loss of bone mass, decreased muscle mass and strength, diminished reflexes, decreased sensory perception and motor responses, decreased production of some hormones, increased incidence of coronary artery disease and congestive heart failure, increased susceptibility to infections and malignancies, decreased lung capacity, less efficient functioning of the digestive system, decreased kidney function, menopause, and enlarged prostate. These and other effects of aging will be discussed in detail in later chapters.

ANATOMICAL TERMINOLOGY

OBJECTIVES

- **Describe the orientation of the body in the anatomical position.**
- **Relate the common names to the corresponding anatomical descriptive terms for various regions of the human body.**
- **Define the anatomical planes and sections used to describe the human body.**
- **Describe the major body cavities, the organs they contain, and their associated linings.**

Scientists and health-care professionals use a common language of special terms when referring to body structures and their functions. The language of anatomy has precisely defined meanings that allow us to communicate without using unneeded or ambiguous words. For example, is it correct to say, "The wrist is above the fingers"? This might be true if your arms are at your sides. But if you hold your hands up above your head, your fingers would be above your wrists. To prevent this kind of confusion, anatomists developed a standard anatomical position and a special vocabulary for relating body parts to one another.

Body Positions

Descriptions of any region or part of the human body assume that it is in a specific stance called the **anatomical position.** In the anatomical position, the subject stands erect facing the observer, with the head level and the eyes facing directly forward. The feet are flat on the floor and directed forward, and the arms are at the sides with the palms turned forward (Figure 1.5). In the anatomical position, the body is upright. Two terms describe a reclining body. If the body is lying face down, it is in the **prone** position. If the body is lying face up, it is in the **supine** position.

Regional Names

The human body is divided into several major regions that can be identified externally. The principal regions are the head, neck, trunk, upper limbs, and lower limbs (Figure 1.5). The **head** consists of the skull and face. Whereas the *skull* encloses and protects the brain, the *face* is the anterior (front) portion of the head that includes the eyes, nose, mouth, forehead, cheeks, and chin. The **neck** supports the head and attaches it to the trunk. The **trunk** consists of the chest, abdomen, and pelvis. Each **upper limb** attaches to the trunk and consists of the shoulder, armpit, arm (portion of the limb from the shoulder to the elbow), forearm (portion of the limb from the elbow to the wrist), wrist, and hand. Each **lower limb** also attaches to the trunk and consists of the buttock, thigh (portion of the limb from the buttock to the knee), leg (portion of the limb from the knee to the ankle), ankle, and foot. The *groin* is the area on the anterior surface of the body marked by a crease on each side, where the trunk attaches to the thighs.

Figure 1.5 shows the common names of major parts of the body. The corresponding anatomical descriptive form (adjective) for each part appears in parentheses next to the common name. For example, if you receive a tetanus shot in your *buttock,* it is a *gluteal* injection. Because the descriptive form of a body part usually is based on a Greek or Latin word, it may look different from the common name for the same part or area. For example, the Latin word for armpit is *axilla* (ak-SIL-a). Thus, one of the nerves passing within the armpit is named the axillary nerve. You will learn more about the Greek and Latin word roots of anatomical and physiological terms as you read this book.

Directional Terms

To locate various body structures, anatomists use specific **directional terms,** words that describe the position of one body part relative to another. Several directional terms are grouped in pairs that have opposite meanings, such as anterior (front) and posterior (back). Exhibit 1.1 and Figure 1.6 on pages 14–15 present the main directional terms.

Planes and Sections

You will also study parts of the body relative to planes, imaginary flat surfaces that pass through the body parts (Figure 1.7 on page 16). A **sagittal plane** (SAJ-i-tal; *sagitt-* = arrow) is a vertical plane that divides the body or an organ into right and left sides. More specifically, when such a plane passes through the midline of the body or an organ and divides it into *equal* right and left sides, it is called a **midsagittal plane** or a **median plane.** If the sagittal plane does not pass through the midline but instead divides the body or an organ into *unequal* right and left sides, it is called a **parasagittal plane** (*para-* = near). A **frontal** or **coronal plane** (kō-RŌ-nal; *corona* = crown) divides the body or an organ into anterior (front) and posterior (back) portions. A **transverse plane** divides the body or an organ into superior (upper) and inferior (lower) portions. Other names for a transverse plane are a cross-sectional or horizontal plane. Sagittal, frontal, and transverse planes are all at right angles to one another. An **oblique plane,** by contrast, passes through the body or an organ at an angle between the transverse plane and either a sagittal or frontal plane.

When you study a body region, you often view it in section. A **section** is one flat surface of a three-dimensional structure. It is important to know the plane of the section so you can understand the anatomical relationship of one part to another. Figure 1.8 on page 16 indicates how three different sections—*transverse, frontal,* and *midsagittal*—provide different views of the brain.

Body Cavities

Body cavities are spaces within the body that help protect, separate, and support internal organs. Bones, muscles, and ligaments

Figure 1.5 **The anatomical position.** The common names and corresponding anatomical terms (in parentheses) are indicated for specific body regions. For example, the head is the cephalic region.

In the anatomical position, the subject stands erect facing the observer with the head level and the eyes facing forward. The feet are flat on the floor and directed forward, and the arms are at the sides with the palms facing forward.

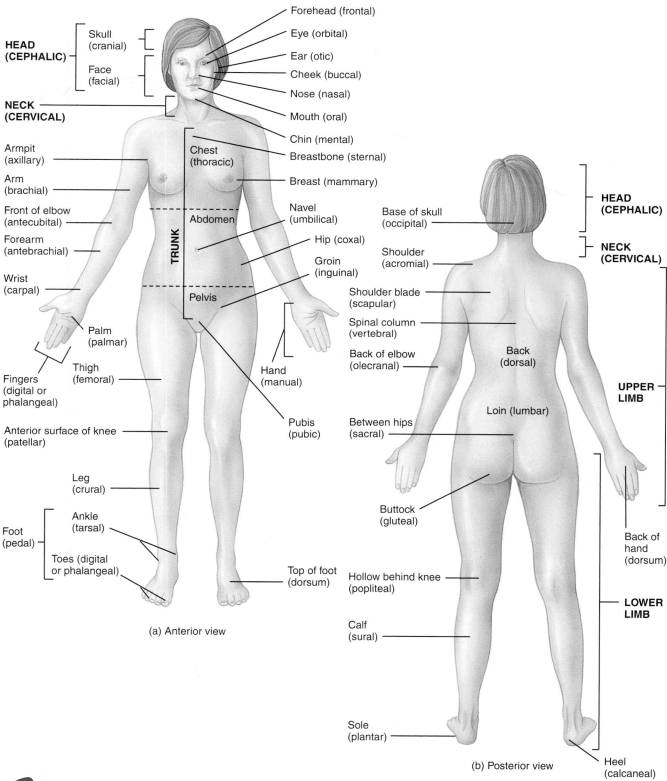

(a) Anterior view

(b) Posterior view

What is the usefulness of defining one standard anatomical position?

Exhibit 1.1 Directional Terms (Figure 1.6)

▶ OBJECTIVE

• Define each directional term used to describe the human body.

Overview

Most of the directional terms used to describe the human body can be grouped into pairs that have opposite meanings. For example, **superior** means toward the upper part of the body, whereas **inferior** means toward the lower part of the body. Moreover, it is important to understand that directional terms have relative meanings; they only make sense when used to describe the position of one structure relative to another.

For example, your knee is superior to your ankle, even though both are located in the inferior half of the body. Study the directional terms below and the example of how each is used. As you read the examples, look at Figure 1.6 to see the location of each structure.

▶ CHECKPOINT

Which directional terms can be used to specify the relationships between (1) the elbow and the shoulder, (2) the left and right shoulders, (3) the sternum and the humerus, and (4) the heart and the diaphragm?

Directional Term	Definition	Example of Use
Superior (soo′-PEER-ē-or) **(cephalic or cranial)**	Toward the head, or the upper part of a structure.	The heart is superior to the liver.
Inferior (in′-FEER-ē-or) **(caudal)**	Away from the head, or the lower part of a structure.	The stomach is inferior to the lungs.
Anterior (an-TEER-ē-or) **(ventral)***	Nearer to or at the front of the body.	The sternum (breastbone) is anterior to the heart.
Posterior (pos-TEER-ē-or) **(dorsal)**	Nearer to or at the back of the body.	The esophagus is posterior to the trachea (windpipe).
Medial (MĒ-dē-al)	Nearer to the midline.†	The ulna is medial to the radius.
Lateral (LAT-er-al)	Farther from the midline.	The lungs are lateral to the heart.
Intermediate (in′-ter-MĒ-dē-at)	Between two structures.	The transverse colon is intermediate between the ascending and descending colons.
Ipsilateral (ip-si-LAT-er-al)	On the same side of the body as another structure.	The gallbladder and ascending colon are ipsilateral.
Contralateral (CON-tra-lat-er-al)	On the opposite side of the body from another structure.	The ascending and descending colons are contralateral.
Proximal (PROK-si-mal)	Nearer to the attachment of a limb to the trunk; nearer to the origination of a structure.	The humerus is proximal to the radius.
Distal (DIS-tal)	Farther from the attachment of a limb to the trunk; farther from the origination of a structure.	The phalanges are distal to the carpals.
Superficial (soo′-per-FISH-al)	Toward or on the surface of the body.	The ribs are superficial to the lungs.
Deep (DĒP)	Away from the surface of the body.	The ribs are deep to the skin of the chest and back.

*Ventral refers to the belly side, whereas dorsal refers to the back side. In four-legged animals, anterior = cephalic (toward the head), ventral = inferior, posterior = caudal (toward the tail), and dorsal = superior.

†The midline is an imaginary vertical line that divides the body into equal right and left sides.

Figure 1.6 Directional terms.

 Directional terms precisely locate various parts of the body relative to one another.

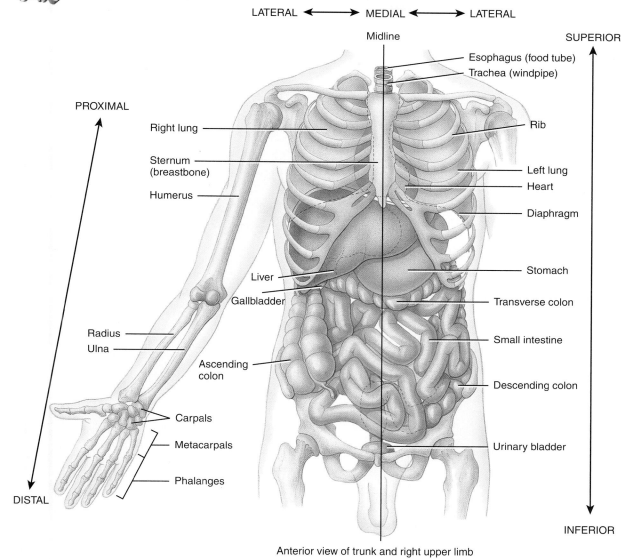

Anterior view of trunk and right upper limb

Is the radius proximal to the humerus? Is the esophagus anterior to the trachea? Are the ribs superficial to the lungs? Is the urinary bladder medial to the ascending colon? Is the sternum lateral to the descending colon?

Figure 1.7 **Planes through the human body.**

Frontal, transverse, sagittal, and oblique planes divide the body in specific ways.

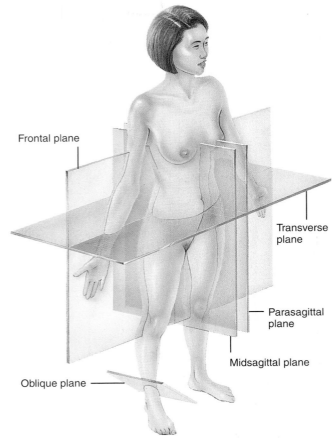

Frontal plane

Transverse plane

Parasagittal plane

Midsagittal plane

Oblique plane

Right anterolateral view

Which plane divides the heart into anterior and posterior portions?

Figure 1.8 **Planes and sections through different parts of the brain.** The diagrams (left) show the planes, and the photographs (right) show the resulting sections. Note: The arrows in the diagrams indicate the direction from which each section is viewed. This aid is used throughout the book to indicate viewing perspectives.

Planes divide the body in various ways to produce sections.

(a) Transverse plane

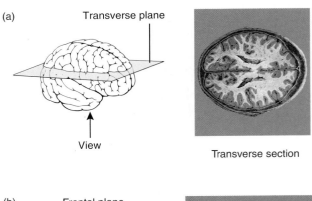

View

Transverse section

(b) Frontal plane

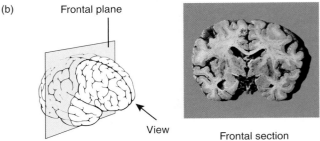

View

Frontal section

(c) Midsagittal plane

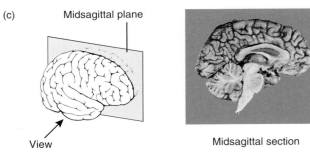

View

Midsagittal section

Which plane divides the brain into unequal right and left portions?

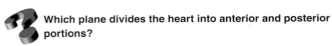

separate the various body cavities from one another. The two major cavities are the dorsal and ventral body cavities (Figure 1.9).

Dorsal Body Cavity

The **dorsal body cavity** is located near the dorsal (posterior) surface of the body and has two subdivisions, the cranial cavity and the vertebral canal. The cranial bones form the **cranial cavity,** which contains the brain. The bones of the vertebral column (backbone) form the **vertebral (spinal) canal,** which contains the spinal cord. Three layers of protective tissue, the **meninges** (me-NIN-jēz), line the dorsal body cavity.

Ventral Body Cavity

The other major body cavity—the **ventral body cavity**—is located on the ventral (anterior) aspect of the body. The ventral

body cavity also has two main subdivisions, the superior thoracic and the inferior abdominopelvic cavities. The **diaphragm** (DĪ-a-fram = partition or wall) is a dome-shaped muscle that separates the thoracic cavity from the abdominopelvic cavity. The organs inside the ventral body cavity are termed the **viscera** (VIS-er-a).

The superior part of the ventral body cavity is the **thoracic cavity** (thor-AS-ik; *thorac-* = chest) or chest cavity (Figure 1.10). The ribs, the muscles of the chest, the sternum (breastbone), and the thoracic portion of the vertebral column (backbone) encircle

Figure 1.9   Body cavities.   The dashed lines in (a) and (b) indicate the border between the abdominal and pelvic cavities. (See Tortora, *A Photographic Atlas of the Human Body,* Figures 6.5, 6.11, and 11.12.)

 The two major body cavities are the dorsal and ventral cavities.

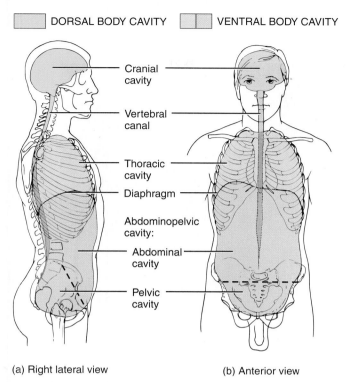

DORSAL BODY CAVITY   VENTRAL BODY CAVITY

Cranial cavity

Vertebral canal

Thoracic cavity

Diaphragm

Abdominopelvic cavity:

Abdominal cavity

Pelvic cavity

(a) Right lateral view   (b) Anterior view

In which cavities are the following organs located: urinary bladder, stomach, heart, small intestine, lungs, internal female reproductive organs, thymus, spleen, liver? Use the following symbols for your response: T = thoracic cavity, A = abdominal cavity, or P = pelvic cavity.

Figure 1.10   The thoracic cavity.   The dashed lines indicate the borders of the mediastinum. Note: When transverse sections are viewed inferiorly (from below), the anterior aspect of the body appears on top and the left side of the body appears on the right side of the illustration. (See Tortora, *A Photographic Atlas of the Human Body,* Figure 6.6.)

The thoracic cavity is the superior part of the ventral body cavity.

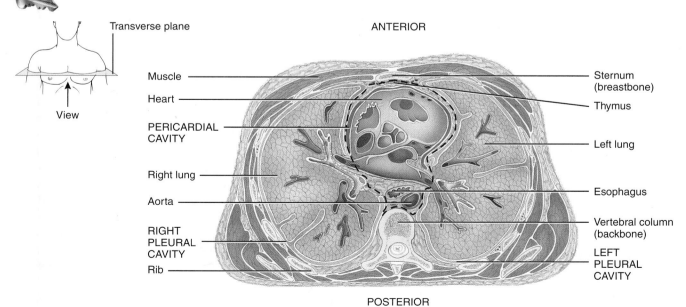

Transverse plane

View

ANTERIOR

Muscle

Heart

PERICARDIAL CAVITY

Right lung

Aorta

RIGHT PLEURAL CAVITY

Rib

Sternum (breastbone)

Thymus

Left lung

Esophagus

Vertebral column (backbone)

LEFT PLEURAL CAVITY

POSTERIOR

Inferior view of transverse section of thoracic cavity

What is the name of the cavity that surrounds the heart? Which cavities surround the lungs?

Figure 1.11 **The mediastinum.** The dashed lines indicate the borders of the mediastinum.

 The mediastinum is medial to the lungs. It extends from the sternum to the vertebral column and from the neck to the diaphragm.

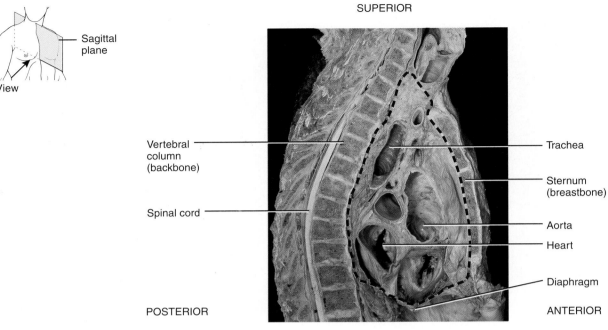

Sagittal section of thoracic cavity

Which of the following structures are contained in the mediastinum: right lung, heart, esophagus, spinal cord, trachea, rib, thymus, left pleural cavity?

the thoracic cavity. The **pericardial cavity** (per′-i-KAR-dē-al; *peri-* = around; *-cardial* = heart), a fluid-filled space that surrounds the heart, and two **pleural cavities** (PLOOR-al; *pleur-* = rib or side) lie within the thoracic cavity. Each pleural cavity surrounds one lung and contains a small amount of fluid. The central part of the thoracic cavity is called the **mediastinum** (mē′-dē-as-TĪ-num; *media-* = middle; *-stinum* = partition). It is between the lungs, extending from the sternum to the vertebral column and from the neck to the diaphragm (Figure 1.11). The mediastinum contains all thoracic viscera except the lungs themselves. Among the structures in the mediastinum are the heart, esophagus, trachea, thymus, and several large blood vessels.

The inferior part of the ventral body cavity is the **abdominopelvic cavity** (ab-dom′-i-nō-PEL-vik; see Figure 1.8), which extends from the diaphragm to the groin and is encircled by the abdominal wall and the bones and muscles of the pelvis. As the name suggests, the abdominopelvic cavity is divided into two portions, even though no wall separates them (Figure 1.12). The superior portion, the **abdominal cavity** (*abdomin-* = belly), contains the stomach, spleen, liver, gallbladder, small intestine, and most of the large intestine. The inferior portion, the **pelvic**

cavity (*pelv-* = basin), contains the urinary bladder, portions of the large intestine, and internal organs of the reproductive system.

Thoracic and Abdominal Cavity Membranes

A thin, slippery **serous membrane** covers the viscera within the thoracic and abdominal cavities and also lines the walls of the thorax and abdomen. The parts of a serous membrane are (1) the *parietal layer* (pa-RĪ-e-tal), which lines the walls of the cavities, and (2) the *visceral layer,* which covers and adheres to the viscera within the cavities. Serous fluid between the two layers reduces friction, allowing the viscera to slide somewhat during movements, for example, when the lungs inflate and deflate during breathing.

The serous membrane of the pleural cavities is called the **pleura** (PLOO-ra). The *visceral pleura* clings to the surface of the lungs, whereas the *parietal pleura* lines the chest wall. In between is the pleural cavity, filled with a small volume of serous fluid. The serous membrane of the pericardial cavity is the **pericardium.** The *visceral pericardium* covers the surface of

Figure 1.12 **The abdominopelvic cavity.** The dashed line shows the approximate boundary between the abdominal and pelvic cavities. (See Tortora, *A Photographic Atlas of the Human Body,* Figure 12.2.)

🔑 **The abdominopelvic cavity extends from the diaphragm to the groin.**

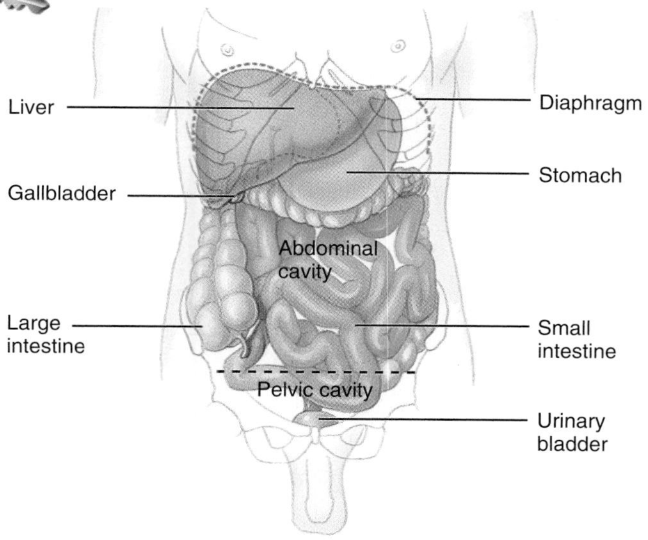

Liver

Gallbladder

Abdominal cavity

Large intestine

Pelvic cavity

Diaphragm

Stomach

Small intestine

Urinary bladder

Anterior view

❓ **To which body systems do the organs shown here within the abdominal and pelvic cavities belong? (*Hint: Refer to Table 1.2.*)**

the heart, whereas the *parietal pericardium* lines the chest wall. Between them is the pericardial cavity. The **peritoneum** (per-i-tō-NĒ-um) is the serous membrane of the abdominal cavity. The *visceral peritoneum* covers the abdominal viscera, whereas the *parietal peritoneum* lines the abdominal wall. Between them is the *peritoneal cavity.* Most abdominal organs are located in the peritoneal cavity. Some are located behind the parietal peritoneum between it and the posterior abdominal wall. Such organs are said to be *retroperitoneal* (re′-trō-per-i-tō-NĒ-al; *retro* = behind). The kidneys, adrenal glands, pancreas, duodenum of the small intestine, ascending and descending colons of the large intestine, and portions of the abdominal aorta and inferior vena cava are retroperitoneal.

Table 1.3 presents a summary of body cavities and their membranes.

Abdominopelvic Regions and Quadrants

To describe the location of the many abdominal and pelvic organs more easily, anatomists and clinicians use two methods of dividing the abdominopelvic cavity into smaller compartments. In the first method, two horizontal and two vertical lines, aligned like a tick-tack-toe grid, partition this cavity into nine **abdominopelvic regions** (Figure 1.13a). The top horizontal line, the *subcostal line,* is drawn just inferior to the rib cage, across the inferior portion of the stomach; the bottom horizontal line, the *transtubercular line,* is drawn just inferior to the tops of the hip bones. Two vertical lines, the left and right *midclavicular lines,* are drawn through the midpoints of the clavicles (collar bones), just medial to the nipples. The four lines divide the abdominopelvic cavity into a larger middle section and smaller left and right sections. The names of the nine abdominopelvic regions are right hypochondriac, epigastric, left hypochondriac, right lumbar, umbilical, left lumbar, right inguinal (iliac), hypogastric (pubic), and left inguinal (iliac).

Table 1.3	Summary of Body Cavities and Their Membranes
Cavity	**Comments**
Dorsal cavity	
Cranial cavity	Formed by cranial bones and contains brain.
Vertebral cavity	Formed by vertebral column and contains spinal cord.
Ventral cavity	
Thoracic cavity	Superior portion of ventral body cavity; contains pleural and pericardial cavities and the mediastinum.
Pleural cavity	Each surrounds a lung; the serous membrane of the pleural cavities is the pleura.
Pericardial cavity	Surrounds the heart; the serous membrane of the pericardial cavity is the pericardium.
Mediastinum	Central portion of thoracic cavity between the lungs; extends from sternum to vertebral column and from neck to diaphragm; contains heart, thymus, esophagus, trachea, and several large blood vessels.
Abdominopelvic cavity	Inferior portion of ventral body cavity; subdivided into abdominal and pelvic cavities.
Abdominal cavity	Contains stomach, spleen, liver, gallbladder, small intestine, and most of large intestine; the serous membrane of the abdominal cavity is the peritoneum.
Pelvic cavity	Contains urinary bladder, portions of large intestine, and internal organs of reproduction.

Figure 1.13 **Regions and quadrants of the abdominopelvic cavity.**

The nine-region designation is used for anatomical studies, whereas the quadrant designation is used to locate the site of pain, tumor, or some other abnormality.

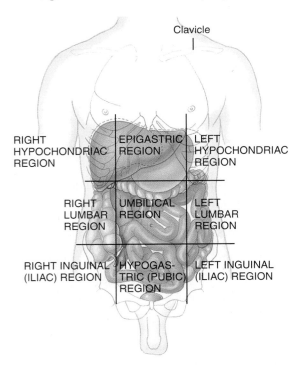

(a) Anterior view showing abdominopelvic regions

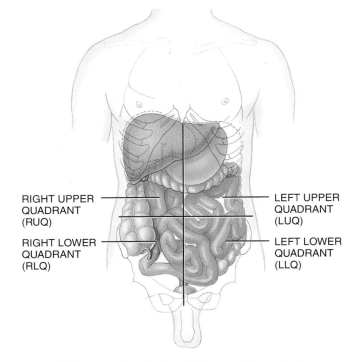

(b) Anterior view showing abdominopelvic quadrants

 In which abdominopelvic region is each of the following found: most of the liver, transverse colon, urinary bladder, spleen? In which abdominopelvic quadrant would pain from appendicitis (inflammation of the appendix) be felt?

The second method is simpler and divides the abdominopelvic cavity into **quadrants** (KWOD-rantz; *quad-* = onefourth), as shown in Figure 1.13b. In this method, a vertical line and a horizontal line are passed through the **umbilicus** (um-bi-LĪ-kus; *umbilic-* = navel) or *belly button*. The names of the abdominopelvic quadrants are right upper quadrant (RUQ), left upper quadrant (LUQ), right lower quadrant (RLQ), and left lower quadrant (LLQ). Whereas the nine-region division is more widely used for anatomical studies, quadrants are more commonly used by clinicians for describing the site of abdominopelvic pain, tumor, or other abnormality.

▶ C H E C K P O I N T

12. Locate each region shown in Figure 1.5 on your own body, and then identify it by its common name and the corresponding anatomical descriptive form.

13. What structures separate the various body cavities from one another?

14. Locate the nine abdominopelvic regions and the four abdominopelvic quadrants on yourself, and list some of the organs found in each.

MEDICAL IMAGING

▶ O B J E C T I V E

• Describe the principles and importance of medical imaging procedures in the evaluation of organ functions and the diagnosis of disease.

Various kinds of **medical imaging** procedures allow visualization of structures inside our bodies and are increasingly helpful for precise diagnosis of a wide range of anatomical and physiological disorders. The grandparent of all medical imaging techniques is conventional radiography, in medical use since the late 1940s. The newer imaging technologies not only contribute to diagnosis of disease, but they also are advancing our understanding of normal physiology. Table 1.4 describes some commonly used medical imaging techniques. Other imaging methods, such as cardiac catheterization, will be discussed in later chapters.

▶ C H E C K P O I N T

15. Which forms of medical imaging use radiation? Which do not?

16. Which imaging modality best reveals the physiology of a structure?

Table 1.4 Common Medical Imaging Procedures

Radiography

Procedure: A single barrage of x rays passes through the body, producing an image of interior structures on x-ray–sensitive film. The resulting two-dimensional image is a *radiograph* (RĀ-dē-ō-graf′), commonly called an *x ray.*

Comments: Produces clear images of bony or dense structures, which appear bright, but poor images of soft tissues or organs, which appear hazy or dark.

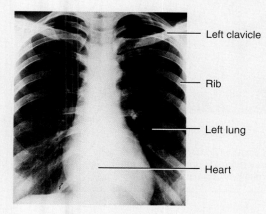

Anterior view of thorax

Computed tomography (CT) [formerly called computerized axial tomography (CAT) scanning]

Procedure: Computer-assisted radiography in which an x-ray beam traces an arc at multiple angles around a section of the body. The resulting transverse section of the body, called a *CT scan,* is reproduced on a video monitor.

Comments: Visualizes soft tissues and organs with much more detail than conventional radiographs. Differing tissue densities show up as various shades of gray. Multiple scans can be assembled to build three-dimensional views of structures.

ANTERIOR

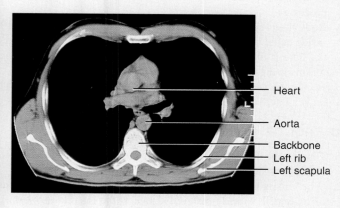

POSTERIOR

Inferior view of transverse section of the thorax

Magnetic resonance imaging (MRI)

Procedure: The body is exposed to a high-energy magnetic field, which causes protons (small positive particles within atoms, such as hydrogen) in body fluids and tissues to arrange themselves in relation to the field. Then a pulse of radiowaves "reads" these ion patterns, and a color-coded image is assembled on a video monitor. The resulting image is a two- or three-dimensional blueprint of cellular chemistry.

Comments: Relatively safe, but can't be used on patients with metal in their bodies. Shows fine details for soft tissues but not for bones. Most useful for differentiating between normal and abnormal tissues. Used to detect tumors and artery-clogging fatty plaques, reveal brain abnormalities, measure blood flow, and detect a variety of musculoskeletal, liver, and kidney disorders.

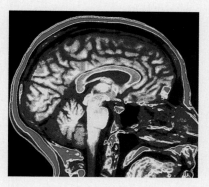

Sagittal section of brain

Sonography

Procedure: High-frequency sound waves produced by a handheld wand reflect off body tissues and are detected by the same instrument. The image, which may be still or moving, is called a *sonogram* (SON-ō-gram) and is reproduced on a video monitor.

Comments: Safe, noninvasive, painless, and uses no dyes. Most commonly used to visualize the fetus during pregnancy. Also used to observe the size, location, and actions of organs and blood flow through blood vessels.

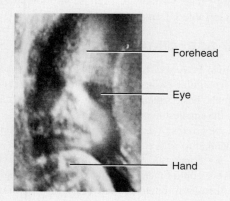

Courtesy of Andrew Joseph Tortora and Damaris Soler

(continues)

Table 1.4   Common Medical Imaging Procedures *(continued)*

Positron emission tomography (PET)

Procedure: A substance that emits positrons (positively charged particles) is injected into the body, where it is taken up by tissues. The collision of positrons with negatively charged electrons in body tissues produces gamma rays (similar to x rays) that are detected by gamma cameras positioned around the subject. A computer receives signals from the gamma cameras and constructs a *PET scan* image, displayed in color on a video monitor. The PET scan shows where the injected substance is being used in the body. In the PET scan image shown here, the black and blue colors indicate minimal activity, whereas the red, orange, yellow, and white colors indicate areas of increasingly greater activity.

Comments: Used to study the physiology of body structures, such as metabolism in the brain or heart.

ANTERIOR

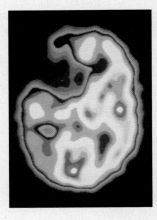

POSTERIOR

Transverse section showing blood flow through brain (darkened area at upper left indicates where a stroke has occurred)

STUDY OUTLINE

ANATOMY AND PHYSIOLOGY DEFINED (p. 2)

1. Anatomy is the science of body structures and the relationships among structures; physiology is the science of body functions.
2. Dissection is the careful cutting apart of body structures to study their relationships.
3. Some subdisciplines of anatomy are embryology, developmental biology, histology, surface anatomy, gross anatomy, systemic anatomy, regional anatomy, radiographic anatomy, and pathological anatomy (see Table 1.1 on page 2).
4. Some subdisciplines of physiology are neurophysiology, endocrinology, cardiovascular physiology, immunology, respiratory physiology, renal physiology, exercise physiology, and pathophysiology (see Table 1.1 on page 2).

LEVELS OF BODY ORGANIZATION (p. 3)

1. The human body consists of six levels of structural organization: chemical, cellular, tissue, organ, system, and organismal levels.
2. Cells are the basic structural and functional living units of an organism and the smallest living units in the human body.
3. Tissues are groups of cells and the materials surrounding them that work together to perform a particular function.
4. Organs are composed of two or more different types of tissues; they have specific functions and usually have recognizable shapes.
5. Systems consist of related organs that have a common function.
6. An organism is any living individual.
7. Table 1.2 on page 16 introduces the eleven systems of the human organism: the integumentary, skeletal, muscular, nervous, endocrine, cardiovascular, lymphatic and immune, respiratory, digestive, urinary, and reproductive systems.

CHARACTERISTICS OF THE LIVING HUMAN ORGANISM (p. 7)

1. All organisms carry on certain processes that distinguish them from nonliving things.
2. The most important life processes of the human body are metabolism, responsiveness, movement, growth, differentiation, and reproduction.
3. Homeostasis is a condition of equilibrium in the body's internal environment produced by the interplay of all the body's regulatory processes.
4. Body fluids are dilute, watery solutions. Intracellular fluid (ICF) is inside cells, and extracellular fluid (ECF) is outside cells. Interstitial fluid is the ECF that fills spaces between tissue cells, whereas plasma is the ECF within blood vessels.
5. Because it surrounds all body cells, interstitial fluid is called the body's internal environment.

CONTROL OF HOMEOSTASIS (p. 8)

1. Disruptions of homeostasis come from external and internal stimuli and psychological stresses.
2. When disruption of homeostasis is mild and temporary, responses of body cells quickly restore balance in the internal environment. If disruption is extreme, regulation of homeostasis may fail.
3. Most often, the nervous and endocrine systems acting together or separately regulate homeostasis. The nervous system detects body changes and sends nerve impulses to counteract changes in controlled conditions. The endocrine system regulates homeostasis by secreting hormones.
4. Feedback systems include three components. (1) Receptors moni-

tor changes in a controlled condition and send input to a control center. (2) The control center sets the value at which a controlled condition should be maintained, evaluates the input it receives from receptors, and generates output commands when they are needed. (3) Effectors receive output from the control center and produce a response (effect) that alters the controlled condition.

5. If a response reverses the original stimulus, the system is operating by negative feedback. If a response enhances the original stimulus, the system is operating by positive feedback.

6. One example of negative feedback is a system that helps regulate blood pressure. If a stimulus causes blood pressure (controlled condition) to rise, baroreceptors (pressure-sensitive nerve cells, the receptors) in blood vessels send impulses (input) to the brain (control center). The brain sends impulses (output) to the heart (effector). As a result, heart rate decreases (response) and blood pressure decreases to normal (restoration of homeostasis).

7. One example of positive feedback occurs during the birth of a baby. When labor begins, the cervix of the uterus is stretched (stimulus), and stretch-sensitive nerve cells in the cervix (receptors) send nerve impulses (input) to the brain (control center). The brain responds by releasing oxytocin (output), which stimulates the uterus (effector) to contract more forcefully (response). Movement of the baby further stretches the cervix, more oxytocin is released, and even more forceful contractions occur. The cycle is broken with the birth of the baby.

8. Disruptions of homeostasis—homeostatic imbalances—can lead to disorders, diseases, and even death.

9. Disorder is a general term for any derangement or abnormality of function. A disease is an illness with a definite set of signs and symptoms.

10. Symptoms are subjective changes in body functions that are not apparent to an observer, whereas signs are objective changes that can be observed and measured.

AGING AND HOMEOSTASIS (p. 11)

1. Aging produces observable changes in structure and function and increases vulnerability to stress and disease.

2. Changes associated with aging occur in all body systems.

ANATOMICAL TERMINOLOGY (p. 12)

1. Descriptions of any region of the body assume the body is in the anatomical position, in which the subject stands erect facing the observer, with the head level and the eyes facing directly forward. The feet are flat on the floor and directed forward, and the arms are at the sides, with the palms turned forward.

2. A body lying face down is prone, whereas a body lying face up is supine.

3. Regional names are terms given to specific regions of the body. The principal regions are the head, neck, trunk, upper limbs, and lower limbs.

4. Within the regions, specific body parts have common names and are specified by corresponding anatomical terms. Examples are chest (thoracic), nose (nasal), and wrist (carpal).

5. Directional terms indicate the relationship of one part of the body to another. Exhibit 1.1 on page 14 summarizes commonly used directional terms.

6. Planes are imaginary flat surfaces that are used to divide the body or organs to visualize interior structures. A midsagittal plane

divides the body or an organ into equal right and left sides. A parasagittal plane divides the body or an organ into unequal right and left sides. A frontal plane divides the body or an organ into anterior and posterior portions. A transverse plane divides the body or an organ into superior and inferior portions. An oblique plane passes through the body or an organ at an angle between a transverse plane and either a midsagittal, parasagittal, or frontal plane.

7. Sections are flat surfaces resulting from cuts through body structures. They are named according to the plane on which the cut is made and include transverse, frontal, and sagittal sections.

8. Body cavities are spaces in the body that help protect, separate, and support internal organs.

9. The dorsal and ventral cavities are the two major body cavities.

10. The dorsal cavity is subdivided into the cranial cavity, which contains the brain, and the vertebral canal, which contains the spinal cord. The meninges are protective tissues that line the dorsal cavity.

11. The diaphragm subdivides the ventral body cavity into a superior thoracic cavity and an inferior abdominopelvic cavity. The viscera are organs within the ventral body cavity. A serous membrane lines the wall of the cavity and adheres to the viscera.

12. The thoracic cavity is subdivided into three smaller cavities: a pericardial cavity, which contains the heart, and two pleural cavities, each of which contains a lung.

13. The central part of the thoracic cavity is the mediastinum. It is located between the pleural cavities, extending from the sternum to the vertebral column and from the neck to the diaphragm. It contains all thoracic viscera except the lungs.

14. The abdominopelvic cavity is divided into a superior abdominal and an inferior pelvic cavity.

15. Viscera of the abdominal cavity include the stomach, spleen, liver, gallbladder, small intestine, and most of the large intestine.

16. Viscera of the pelvic cavity include the urinary bladder, portions of the large intestine, and internal organs of the reproductive system.

17. Serous membranes line the walls of the thoracic and abdominal cavities and cover the organs within them. They include the pleura, associated with the lungs; the pericardium, associated with the heart; and the peritoneum, associated with the abdominal cavity.

18. Table 1.3 on page 19 summarizes body cavities and their membranes.

19. To describe the location of organs more easily, the abdominopelvic cavity is divided into abdominopelvic regions: right hypochondriac, epigastric, left hypochondriac, right lumbar, umbilical, left lumbar, right inguinal (iliac), hypogastric (pubic), and left inguinal (iliac).

20. To locate the site of an abdominopelvic abnormality in clinical studies, the abdominopelvic cavity is divided into quadrants: right upper quadrant (RUQ), left upper quadrant (LUQ), right lower quadrant (RLQ), and left lower quadrant (LLQ).

MEDICAL IMAGING (p. 20)

1. Medical imaging techniques allow visualization of internal structures to diagnose abnormal anatomy and deviations from normal physiology.

2. Table 1.4 on pages 21–22 summarizes several medical imaging techniques.

 **SELF-QUIZ QUESTIONS**

Fill in the blanks in the following statements.

1. The basic structural and functional unit of the human organism is the _____.

2. The phase of metabolism that involves breaking down large, complex molecules into smaller, simpler ones is called _____.

3. The condition in which the body's internal environment remains within certain physiological limits is termed _____.

Indicate whether the following statements are true or false.

4. In a negative feedback system, the response enhances or intensifies the original stimulus.

5. The ventral body cavity contains the heart, lungs, and abdominal viscera.

6. The parietal layer is the part of a serous membrane that lines the walls of the thoracic and abdominal cavities.

Choose the one best answer to the following questions.

7. Which of the following statements are *true* of a feedback system? (1) A feedback system consists of a control center, a receptor, and an effector. (2) The control center receives input from the effector. (3) The receptor monitors environmental changes that affect the body. (4) The effector produces a response to return the controlled condition to normal. (5) A feedback system involves a cycle of events in which information about the status of a condition is monitored and fed back to a central control region. (a) 1 and 2, (b) 1, 2, and 3, (c) 1, 3, and 5, (d) 1, 3, 4, and 5, (e) 2 and 4.

8. Which of the following statements are *true* of a serous membrane? (1) It lines body cavities that do not open directly to the exterior. (2) It lines body cavities that open directly to the exterior. (3) It is a double-layered membrane. (4) Examples include the pleura, pericardium, and peritoneum. (5) It lines the cavity only and does not cover the organs within the cavity. (a) 1, 3, 4, and 5, (b) 1, 3, and 4, (c) 1, 3 and 5, (d) 1 and 3, (e) 1 and 5.

9. A vertical plane that divides the body or an organ into right and left sides is termed a (an) (a) frontal plane, (b) sagittal plane, (c) transverse plane, (d) oblique plane, (e) coronal plane.

10. The two systems that regulate most homeostatic responses of the body are the (a) nervous and cardiovascular systems, (b) respiratory and cardiovascular systems, (c) endocrine and cardiovascular systems, (d) cardiovascular and urinary systems, (e) endocrine and nervous systems.

11. The fluid found within body cells is (a) blood plasma, (b) extracellular fluid, (c) interstitial fluid, (d) cerebrospinal fluid, (e) intracellular fluid.

12. Match the following life processes to their definitions:
_____ (a) responsiveness
_____ (b) reproduction
_____ (c) growth
_____ (d) metabolism
_____ (e) differentiation
_____ (f) movement

(1) development of a cell from an unspecialized to specialized state
(2) motion of body or body parts
(3) ability to detect and respond to change
(4) formation of new cells or a new individual
(5) sum of all of the body's chemical processes
(6) increase in size

13. Match the following:
_____ (a) nervous system
_____ (b) endocrine system
_____ (c) urinary system
_____ (d) cardiovascular system
_____ (e) muscular system
_____ (f) respiratory system
_____ (g) digestive system
_____ (h) skeletal system
_____ (i) integumentary system
_____ (j) lymphatic & immune system
_____ (k) reproductive system

(1) regulates body activities through chemicals transported in blood to various target organs of the body
(2) produces gametes; releases hormones from gonads
(3) protects against disease; returns fluids to blood
(4) protects body by forming a barrier between the outside environment and internal organs
(5) transports oxygen and nutrients to cells; protects against disease; carries wastes away from cells
(6) regulates body activities through action potentials; receives sensory information, interprets the information, and responds to the information
(7) carries out the physical and chemical breakdown of food and absorption of nutrients
(8) transfers oxygen and carbon dioxide between air and blood
(9) supports and protects the body; provides internal framework
(10) powers movements of the body and stabilizes body position
(11) eliminates wastes; regulates the volume and chemical composition of blood

14. Match the following common names and anatomical descriptive adjectives:
_____ (a) axillary
_____ (b) inguinal
_____ (c) cervical
_____ (d) cranial
_____ (e) brachial
_____ (f) orbital
_____ (g) gluteal
_____ (h) buccal

(1) skull
(2) eye
(3) cheek
(4) armpit
(5) arm
(6) groin
(7) buttock
(8) neck

15. Match each of the following directional terms to its definition:
_____ (a) at the front of the body
_____ (b) closer to the trunk
_____ (c) toward the upper part of a structure
_____ (d) nearer to the midline of the body
_____ (e) farther from the midline of the body
_____ (f) at the back of the body
_____ (g) farther from the trunk
_____ (h) toward the lower part of a structure

(1) superior
(2) inferior
(3) anterior (ventral)
(4) posterior (dorsal)
(5) medial
(6) lateral
(7) proximal
(8) distal

CRITICAL THINKING QUESTIONS

1. On her first anatomy and physiology exam, Heather defined homeostasis as "the condition in which the body approaches room temperature and stays there." Do you agree with Heather's definition?
   HINT *Does an oral thermometer measure down to 72°F (25°C)?*

2. Elena complained of numbness and tingling sensations in both of her hands. Her physician recommended bilateral splints for carpal tunnel syndrome. Where would she wear these splints?
   HINT *The splints somewhat restrict movement of the hand.*

3. Eighty-year-old Harold fell off a ladder and may have broken his arm. The emergency room physician ordered some x rays, but Harold said that he "wants one of those fancy, new x rays—an MRI. And while you're taking my picture," he continued, "check out that pacemaker near my heart, too." The physician refused to order an MRI. Why?
   HINT *Pacemakers contain metal.*

ANSWERS TO FIGURE QUESTIONS

1.1 Organs have two or more different types of tissues that work together to perform a specific function.

1.2 The difference between negative and positive feedback systems is that in negative feedback systems, the response reverses the original stimulus, whereas in positive feedback systems, the response enhances the original stimulus.

1.3 When something causes blood pressure to decrease, then heart rate increases due to operation of this negative feedback system.

1.4 Because positive feedback systems continually intensify or reinforce the original stimulus, a termination mechanism is needed to end the response.

1.5 Having one standard anatomical position allows directional terms to be clear, and any body part can be described in relation to any other part.

1.6 No, No, Yes, Yes, No.

1.7 The frontal plane divides the heart into anterior and posterior portions.

1.8 The parasagittal plane divides the brain into unequal right and left portions.

1.9 P, A, T, A, T, P, T, A, A.

1.10 The pericardial cavity surrounds the heart, and the pleural cavities surround the lungs.

1.11 Structures in the mediastinum are the heart, esophagus, trachea, thymus, and several large blood vessels.

1.12 The illustrated abdominal cavity organs all belong to the digestive system (liver, gallbladder, stomach, appendix, small intestine, and most of the large intestine). Illustrated pelvic cavity organs belong to the urinary system (the urinary bladder) and the digestive system (part of the large intestine).

1.13 The liver is mostly in the epigastric region; the transverse colon is in the umbilical region; the urinary bladder is in the hypogastric region; the spleen is in the left hypochondriac region. The pain associated with appendicitis would be felt in the right lower quadrant (RLQ).

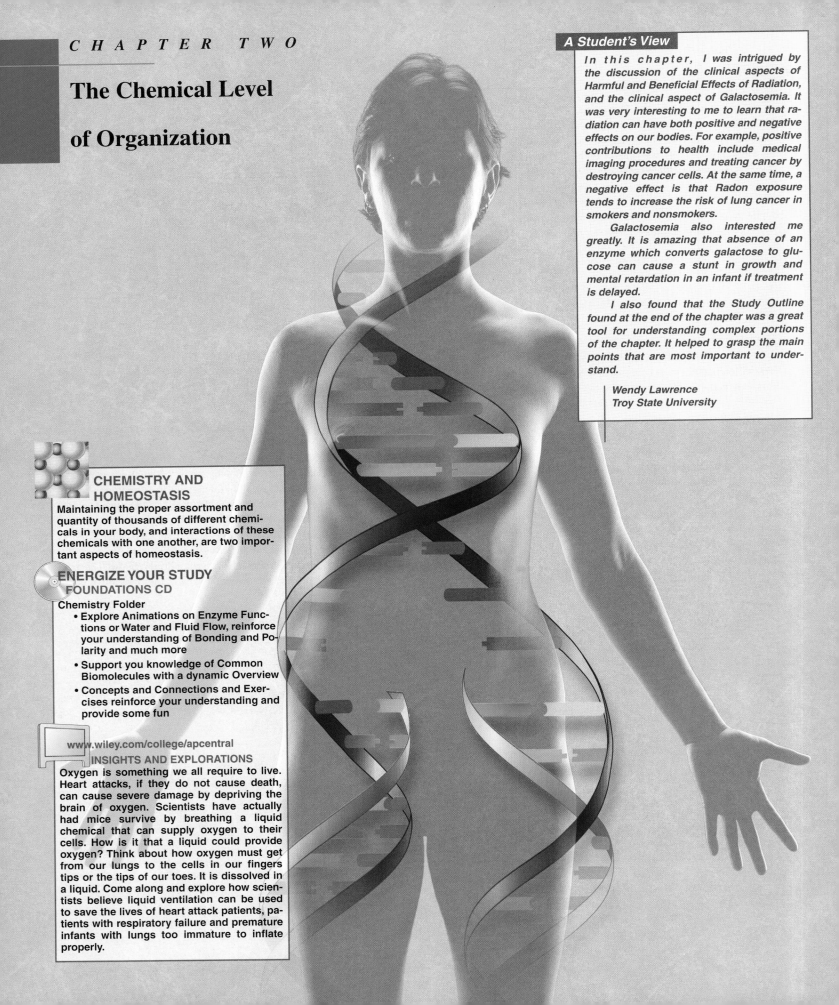

The Chemical Level of Organization

CHEMISTRY AND HOMEOSTASIS

Maintaining the proper assortment and quantity of thousands of different chemicals in your body, and interactions of these chemicals with one another, are two important aspects of homeostasis.

ENERGIZE YOUR STUDY
FOUNDATIONS CD

Chemistry Folder

- **Explore Animations on Enzyme Functions or Water and Fluid Flow, reinforce your understanding of Bonding and Polarity and much more**
- **Support you knowledge of Common Biomolecules with a dynamic Overview**
- **Concepts and Connections and Exercises reinforce your understanding and provide some fun**

www.wiley.com/college/apcentral

INSIGHTS AND EXPLORATIONS

Oxygen is something we all require to live. Heart attacks, if they do not cause death, can cause severe damage by depriving the brain of oxygen. Scientists have actually had mice survive by breathing a liquid chemical that can supply oxygen to their cells. How is it that a liquid could provide oxygen? Think about how oxygen must get from our lungs to the cells in our fingers tips or the tips of our toes. It is dissolved in a liquid. Come along and explore how scientists believe liquid ventilation can be used to save the lives of heart attack patients, patients with respiratory failure and premature infants with lungs too immature to inflate properly.

ou learned in Chapter 1 that the chemical level of organization, the lowest level of structural organization, consists of atoms and molecules. These small particles combine to form body structures and systems of astonishing size and complexity. In this chapter, we consider how atoms bond together to form molecules and how atoms and molecules release or store energy in processes known as chemical reactions. Water, which accounts for nearly two-thirds of body weight, is vitally important for chemical reactions and homeostasis in the body. Finally, you will learn about five families of molecules whose unique proper-

ties contribute to assembly of the body's structures or to powering the processes that characterize life.

Chemistry (KEM-is-trē) is the science of the structure and interactions of matter. All living and nonliving things consist of **matter,** which is anything that occupies space and has mass. The amount of matter in any object is its **mass.** On Earth, *weight* is the force of gravity acting on matter. Things weigh less when they are farther from Earth because the pull of gravity is weaker. In outer space, weight is close to zero but mass remains the same as it is on Earth.

HOW MATTER IS ORGANIZED

OBJECTIVES

- **Identify the main chemical elments of the human body.**
- **Describe the structures of atoms, ions, molecules, free radicals, and compounds.**

Chemical Elements

All forms of matter—both living and nonliving—are made up of a limited number of building blocks called **chemical elements.** Each element is a substance that cannot be split into a simpler substance by ordinary chemical means. Scientists now recognize 112 elements. Of these, 92 occur naturally on Earth. The rest have been produced from the natural elements using particle accelerators or nuclear reactors. Each element is designated by a **chemical symbol,** one or two letters of the element's name in English, Latin, or another language. Examples of chemical symbols are H for hydrogen, C for carbon, O for oxygen, N for nitrogen, Ca for calcium, and Na for sodium (*natrium* = sodium).*

Twenty-six different elements normally are present in your body. Just four elements—oxygen, carbon, hydrogen, and nitrogen—constitute about 96% of the body's mass. Eight others—calcium, phosphorus (P), potassium (K), sulfur (S), sodium, chlorine (Cl), magnesium (Mg), and iron (Fe)—contribute 3.8% of the body's mass. An additional 14 elements—the *trace elements*—are present in tiny amounts. Together, they account for the remaining 0.2% of the body's mass. Several trace elements have important functions in the body. For example, iodine is needed to make thyroid hormones. The functions of some trace elements are unknown. Table 2.1 lists the main chemical elements that compose the human body.

---

*The periodic table of elements, which lists all of the known chemical elements, can be found in Appendix B.

Structure of Atoms

Each element is made up of **atoms,** the smallest units of matter that retain the properties and characteristics of an element. Atoms are extremely small. Two hundred thousand of the largest atoms would fit on the period at the end of this sentence. Hydrogen atoms, the smallest atoms, have a diameter less than 0.1 nanometer (0.1×10^{-9} m = 0.0000000001 m), and the largest atoms are only five times larger.

Dozens of different **subatomic particles** compose individual atoms. However, only three types of subatomic particles are important for understanding chemical reactions: protons, neutrons, and electrons (Figure 2.1). The dense central core of an atom is its **nucleus.** Within the nucleus are positively charged **protons (p^+)** and uncharged (neutral) **neutrons (n^0).** The tiny negatively charged **electrons (e^-)** move about in a large space surrounding the nucleus. They do not follow a fixed path or orbit but instead form a negatively charged "cloud" that envelopes the nucleus (Figure 2.1a).

Even though their exact positions cannot be predicted, specific groups of electrons are most likely to move about within certain regions around the nucleus. These regions are called **electron shells.** We depict the electron shells as simple circles around the nucleus even though some of their shapes are not spherical. Because each electron shell can hold a specific number of electrons, the electron shell model best conveys this aspect of atomic structure (Figure 2.1b). For instance, the first electron shell (nearest the nucleus) never holds more than 2 electrons. The second shell holds a maximum of 8 electrons, whereas the third can hold up to 18 electrons. The electron shells fill with electrons in a specific order, beginning with the first shell. For example, notice in Figure 2.2 that sodium (Na), which has 11 electrons total, contains 2 electrons in the first shell, 8 electrons in the second shell, and 1 electron in the third shell. The most massive element present in the human body is iodine, which has a total of 53 electrons: 2 in the first shell, 8 in the second shell, 18 in the third shell, 18 in the fourth shell, and 7 in the fifth shell.

Table 2.1 Main Chemical Elements in the Body

Chemical Element (Symbol)	% Of Total Body Mass	Significance
Oxygen (O)	65.0	Part of water and many organic (carbon-containing molecules); used to generate ATP, a molecule used by cells to temporarily store chemical energy.
Carbon (C)	18.5	Forms backbone chains and rings of all organic molecules: carbohydrates, lipids (fats), proteins, and nucleic acids (DNA and RNA).
Hydrogen (H)	9.5	Constituent of water and most organic molecules; ionized form (H^+) makes body fluids more acidic.
Nitrogen (N)	3.2	Component of all proteins and nucleic acids.
Calcium (Ca)	1.5	Contributes to hardness of bones and teeth; ionized form (Ca^{2+}) needed for blood clotting, release of hormones, contraction of muscle, and many other processes.
Phosphorus (P)	1.0	Component of nucleic acids and ATP; required for normal bone and tooth structure.
Potassium (K)	0.35	Ionized form (K^+) is the most plentiful cation (positively charged particle) in intracellular fluid; needed to generate action potentials.
Sulfur (S)	0.25	Component of some vitamins and many proteins.
Sodium (Na)	0.2	Ionized form (Na^+) is the most plentiful cation in extracellular fluid; essential for maintaining water balance; needed to generate action potentials.
Chlorine (Cl)	0.2	Ionized form (Cl^-) is the most plentiful anion (negatively charged particle) in extracellular fluid; essential for maintaining water balance.
Magnesium (Mg)	0.1	Ionized form (Mg^{2+}) needed for action of many enzymes, molecules that increase the rate of chemical reactions in organisms.
Iron (Fe)	0.005	Ionized forms (Fe^{2+} and Fe^{3+}) are part of hemoglobin (oxygen-carrying protein in red blood cells) and some enzymes (proteins that catalyze chemical reactions in living cells).
Trace elements	0.2	Aluminum (Al), Boron (B), Chromium (Cr), Cobalt (Co), Copper (Cu), Fluorine (F), Iodine (I), Manganese (Mn), Molybdenum (Mo), Selenium (Se), Silicon (Si), Tin (Sn), Vanadium (V), and Zinc (Zn).

Figure 2.1 Two representations of the structure of an atom. Electrons move about the nucleus, which contains neutrons and protons. (a) In the electron cloud model of an atom, the shading represents the chance of finding an electron in regions outside the nucleus. (b) In the electron shell model, filled circles represent individual electrons, which are grouped into concentric circles according to the shells they occupy. Both models depict a carbon atom, with six protons, six neutrons, and six electrons.

An atom is the smallest unit of matter that retains the properties and characteristics of its element.

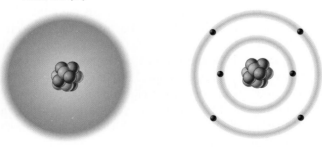

- Protons (p^+) ⎤
- Neutrons (n^0) ⎦ Nucleus
- Electrons (e^-)

(a) Electron cloud model (b) Electron shell model

 How are the electrons of carbon distributed between the first and second electron shells?

The number of electrons in an atom of an element always equals the number of protons. Because each electron and proton carries one charge, the negatively charged electrons and the positively charged protons balance each other. Thus, each atom is electrically neutral; its total charge is zero.

Atomic Number and Mass Number

The *number of protons* in the nucleus of an atom, designated by the atom's **atomic number,** distinguishes the atoms of one element from those of another. Figure 2.2 shows that atoms of different elements have different atomic numbers because they have different numbers of protons. For example, oxygen has an atomic number of 8 because its nucleus has eight protons, whereas sodium has an atomic number of 11 because its nucleus has 11 protons.

The **mass number** of an atom is the sum of its protons and neutrons. For sodium, which has 11 protons and 12 neutrons, the mass number is 23 (Figure 2.2). Although all atoms of one element have the same number of protons, they may have different numbers of neutrons and thus different mass numbers. **Isotopes** are atoms of an element that have different numbers of neutrons and therefore different mass numbers. In a sample of oxygen, for example, most atoms have 8 neutrons, but a few have 9 or 10, even though all have 8 protons and 8 electrons. Most isotopes are stable, which means that their nuclear structure does not

Figure 2.2   Atomic structures of several stable atoms.

 The atoms of different elements have different atomic numbers because they have different numbers of protons.

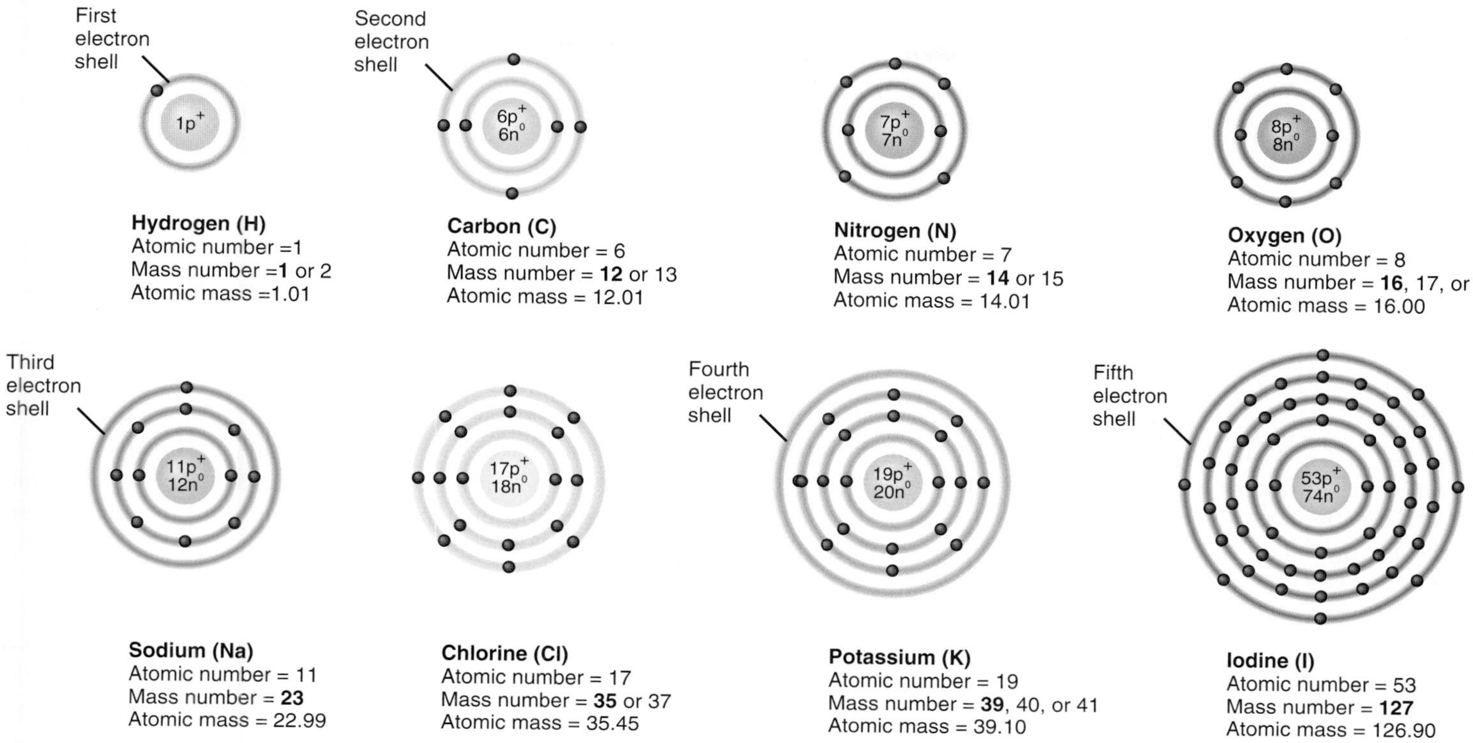

First electron shell

Second electron shell

Hydrogen (H)
Atomic number =1
Mass number =**1** or 2
Atomic mass =1.01

Carbon (C)
Atomic number = 6
Mass number = **12** or 13
Atomic mass = 12.01

Nitrogen (N)
Atomic number = 7
Mass number = **14** or 15
Atomic mass = 14.01

Oxygen (O)
Atomic number = 8
Mass number = **16**, 17, or 18
Atomic mass = 16.00

Third electron shell

Fourth electron shell

Fifth electron shell

Sodium (Na)
Atomic number = 11
Mass number = **23**
Atomic mass = 22.99

Chlorine (Cl)
Atomic number = 17
Mass number = **35** or 37
Atomic mass = 35.45

Potassium (K)
Atomic number = 19
Mass number = **39**, 40, or 41
Atomic mass = 39.10

Iodine (I)
Atomic number = 53
Mass number = **127**
Atomic mass = 126.90

Atomic number = number of protons in an atom
Mass number = number of protons and neutrons in an atom (boldface indicates most common isotope)
Atomic mass = average mass of all stable atoms of a given element in daltons

Which four of these elements are present most abundantly in living organisms?

change over time. The stable isotopes of oxygen are designated ^{16}O, ^{17}O, and ^{18}O (or O-16, O-17, and O-18). The numbers indicate the mass number (total number of protons and neutrons) of each isotope. As you will discover, the electrons of an atom determine the atom's chemical properties. Although the isotopes of an element have different numbers of neutrons, they have the same number of electrons and thus identical chemical properties.

Certain isotopes called **radioactive isotopes** are unstable; their nuclei decay (spontaneously change) into a stable configuration. Examples are H-3, C-14, O-15, and O-19. As they decay, these atoms emit radiation—either subatomic particles or packets of energy—and in the process often transform into a different element. For example, the radioactive isotope of carbon, C-14, decays to N-14. The decay of a radioisotope may be fast, occurring in a fraction of a second, or slow, taking millions of years. The **half-life** of an isotope is the time required for half of the radioactive atoms in a sample of that isotope to decay into a more stable form. For instance, the half-life of C-14 is 5600 years, whereas the half-life of I-131 is 8 days.

Harmful and Beneficial Effects of Radiation

Radioactive isotopes have both harmful and helpful effects. They can pose a serious threat to the human body because their radiations can break apart molecules, which may produce tissue damage as well as cause various types of cancer. Although the decay of naturally occurring radioactive isotopes typically releases just a small amount of radiation into the environment, localized accumulations can occur. One example of this is radon-222, a colorless and odorless gas that is a naturally occurring radioactive breakdown product of uranium. Radon may seep out of the soil and accumulate in buildings. Radon exposure not only greatly increases the risk of lung cancer in smokers but also causes many cases of lung cancer in nonsmokers. Beneficial uses of certain radioisotopes include their use in medical imaging procedures, for example, to diagnose heart disease, and to treat cancer by killing cancerous cells. ■

Atomic Mass

The standard unit for measuring the mass of atoms and their subatomic particles is a **dalton,** also known as an *atomic mass unit (amu).* A neutron has a mass of 1.008 daltons, and a proton has a mass of 1.007 daltons. The mass of an electron, however, is 0.0005 dalton, almost 2000 times smaller than the mass of a neutron or proton. The **atomic mass** (also called the *atomic weight*) of an element is the average mass of all its naturally occurring isotopes. The atomic mass of chlorine, for example, is 35.45 daltons. About 76% of all chlorine atoms have 18 neutrons (mass number = 35), whereas 24% have 20 neutrons (mass number = 37). Typically, the atomic mass of an element is close to the mass number of its most abundant isotope.

The mass of a single atom is slightly less than the sum of the masses of its neutrons, protons, and electrons because some mass (less than 1%) was lost when the atom's subatomic particles came together to form an atom. This aspect of atom formation explains why an element's atomic mass can be slightly less than the mass number of its smallest stable isotope. For example, the atomic mass of sodium is 22.99 daltons although its mass number is 23.

Ions, Molecules, Free Radicals, and Compounds

As we discussed, atoms of the same element have the same number of protons. In addition, the atoms of each element have a characteristic way of losing, gaining, or sharing their electrons when interacting with other atoms to achieve stability. The way that electrons behave enables atoms in the body to exist in electrically charged forms called ions, or to join with each other into the complex combinations called molecules. If an atom either *gives up* or *gains* electrons, it becomes an ion. An **ion** is an atom that has a positive or negative charge because it has unequal numbers of protons and electrons. *Ionization* is the process of giving up or gaining electrons. An ion of an atom is symbolized by writing its chemical symbol followed by the number of its positive (+) or negative (−) charges. Thus, Ca^{2+} stands for a calcium ion that has two positive charges because it has lost two electrons.

By contrast, when two or more atoms *share* electrons, the resulting combination is called a **molecule** (MOL-e-kūl). A molecule may consist of two atoms of the same kind, such as an oxygen molecule (Figure 2.3a). A *molecular formula* indicates the elements and the number of atoms of each element that make up a molecule. The molecular formula for a molecule of oxygen is O_2. The subscript 2 indicates that the molecule contains two atoms. Two or more different kinds of atoms may also form a molecule, as in a water molecule (H_2O). In H_2O one atom of oxygen shares electrons with two atoms of hydrogen.

A **free radical** is an electrically charged atom or group of atoms with an unpaired electron in its outermost shell. A common example is superoxide, which is formed by the addition of

Figure 2.3 Atomic structures of an oxygen molecule and a superoxide free radical.

A free radical has an unpaired electron in its outermost electron shell.

Unpaired electron

(a) Oxygen molecule   (b) Superoxide free radical

What substances in the body can inactivate oxygen-derived free radicals?

an electron to an oxygen molecule (Figure 2.3b). Having an unpaired electron makes a free radical unstable, highly reactive, and destructive to nearby molecules. Free radicals become stable by either giving up their unpaired electron to, or taking on an electron from, another molecule. In so doing, free radicals may break apart important body molecules.

A **compound** is a substance that contains atoms of two or more different elements. Most of the atoms in the body are joined into compounds. Water (H_2O) and sodium chloride (NaCl), common table salt, are compounds. A molecule of oxygen (O_2), however, is not a compound because it consists of atoms of only one element. Thus, while all compounds are molecules, not all molecules are compounds.

Free Radicals and Their Effects on Health

In our bodies, several processes can generate free radicals. They may result from exposure to ultraviolet radiation in sunlight or to x rays. Some reactions that occur during normal metabolic processes produce free radicals. Moreover, certain harmful substances, such as carbon tetrachloride (a solvent used in dry cleaning), give rise to free radicals when they participate in metabolic reactions in the body. Among the many disorders and diseases linked to oxygen-derived free radicals are cancer, atherosclerosis, Alzheimer disease, emphysema, diabetes mellitus, cataracts, macular degeneration, rheumatoid arthritis, and deterioration associated with aging. Consuming more *antioxidants*— substances that inactivate oxygen-derived free radicals—is thought to slow the pace of damage caused by free radicals. Important dietary antioxidants include selenium, beta-carotene, and vitamins C and E. ■

▶ C H E C K P O I N T

1. What are the names and chemical symbols of the 12 most abundant chemical elements in the human body?

2. Compare the meanings of atomic number, mass number, and atomic mass.

3. What are isotopes and radioactive isotopes of chemical elements?

CHEMICAL BONDS

O B J E C T I V E S

- **Describe how valence electrons form chemical bonds.**
- **Distinguish among ionic, covalent, and hydrogen bonds.**

The forces that hold together the atoms of a molecule or a compound are **chemical bonds.** The likelihood that an atom will form a chemical bond with another atom depends on the number of electrons in its outermost shell, also called the **valence shell.** An atom with a valence shell holding eight electrons is *chemically stable,* which means it is unlikely to form chemical bonds with other atoms. Neon, for example, has eight electrons in its valence shell, and for this reason it does not easily bond with other atoms. Hydrogen and helium are exceptions; their valence shell is the first electron shell, which holds a maximum of two electrons. Because helium has two valence electrons (see Figure 2.2), it too is stable and seldom bonds with other atoms.

The atoms of most biologically important elements do not have eight electrons in their valence shells. Under the right conditions, two or more atoms can interact in ways that produce a chemically stable arrangement of eight valence electrons for each atom. This chemical principle, called the **octet rule** (*octet* = set of eight), helps explain why atoms interact in predictable ways. One atom is more likely to interact with another atom if doing so will leave both with eight valence electrons. For this to happen, an atom either empties its partially filled valence shell, fills it with donated electrons, or shares electrons with other atoms. The way that valence electrons are distributed determines what kind of chemical bond results. We will consider three kinds of chemical bonds: ionic bonds, covalent bonds, and hydrogen bonds.

Ionic Bonds

When atoms lose or gain one or more valence electrons, ions are formed. Positively and negatively charged ions are attracted to one another—opposites attract. The force of attraction that holds ions having opposite charges together is an **ionic bond.** Consider sodium and chlorine atoms to see how this happens. Sodium has one valence electron (Figure 2.4a). If sodium *loses* this electron, it is left with the eight electrons in its second shell, a complete octet. As a result, however, the total number of protons (11) exceeds the number of electrons (10). Thus, the sodium atom has become a **cation,** or positively charged ion. A sodium ion has a charge of $1+$ and is written Na^+. By contrast, chlorine has seven valence electrons (Figure 2.4b). If chlorine *gains* an electron from a neighboring atom, it will have a complete octet in its third electron shell. After gaining an electron, the total number of electrons (18) exceeds the number of protons (17), and the chlorine atom has become an **anion,** a negatively charged ion. The ionic form of chlorine is called a chloride ion. It has a charge of $1-$ and is written Cl^-. When an atom of

Figure 2.4 Ions and ionic bond formation. (a) A sodium atom can have a complete octet of electrons in its outermost shell by losing one electron. (b) A chlorine atom can have a complete octet by gaining one electron. (c) An ionic bond may form between oppositely charged ions. (d) In a crystal of NaCl, each Na^+ is surrounded by six Cl^-. In (a), (b), and (c), the electron that is lost or accepted is colored red.

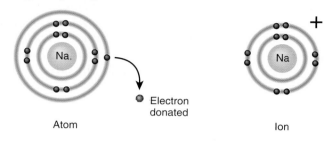

An ionic bond is the force of attraction that holds together oppositely charged ions.

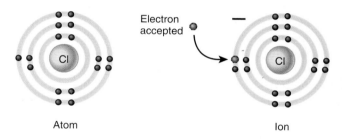

Atom   Electron donated   Ion

(a) Sodium: 1 valence electron

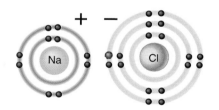

Electron accepted

Atom   Ion

(b) Chlorine: 7 valence electrons

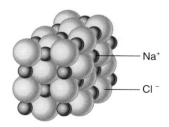

(c) Ionic bond in sodium chloride (NaCl)

Na$^+$

Cl$^-$

(d) Packing of ions in a crystal of sodium chloride

What are cations and anions?

sodium donates its sole valence electron to an atom of chlorine, the resulting positive and negative charges pull both ions tightly together into an ionic bond (Figure 2.4c). The resulting compound is sodium chloride, written NaCl.

In general, ionic compounds exist as solids, with an orderly, repeating arrangement of the ions, as in a crystal of NaCl (Figure 2.4d). A crystal of NaCl may be large or small—the total number of ions can vary—but the ratio of Na^+ to Cl^- is always $1:1$. In the body, ionic bonds are found mainly in teeth and bones, where they give great strength to the tissue. Most other ions in the body are dissolved in body fluids. An ionic compound that dissociates (breaks apart) into positive and negative ions in solution is called an **electrolyte** (e-LEK-trō-līt). Electrolytes are so-named because their solutions can conduct an electric current. (In Chapter 27 we will discuss the chemistry and importance of electrolytes.)

When a single positively charged ion of sodium and a single negatively charged ion of chlorine form an ionic bond, the net charge of the resulting compound, NaCl, is zero. The same is true for every compound with ionic bonds—the net charge of the compound is zero. In your study of the human body, however, you will discover many compounds that are positively or negatively charged. For example, ammonium (NH_4^+) and hydroxide (OH^-) are very common ionic compounds found in the body. The reason these compounds can carry a charge is that they are the result of covalent bonds (discussed next), not ionic bonds. Table 2.2 lists the names and symbols of the most common ions in the body.

Covalent Bonds

When a **covalent bond** forms, two or more atoms *share* electrons rather than gaining or losing whole electrons. Atoms form a covalently bonded molecule by sharing one, two, or three pairs

of their valence electrons. The larger the number of electron pairs shared between two atoms, the stronger the covalent bond. The electrons in each pair are said to be shared because they spend most of their time in the region between the nuclei of the two atoms. Covalent bonds may form between atoms of the same element or between atoms of different elements. They are the most common chemical bonds in the body, and the compounds that result from them form most of the body's structures.

A **single covalent bond** results when two atoms share one electron pair. For example, a molecule of hydrogen forms when two hydrogen atoms share their single valence electrons (Figure 2.5a), which allows both atoms to have a full valence shell at least part of the time. A **double covalent bond** results when two atoms share two pairs of electrons, as happens in an oxygen molecule (Figure 2.5b). A **triple covalent bond** occurs when two atoms share three pairs of electrons, as in a molecule of nitrogen (Figure 2.5c). Notice in the *structural formulas* for covalently bonded molecules in Figure 2.5 that the number of lines between the chemical symbols for two atoms indicates whether the bond is a single ($-$), double ($=$), or triple ($\equiv$) covalent bond.

The same principles of covalent bonding that apply to atoms of the same element also apply to covalent bonds between atoms of different elements. The gas methane (CH_4) contains covalent bonds formed between the atoms of two different elements (Figure 2.5d). The valence shell of the carbon atom can hold eight electrons but has only four of its own. The single electron shell of a hydrogen atom can hold two electrons, but each hydrogen atom has only one of its own. A methane molecule contains four separate single covalent bonds. Each hydrogen atom shares one pair of electrons with the carbon atom.

In some covalent bonds, two atoms share the electrons equally—one atom does not attract the shared electrons more strongly than the other atom. This type of bond is a **nonpolar covalent bond**. The bonds between two identical atoms are always nonpolar covalent bonds (Figure 2.5a–c). Also, the bonds between carbon and hydrogen atoms are nonpolar, for example, each of the four C–H bonds in a methane molecule (Figure 2.5d).

In a **polar covalent bond,** the sharing of electrons between two atoms is unequal—one atom attracts the shared electrons more strongly than the other. When polar covalent bonds form, the resulting molecule has a partial negative charge near the atom that attracts electrons more strongly. This atom has greater **electronegativity,** the power to attract electrons to itself. At least one other atom in the molecule then will have a partial positive charge. The partial charges are indicated by a lowercase Greek delta with a minus or plus sign: δ^- and δ^+. A very important example of a polar covalent bond in living systems is the bond between oxygen and hydrogen, for example, in a molecule of water (Figure 2.6 on page 34). Later in the chapter, we will see how polar covalent bonds allow water to dissolve many molecules that are important to life. The N$-$H and O$-$C bonds are also polar bonds.

Table 2.2	Common Ions and Ionic Compounds in the Body		
Cations		**Anions**	
Name	**Symbol**	**Name**	**Symbol**
Hydrogen ion	H^+	Fluoride ion	F^-
Sodium ion	Na^+	Chloride ion	Cl^-
Potassium ion	K^+	Iodide ion	I^-
Ammonium ion	NH_4^+	Hydroxide ion	OH^-
Hydronium ion	H_3O^+	Nitrate ion	NO_3^-
Magnesium ion	Mg^{2+}	Bicarbonate ion	HCO_3^-
Calcium ion	Ca^{2+}	Oxide ion	O^{2-}
Iron (II) ion	Fe^{2+}	Sulfide ion	S^{2-}
Iron (III) ion	Fe^{3+}	Phosphate ion	PO_4^{3-}

Figure 2.5   Covalent bond formation. The red electrons are shared equally. In writing the structural formula of a covalently bonded molecule, each straight line between the chemical symbols for two atoms denotes a pair of shared electrons. In molecular formulas, the number of atoms in each molecule is noted by subscripts.

🔑 **In a covalent bond, two atoms share one, two, or three pairs of valence electrons.**

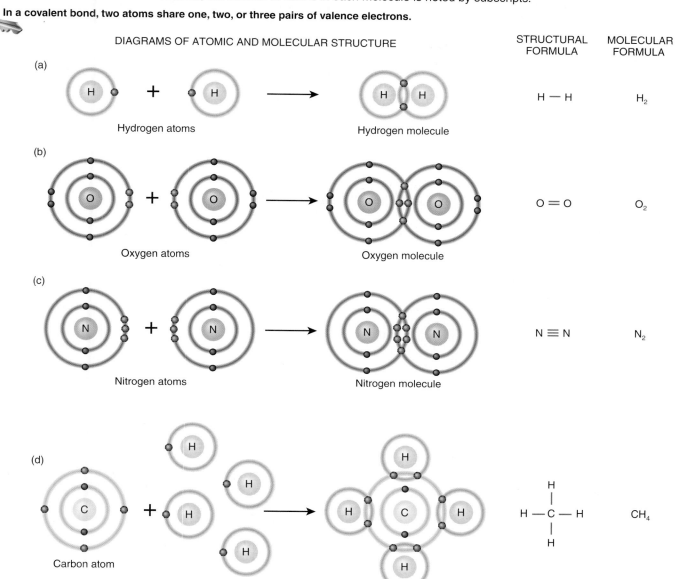

? **What is the principal difference between an ionic bond and a covalent bond?**

Hydrogen Bonds

The polar covalent bonds that form between hydrogen atoms and other atoms can give rise to a third type of chemical bond, a hydrogen bond (Figure 2.7). A **hydrogen bond** forms when a hydrogen atom with a partial positive charge (δ^+) attracts the partial negative charge (δ^-) of neighboring electronegative atoms, most often oxygen or nitrogen. Thus, hydrogen bonds result from attraction of oppositely charged parts of molecules rather than from sharing of electrons as in covalent bonds. Hydrogen bonds are weak—only about 5% as strong as covalent bonds. Thus, they cannot bind atoms into molecules. However, hydrogen bonds do establish important links between molecules or between different parts of a large molecule, such as a protein or nucleic acid (both discussed later in this chapter).

Figure 2.6 Polar covalent bonds between oxygen and hydrogen atoms in a water molecule.   The red electrons are shared unequally. Because the oxygen nucleus attracts the shared electrons more strongly, the oxygen end of a water molecule has a partial negative charge, written δ^-, and the hydrogen ends have partial positive charges, written δ^+.

A polar covalent bond occurs when one atomic nucleus attracts the shared electrons more strongly than does the nucleus of another atom in the molecule.

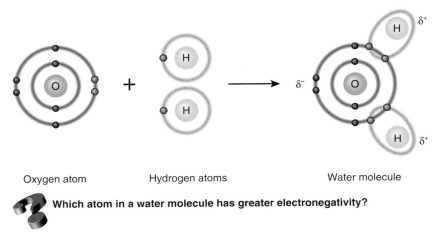

Oxygen atom   Hydrogen atoms   Water molecule

Which atom in a water molecule has greater electronegativity?

The hydrogen bonds that link neighboring water molecules give water considerable *cohesion,* the tendency of like particles to stay together. The cohesion of water molecules creates a very high **surface tension,** a measure of the difficulty of stretching or breaking the surface of a liquid. At the boundary between water and air, water's surface tension is very high because the water molecules are much more attracted to one another than they are attracted to molecules in the air. The influence of water's surface tension on the body can be seen in the way it increases the work required for breathing. A thin film of watery fluid coats the air sacs of the lungs. So, each inhalation must have enough force to overcome the opposing effect of surface tension as the air sacs stretch and enlarge when taking in air.

Even though single hydrogen bonds are weak, very large molecules may contain hundreds of these bonds. Acting collectively, hydrogen bonds provide considerable strength and stability and help determine the three-dimensional shape of large molecules. As you will see later in this chapter, a large molecule's shape determines how it functions.

▶ CHECKPOINT

4. Which electron shell is the valence shell of an atom, and what is its significance?

5. How does valence relate to the octet rule?

6. What information is conveyed when you write the molecular or structural formula for a molecule?

Figure 2.7 Hydrogen bonding among water molecules.   Each water molecule forms hydrogen bonds, indicated by dotted lines, with three to four neighboring water molecules.

Hydrogen bonds occur because hydrogen atoms in one water molecule are attracted to the partial negative charge of the oxygen atom in another water molecule.

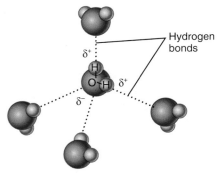

Hydrogen bonds

Why would you expect ammonia (NH_3) to form hydrogen bonds with water molecules?

CHEMICAL REACTIONS

▶ OBJECTIVES

• **Define a chemical reaction.**
• **Describe the various forms of energy.**
• **Compare exergonic and endergonic chemical reactions.**
• **Describe the role of activation energy and catalysts in chemical reactions.**
• **Describe synthesis, decomposition, exchange, and reversible reactions.**

A **chemical reaction** occurs when new bonds form or old bonds break between atoms. Chemical reactions are the foundation of all life processes, and as we have seen, the interactions of valence electrons are the basis of all chemical reactions. Consider how hydrogen and oxygen molecules react to form water molecules (Figure 2.8). The starting substances—two H_2 and one O_2—are known as the **reactants.** The ending substances—two molecules of H_2O—are the **products.** The arrow in the figure indicates the direction in which the reaction proceeds. In a chemical reaction, the total mass of the reactants equals the total mass of the products, a relationship known as the **law of conservation of mass.** Because of this law, the number of atoms of each element is the same before and after the reaction. However, because the atoms are rearranged, the reactants and products have different chemical properties. Through thousands of different chemical reactions, body structures are built and body functions are carried out. The term **metabolism** refers to all the chemical reactions occurring in the body.

Forms of Energy and Chemical Reactions

Each chemical reaction involves energy changes. **Energy** (*en-* = in; *-ergy* = work) is the capacity to do work. Two principal forms of energy are **potential energy,** energy stored by matter due to its position, and **kinetic energy,** the energy associated with matter in motion. For example, the energy stored in a battery, in water behind a dam, or in a person poised to jump down some steps is potential energy. When the battery is used to run a clock, or the gates of the dam are opened and the falling water turns a generator, or the person jumps, potential energy is converted into kinetic energy. **Chemical energy** is a form of potential energy that is stored in the bonds of compounds and molecules. The total amount of energy present at the beginning and end of a chemical reaction is the same. Although energy can be neither created nor destroyed, it may be converted from one form to another. This principle is known as the **law of conservation of energy.** For example, some of the chemical energy in the foods we eat is eventually converted into various forms of kinetic energy, such as mechanical energy used to walk and talk. Conversion of energy from one form to another generally releases heat, some of which is used to maintain normal body temperature.

Energy Transfer in Chemical Reactions

In chemical reactions, breaking old bonds requires energy, and forming new bonds releases energy. Because most chemical reactions involve both breaking bonds in the reactants and forming bonds in the products, the *overall reaction* may either release energy or absorb energy. **Exergonic reactions** (*ex-* = out) release more energy than they absorb. In an exergonic reaction, the energy released when new bonds form is *greater* than the energy needed to break apart old bonds. As a result, surplus energy is released as the reaction occurs. At the conclusion of an exergonic reaction, the products have *less* potential energy than the reactants. By contrast, **endergonic reactions** (*end-* = within) absorb more energy than they release. In an endergonic reaction, the energy released when new bonds form is *less* than the energy needed to break apart old bonds, so energy must be absorbed

Figure 2.8 The chemical reaction between two hydrogen molecules (H_2) and one oxygen molecule (O_2) to form two molecules of water (H_2O). Note that the reaction occurs by breaking old bonds and making new bonds.

 The number of atoms of each element is the same before and after a chemical reaction.

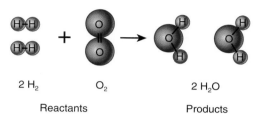

2 H_2	O_2	2 H_2O
Reactants		Products

Why does this reaction require two molecules of H_2?

from sources outside the reactants for the reaction to take place. At the conclusion of an endergonic reaction, the products have *more* potential energy than the reactants.

A key feature of the body's metabolism is the coupling of exergonic reactions and endergonic reactions. Energy released from an exergonic reaction often is used to drive an endergonic one. In general, exergonic reactions occur as nutrients, such as glucose, are broken down. Some of the energy released may be trapped in the covalent bonds of *adenosine triphosphate (ATP),* which we describe more fully later in this chapter. If a molecule of glucose is completely broken down, the chemical energy in its bonds can be used to produce as many as 38 molecules of ATP. The energy transferred to the ATP molecules is then used to drive endergonic reactions needed to build body structures, such as muscles and bones. The energy in ATP is also used to do mechanical work, such as occurs during the contraction of muscle or the movement of substances into or out of cells.

Activation Energy

Because particles of matter such as atoms, ions, and molecules have kinetic energy, they are continuously moving and colliding with one another. A sufficiently forceful collision can disrupt the movement of valence electrons, causing an existing chemical bond to break or a new one to form. The collision energy needed to break chemical bonds in the reactants is called the **activation energy** (Figure 2.9). This initial energy "investment" is needed to start a reaction. The reactants must absorb enough energy for their chemical bonds to become unstable and their valence electrons to form new combinations. Then, as new bonds form, energy is released to the surroundings.

Both the concentration of particles and the temperature influence the chance that a collision will occur and cause a chemical reaction.

- *Concentration.* The more particles of matter present in a confined space, the greater the chance that they will collide. The concentration of particles increases when more are added to a given space or when the pressure on the space

Figure 2.9 Activation energy.

 Activation energy is the energy needed to break chemical bonds in the reactant molecules so a reaction can start.

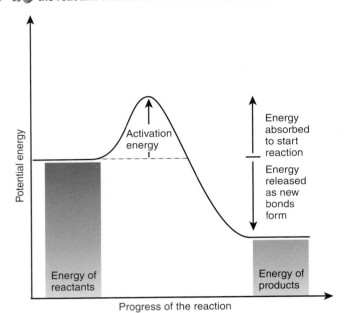

Why is the reaction illustrated here exergonic?

Figure 2.10 Comparison of energy needed for a chemical reaction to proceed with a catalyst (green curve) and without a catalyst (red curve).

 Catalysts speed up chemical reactions by lowering the activation energy.

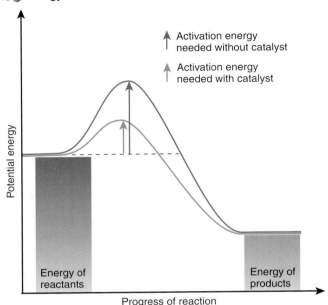

Does a catalyst change the potential energies of the products and reactants?

increases, which forces the particles closer together so that they collide more often.

- *Temperature.* As temperature rises, particles of matter move about more rapidly. Thus, the higher the temperature of matter, the more forcefully particles will collide, and the greater the chance that a collision will produce a reaction.

Catalysts

As we have seen, chemical reactions occur when chemical bonds break or form after atoms, ions, or molecules collide with one another. Body temperature and the concentrations of molecules in body fluids, however, are far too low for most chemical reactions to occur rapidly enough to maintain life. Raising the temperature and the number of reacting particles of matter in the body could increase the frequency of collisions and thus increase the rate of chemical reactions, but doing so could also damage or kill the body's cells.

Substances called catalysts solve this problem. **Catalysts** are chemical compounds that speed up chemical reactions by lowering the activation energy needed for a reaction to occur (Figure 2.10). A catalyst does not alter the difference in potential energy between the reactants and the products. Rather, it lowers the amount of energy needed to start the reaction.

For chemical reactions to occur, some particles of matter—especially large molecules—must not only collide with sufficient force, but they must "hit" one another at precise spots. A catalyst helps to properly orient the colliding particles. Thus,

they interact at the spots that make the reaction happen. Although the action of a catalyst helps to speed up a chemical reaction, the catalyst itself is unchanged at the end of the reaction. A single catalyst molecule can repeatedly assist one chemical reaction after another. The most important catalysts in the body are enzymes, which we will discuss later in this chapter.

Types of Chemical Reactions

After a chemical reaction takes place, the atoms of the reactants are rearranged to yield products with new chemical properties. In this section we will look at the types of chemical reactions common to all living cells. Once you have learned them, you will be able to understand the chemical reactions discussed later in the book.

Synthesis Reactions—Anabolism

When two or more atoms, ions, or molecules combine to form new and larger molecules, the processes are **synthesis reactions.** The word *synthesis* means "to put together." A synthesis reaction can be expressed as follows:

$$A \quad + \quad B \xrightarrow{\text{Combine to form}} AB$$

Atom, ion, or molecule A   Atom, ion, or molecule B   New molecule AB

One example of a synthesis reaction is the reaction between two hydrogen molecules and one oxygen molecule to form two molecules of water (see Figure 2.8). Another example of a synthesis reaction is:

$$N_2 \quad + \quad 3\,H_2 \quad \xrightarrow{\text{Combine to form}} \quad 2\,NH_3$$

One nitrogen molecule   Three hydrogen molecules   Two ammonia molecules

All the synthesis reactions that occur in your body are collectively referred to as **anabolism** (a-NAB-ō-lizm). Overall, anabolic reactions are usually endergonic because they absorb more energy than they release. Combining simple molecules like amino acids (discussed shortly) to form large molecules such as proteins is an example of anabolism.

Decomposition Reactions — Catabolism

Decomposition reactions split up large molecules into smaller atoms, ions, or molecules. A decomposition reaction is expressed as follows:

$$AB \quad \xrightarrow{\text{Breaks down into}} \quad A \quad + \quad B$$

Molecule AB   Atom, ion, or molecule A   Atom, ion, or molecule B

For example, under the proper conditions, a molecule of methane can decompose into a carbon atom and two hydrogen molecules:

$$CH_4 \quad \xrightarrow{\text{Breaks down into}} \quad C \quad + \quad 2\,H_2$$

One methane molecule   One carbon atom   Two hydrogen molecules

The decomposition reactions that occur in your body are collectively referred to as **catabolism** (ka-TAB-ō-lizm). Overall, catabolic reactions are usually exergonic because they release more energy than they absorb. For instance, the series of reactions that break down glucose to pyruvic acid, with the net production of two molecules of ATP, are important catabolic reactions in the body. These reactions will be discussed in Chapter 25.

Exchange Reactions

Many reactions in the body are **exchange reactions;** they consist of both synthesis and decomposition reactions. One type of exchange reaction works like this:

$$AB + CD \longrightarrow AD + BC$$

The bonds between A and B and between C and D break (decomposition), and new bonds then form (synthesis) between A and D and between B and C. An example of an exchange reaction is:

$$HCl \quad + \quad NaHCO_3 \quad \longrightarrow \quad H_2CO_3 \quad + \quad NaCl$$

Hydrochloric acid   Sodium bicarbonate   Carbonic acid   Sodium chloride

Notice that the ions in both compounds have "switched partners": The hydrogen ion (H^+) from HCl has combined with the bicarbonate ion (HCO_3^-) from $NaHCO_3$, and the sodium ion (Na^+) from $NaHCO_3$ has combined with the chloride ion (Cl^-) from HCl.

Reversible Reactions

Some chemical reactions proceed in only one direction, from reactants to products, as previously indicated by the single arrows. Other chemical reactions may be reversible. In a **reversible reaction,** the products can revert to the original reactants. A reversible reaction is indicated by two half arrows pointing in opposite directions:

$$AB \; \underset{\text{Combine to form}}{\overset{\text{Breaks down into}}{\rightleftharpoons}} \; A + B$$

Some reactions are reversible only under special conditions:

$$AB \; \underset{\text{Heat}}{\overset{\text{Water}}{\rightleftharpoons}} \; A + B$$

In that case, whatever is written above or below the arrows indicates the condition needed for the reaction to occur. In these reactions, AB breaks down into A and B only when water is added, and A and B react to produce AB only when heat is applied. Many reversible reactions in the body require enzymes. Often, different enzymes guide the reactions in opposite directions.

▶ CHECKPOINT

7. What is the relationship between reactants and products in a chemical reaction?

8. Compare potential energy and kinetic energy.

9. Describe the law of conservation of energy.

10. How do catalysts affect activation energy?

11. How are anabolism and catabolism related to synthesis and decomposition reactions, respectively?

INORGANIC COMPOUNDS AND SOLUTIONS

▶ OBJECTIVES

• **Describe the properties of water and inorganic acids, bases, and salts.**

• **Distinguish among solutions, colloids, and suspensions.**

• **Define pH and explain the role of buffer systems in homeostasis.**

Most of the chemicals in your body exist in the form of compounds. Biologists and chemists divide these compounds into two principal classes: inorganic compounds and organic compounds. In general, **inorganic compounds** lack carbon and are structurally simple. They include water and many salts, acids, and bases. Inorganic compounds may have either ionic or covalent bonds. Water makes up 55–60% of a lean adult's total body mass; all other inorganic compounds add an additional 1–2%. **Organic compounds,** by contrast, always contain carbon, usually contain hydrogen, and always have covalent bonds. Two carbon-containing compounds—carbon dioxide (CO_2) and bicarbonate ion (HCO_3^-)—are classified as inorganic.

Water

Water is the most important and abundant inorganic compound in all living systems. Although you might be able to survive for weeks without food, without water you would die in a matter of days. Nearly all the body's chemical reactions occur in a watery medium. Water has many properties that make it such an indispensable compound for life. The most important property of water is its polarity—the uneven sharing of valence electrons that confers a partial negative charge near the one oxygen atom and two partial positive charges near the two hydrogen atoms in a water molecule (see Figure 2.6). This property alone makes water an excellent solvent for other ionic or polar substances, gives water molecules cohesion (the tendency to stick together), and allows water to minimize temperature changes.

Water as a Solvent

In medieval times people searched in vain for a "universal solvent," a substance that would dissolve all other materials. They found nothing that worked as well as water. Although it is the most versatile solvent known, however, it is not a universal solvent. If it were, no container could hold it because the container would dissolve! In a **solution,** a substance called the **solvent** dissolves another substance called the **solute.** Usually there is more solvent than solute in a solution. For example, your sweat is a dilute solution of water (the solvent) plus small amounts of salts (the solutes).

The versatility of water as a solvent for ionized or polar substances is due to its polar covalent bonds and its bent shape, which allows each water molecule to interact with several neighboring ions or molecules. Solutes that contain polar covalent bonds are **hydrophilic** (*hydro-* = water; *-philic* = loving), which means they dissolve easily in water. Common examples of hydrophilic solutes are sugar and salt. Molecules that contain mainly nonpolar covalent bonds, by contrast, are **hydrophobic** (*-phobic* = fearing). They are not very water soluble. Examples of hydrophobic compounds include animal fats and vegetable oils.

To understand the dissolving power of water, consider what happens when a crystal of a salt such as sodium chloride (NaCl) is placed in water (Figure 2.11). The electronegative oxygen atom in water molecules attracts the sodium ions (Na^+), and the electropositive hydrogen atoms in water molecules attracts the chloride ions (Cl^-). Soon, water molecules surround and separate some Na^+ and Cl^- from each other at the surface of the crystal, breaking the ionic bonds that held NaCl together. The

Figure 2.11 How polar water molecules dissolve salts and polar substances. When a crystal of sodium chloride is placed in water, the slightly negative oxygen end (red) of water molecules is attracted to the positive sodium ions (Na^+), and the slightly positive hydrogen portions (gray) of water molecules are attracted to the negative chloride ions (Cl^-).

 Water is a versatile solvent because its polar covalent bonds, in which electrons are shared unequally, create positive and negative regions.

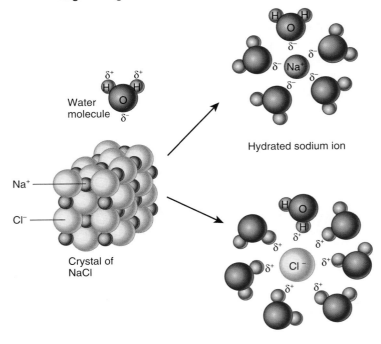

Table sugar (sucrose) easily dissolves in water but is not an electrolyte. Is it likely that all the covalent bonds between atoms in table sugar are nonpolar bonds?

water molecules surrounding the ions also lessen the chance that ions with opposite charges will come together and reform an ionic bond.

The ability of water to form solutions and suspensions is essential to health and survival. Because water can dissolve or suspend so many different substances, it is an ideal medium for metabolic reactions. Water enables dissolved reactants to collide and form products. Water also dissolves waste products, which allows them to be flushed out of the body in the urine.

Water in Chemical Reactions

Water serves as the medium for most chemical reactions in the body, and it also participates as a reactant or product in certain reactions. During digestion, for example, decomposition reactions break down large nutrient molecules into smaller molecules by the addition of water molecules. This type of reaction is called **hydrolysis** (hī-DROL-i-sis; *-lysis* = to loosen or break apart).

Hydrolysis reactions enable dietary nutrients to be absorbed into the body. When two smaller molecules join to form a larger molecule in a **dehydration synthesis reaction** (*de-* = from, down, or out; *hydra-* = water), by contrast, a water molecule is formed and removed from the reactants. As you will see later in the chapter, such reactions occur during synthesis of proteins and other large molecules (for example, see Figure 2.21).

High Heat Capacity of Water

In comparison to most substances, water can absorb or release a relatively large amount of heat with only a modest change in its own temperature. For this reason, water is said to have a high *heat capacity*. The reason for this property is the large number of hydrogen bonds in water. As water absorbs heat energy, some of the energy is used to break hydrogen bonds. Less energy is then left over to increase the motion of water molecules, which would increase the water's temperature. The high heat capacity of water is the reason it is used in automobile radiators; it cools the engine by absorbing heat without its own temperature rising to an unacceptably high level. The large amount of water in the body has a similar effect: It lessens the impact of environmental temperature changes, helping to maintain the homeostasis of body temperature.

Water also requires a large amount of heat to change from a liquid to a gas. Its *heat of vaporization* is high. As water evaporates from the surface of the skin, it removes a large quantity of heat, providing an important cooling mechanism.

Water as a Lubricant

Water is a major part of mucus and other lubricating fluids. Lubrication is especially necessary in the chest (pleural and pericardial cavities) and abdomen (peritoneal cavity), where internal organs touch and slide over one another. It is also needed at joints, where bones, ligaments, and tendons rub against one another. Inside the gastrointestinal tract, mucus and other watery secretions moisten foods, which aids their smooth passage.

Solutions, Colloids, and Suspensions

A **mixture** is a combination of elements or compounds that are physically blended together but not bound by chemical bonds. For example, the air you are breathing is a mixture of gases that includes nitrogen, oxygen, argon, and carbon dioxide. Three common liquid mixtures are solutions, colloids, and suspensions.

Once mixed together, solutes in a solution remain evenly dispersed among the solvent molecules. Because the solute particles in a solution are very small, a solution looks clear and transparent.

A **colloid** differs from a solution mainly because of the size of its particles. The solute particles in a colloid are large enough

to scatter light, just as water droplets in fog scatter light from a car's headlight beams. For this reason, colloids usually look translucent or opaque. Milk is an example of a liquid that is both a colloid and a solution: The large milk proteins make it a colloid, whereas calcium salts, milk sugar (lactose), ions, and other small particles are in solution.

The solutes in both solutions and colloids do not settle out and accumulate on the bottom of the container. In a **suspension,** by contrast, the suspended material may mix with the liquid or suspending medium for some time, but eventually it will settle out. Blood is an example of a suspension. When freshly drawn from the body, blood has an even, reddish color. After blood sits for a while in a test tube, red blood cells settle out of the suspension and drift to the bottom of the tube (see Figure 19.1a on page 635). The upper layer, the liquid portion of blood, appears pale yellow and is called blood plasma. Blood plasma is both a solution of ions and other small solutes and a colloid due to the presence of plasma proteins.

The **concentration** of a solution may be expressed in several ways. One common way is by a mass per volume **percentage,** which gives the relative mass of a solute found in a given volume of solution. Another way expresses concentration in units of **moles per liter (mol/L),** which relate to the total number of molecules in a given volume of solution. A **mole** is the amount of any substance that has a mass in grams equal to the sum of the atomic masses of all its atoms. For example, one mole of the element chlorine (atomic mass = 35.45) is 35.45 grams and one mole of the salt sodium chloride (NaCl) is 58.44 grams (22.99 for Na + 35.45 for Cl). Just as a dozen always means 12 of something, a mole of anything has the same number of particles: 6.023×10^{23}. This huge number is called *Avogadro's number.* Thus, measurements of substances that are stated in moles tell us about the numbers of atoms, ions, or molecules present. This is important when chemical reactions are occurring because each reaction requires a set number of atoms of specific elements. Table 2.3 describes these ways of expressing concentration.

Table 2.3	Percentage and Molarity
Definition	**Example**
Percentage (mass per volume) Number of grams of a substance per 100 milliliters (mL) of solution.	To make a 10% NaCl solution, take 10 gm of NaCl and add enough water to make a total of 100 mL of solution.
Molarity = moles (mol) per liter A 1 molar (1 M) solution = 1 mole of a solute in 1 liter of solution.	To make a 1 molar (1 M) solution of NaCl, Dissolve 1 mole of NaCl (58.44 gm) in enough water to make a total of 1 liter of solution.

Inorganic Acids, Bases, and Salts

When inorganic acids, bases, or salts dissolve in water, they **dissociate** (dis′-sō-sē-ĀT); they separate into ions and become surrounded by water molecules. An **acid** (Figure 2.12a) is a substance that dissociates into one or more **hydrogen ions** (H⁺) and one or more anions. Because H⁺ is a single proton with one positive charge, an acid also is a **proton donor.** A **base,** by contrast (Figure 2.12b), dissociates into one or more **hydroxide ions (OH⁻)** and one or more cations. Because hydroxide ions have a strong attraction for protons, a base is a **proton acceptor.**

A **salt,** when dissolved in water, dissociates into cations and anions, neither of which is H⁺ or OH⁻ (Figure 2.12c). In the body, salts are electrolytes that are important for carrying electrical currents (ions flowing from one place to another), especially in nerve and muscle tissues. The ions of salts also provide many essential chemical elements in intracellular and extracellular fluids such as blood, lymph, and the interstitial fluid of tissues.

Acids and bases react with one another to form salts. For example, the reaction of hydrochloric acid (HCl) and potassium hydroxide (KOH), a base, produces the salt potassium chloride (KCl) and water (H₂O). This exchange reaction can be written as follows:

$$\underset{\text{Acid}}{\text{HCl}} + \underset{\text{Base}}{\text{KOH}} \longrightarrow \underset{\text{Dissociated ions}}{\text{H}^+ + \text{Cl}^- + \text{K}^+ + \text{OH}^-} \longrightarrow \underset{\text{Salt}}{\text{KCl}} + \underset{\text{Water}}{\text{H}_2\text{O}}$$

Acid–Base Balance: The Concept of pH

To ensure homeostasis, intracellular and extracellular fluids must contain almost balanced quantities of acids and bases. The more hydrogen ions (H⁺) dissolved in a solution, the more acidic the solution; conversely, the more hydroxide ions (OH⁻), the more basic (alkaline) the solution. The chemical reactions that take place in the body are very sensitive to even small changes in the acidity or alkalinity of the body fluids in which they occur. Any departure from the narrow limits of normal H⁺ and OH⁻ concentrations greatly disrupts body functions.

A solution's acidity or alkalinity is expressed on the **pH scale,** which extends from 0 to 14 (Figure 2.13). This scale is based on the concentration of H⁺ in moles per liter. A pH of 7 means that a solution contains one ten-millionth (0.0000001) of a mole of hydrogen ions per liter. The number 0.0000001 is written as 1×10^{-7} in scientific notation, which indicates that the number is 1 with the decimal point moved seven places to the left. To convert this value to pH, the negative exponent (−7) is changed to a positive number (7). A solution with a H⁺ concentration of 0.0001 (10^{-4}) moles per liter has a pH of 4; a solution with a H⁺ concentration of 0.000000001 (10^{-9}) moles per liter has a pH of 9; and so on.

The midpoint of the pH scale is 7, where the concentrations of H⁺ and OH⁻ are equal. A substance with a pH of 7, such as

Figure 2.12 **Dissociation of inorganic acids, bases, and salts.**

 Dissociation is the separation of inorganic acids, bases, and salts into ions in a solution.

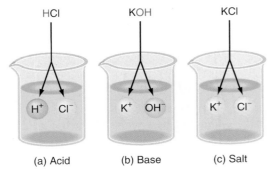

| (a) Acid | (b) Base | (c) Salt |

The compound CaCO₃ (calcium carbonate) dissociates into a calcium ion Ca²⁺ and a carbonate ion CO₃²⁻. Is it an acid, a base, or a salt? What about H₂SO₄, which dissociates into two H⁺ and one SO₄²⁻?

pure water, is neutral. A solution that has more H⁺ than OH⁻ is an **acidic solution** and has a pH below 7. A solution that has more OH⁻ than H⁺ is a **basic (alkaline) solution** and has a pH above 7. It is important to realize that a change of one whole number on the pH scale represents a *tenfold* change in the number of H⁺. A pH of 6 denotes 10 times more H⁺ than a pH of 7, whereas a pH of 8 indicates 10 times fewer H⁺ than a pH of 7 and 100 times fewer H⁺ than a pH of 6.

Maintaining pH: Buffer Systems

Although the pH of body fluids may differ, the normal limits for each fluid are quite narrow. Table 2.4 shows the pH values for certain body fluids together with those of some common substances. Homeostatic mechanisms maintain the pH of blood between 7.35 and 7.45, which is slightly more basic than pure water. Saliva is slightly acidic, and semen is slightly basic. Because the kidneys help remove excess acid from the body, urine can be quite acidic.

Even though strong acids and bases are continually taken into and formed by the body, the pH of fluids inside and outside cells remains almost constant. One important reason is the presence of **buffer systems,** which function to convert strong acids or bases into weak acids or bases. Strong acids (or bases) ionize easily and contribute many H⁺ or (OH⁻) to a solution. Therefore, they can change pH drastically, which can disrupt the body's metabolism. Weak acids (or bases) do not ionize as much and contribute fewer H⁺ (or OH⁻). Hence, they have less effect on the pH. The chemical compounds that can convert strong acids or bases into weak ones are called **buffers.**

One important buffer system in the body is the **carbonic acid–bicarbonate buffer system.** Carbonic acid (H₂CO₃) can

Figure 2.13   The pH scale.   A pH below 7 indicates an acidic solution — more H⁺ than OH⁻. [H⁺] = hydrogen ion concentration; [OH⁻] = hydroxide ion concentration.

The lower the numerical value of the pH, the more acidic is the solution because the H⁺ concentration becomes progressively greater. A pH above 7 indicates a basic (alkaline) solution; that is, there are more OH⁻ than H⁺. The higher the pH, the more basic the solution.

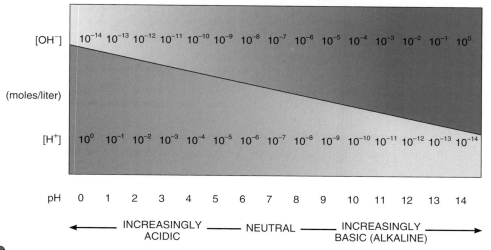

At pH 7 (neutrality), the concentrations of H⁺ and OH⁻ are equal (10⁻⁷ mol/liter). What is the concentration of H⁺ and OH⁻ at pH 6? Which pH is more acidic, 6.82 or 6.91? Which pH is closer to neutral, 8.41 or 5.59?

Table 2.4	pH Values of Selected Substances
Substance	**pH Value**
• Gastric juice (found in the stomach)	1.2–3.0
Lemon juice	2.3
Vinegar	3.0
Carbonated soft drink	3.0–3.5
Orange juice	3.5
• Vaginal fluid	3.5–4.5
Tomato juice	4.2
Coffee	5.0
• Urine	4.6–8.0
• Saliva	6.35–6.85
Milk	6.8
Distilled (pure) water	7.0
• Blood	7.35–7.45
• Semen (fluid containing sperm)	7.20–7.60
• Cerebrospinal fluid (fluid associated with nervous system)	7.4
• Pancreatic juice (digestive juice of the pancreas)	7.1–8.2
• Bile (liver secretion that aids fat digestion)	7.6–8.6
Milk of magnesia	10.5
Lye	14.0

• Denotes substances in the human body.

act as a weak acid, and the bicarbonate ion (HCO_3^-) can act as a weak base. Hence, this buffer system can compensate for either an excess or a shortage of H⁺. For example, if there is an excess of H⁺ (an acidic condition), HCO_3^- can function as a weak base and remove the excess H⁺, as follows:

$$H^+ \ + \ HCO_3^- \ \longrightarrow \ H_2CO_3 \ \longrightarrow \ H_2O \ + \ CO_2$$

Hydrogen ion   Bicarbonate ion (weak base)   Carbonic acid   Water   Carbon dioxide

If there is a shortage of H⁺ (an alkaline condition), by contrast, H_2CO_3 can function as a weak acid and provide needed H⁺ as follows:

$$H_2CO_3 \ \longrightarrow \ H^+ \ + \ HCO_3^-$$

Carbonic acid (weak acid)   Hydrogen ion   Bicarbonate ion

Chapter 27 describes buffers and their roles in maintaining acid–base balance in more detail.

▶ C H E C K P O I N T

12. How do inorganic compounds differ from organic compounds?

13. Describe two ways to express the concentration of a solution.

14. What functions does water perform in the body?

15. What is the chemical reaction whereby bicarbonate ions prevent buildup of excess H⁺?

ORGANIC COMPOUNDS

▶ O B J E C T I V E S

- **Describe the functional groups of organic molecules.**
- **Identify the building blocks and functions of carbohydrates, lipids, proteins, enzymes, deoxyribonucleic acid (DNA), ribonucleic acid (RNA), and adenosine triphosphate (ATP).**

Inorganic compounds are relatively simple. Their molecules have only a few atoms and cannot be used by cells to perform complicated biological functions. Many organic molecules, by contrast, are relatively large and have unique characteristics that allow them to carry out complex functions. Important categories of organic compounds, which make up about 40% of body mass, include carbohydrates, lipids, proteins, nucleic acids, and adenosine triphosphate (ATP).

Carbon and Its Functional Groups

Carbon has several properties that make it particularly useful to living organisms. For one thing, it can form bonds with one to thousands of other carbon atoms to produce large molecules that can have many different shapes. Due to this property of carbon,

the body can build many different organic compounds, each of which has a unique structure and function. Moreover, the large size of most carbon-containing molecules and the fact that some do not dissolve easily in water make them useful materials for building body structures.

Organic compounds are held together mostly or entirely by covalent bonds. Carbon has four electrons in its outermost (valence) shell. It can bond covalently with a variety of atoms, including other carbon atoms, to form rings and straight or branched chains. Other elements that most often bond with carbon in organic compounds are hydrogen (one bond), oxygen (two bonds), and nitrogen (three bonds). Sulfur (two bonds) and phosphorus (five bonds) are also present in organic compounds. The other elements listed in Table 2.1 are present in a smaller number of organic compounds.

The chain of carbon atoms in an organic molecule is called the **carbon skeleton.** Many of the carbons are bonded to hydrogen atoms. Also attached to the carbon skeleton are distinctive **functional groups,** in which other elements form bonds with carbon and hydrogen atoms. Each type of functional group has a specific arrangement of atoms that confers characteristic chemical properties upon organic molecules. Table 2.5 lists the most common functional groups of organic molecules and describes

---

Table 2.5	Major Functional Groups		
Name and Structural Formula*	**Occurrence and Significance**	**Name and Structural Formula***	**Occurrence and Significance**
Hydroxyl R—O—H	*Alcohols* contain an —OH group, which is polar and hydrophilic due to its electronegative O atom. Molecules with many —OH groups dissolve easily in water.	**Ester** O ‖ R—C—O—R	*Esters* predominate in dietary fats and oils and also occur in our body triglycerides. Aspirin is an ester of salicylic acid, a pain-relieving molecule found in bark of the willow tree.
Sulfhydryl R—S—H	*Thiols* have an —SH group, which is polar and hydrophilic due to its electronegative S atom. Certain amino acids, the building blocks of proteins, contain —SH groups, which help stabilize the shape of proteins. An example is the amino acid cysteine.	**Phosphate** O ‖ R—O—P—O⁻ │ O⁻	*Phosphates* contain a phosphate group $(-PO_4^{2-})$, which is very hydrophilic due to the dual negative charges. An important example is adenosine triphosphate (ATP), which transfers chemical energy between organic molecules during chemical reactions.
Carbonyl O ‖ R—C—R or O ‖ R—C—H	*Ketones* contain a carbonyl group within the carbon skeleton. The carbonyl group is polar and hydrophilic due to its electronegative O atom. *Aldehydes* have a carbonyl group at the end of the carbon skeleton.	**Amino** H │ R—N ＼ H or H │ R—N⁺—H │ H	*Amines* have an —NH₂ group, which can act as a base and pick up a hydrogen ion, giving the amino group a positive charge. At the pH of body fluids, most amino groups have a charge of 1+. All amino acids have an amino group at one end.
Carboxyl O ‖ R—C—OH or O ‖ R—C—O⁻	*Carboxylic acids* contain a carboxyl group at the end of the carbon skeleton. All amino acids have a —COOH group at one end. The negatively charged form predominates at the pH of body cells and is hydrophilic.		

*The letter R represents the carbon skeleton of the molecule.

some of their properties. Because organic molecules often are big, there are shorthand methods for representing their structural formulas. Figure 2.14 shows two ways to indicate the structure of the sugar glucose, a molecule with a ring-shaped carbon skeleton that has several hydroxyl groups attached.

Small organic molecules can combine into very large molecules that are called **macromolecules** (*macro-* = large). Macromolecules are usually **polymers** (*poly-* = many; *-mers* = parts). A polymer is a large molecule formed by the covalent bonding of many identical or similar small building-block molecules called **monomers** (*mono-* = one). Usually, the reaction that joins two monomers is a dehydration synthesis. In this type of reaction, a hydrogen atom is removed from one monomer and a hydroxyl group is removed from the other to form a molecule of water (see Figure 2.15). Macromolecules such as carbohydrates, lipids, proteins, and nucleic acids are assembled in cells via dehydration synthesis reactions.

Molecules that have the same molecular formula but different structures are called **isomers** (Ī-so-merz; *iso-* = equal or the same). For example, the molecular formulas for the sugars glucose and fructose are both $C_6H_{12}O_6$. The individual atoms, however, are positioned differently along the carbon skeleton (see Figure 2.15).

Carbohydrates

Carbohydrates include sugars, starches, glycogen, and cellulose. Even though they are a large and diverse group of organic compounds and have several functions, carbohydrates represent only 2 – 3% of your total body mass. Plants store carbohydrate as starch and use the carbohydrate cellulose to build plant cell walls. Although humans eat cellulose, the most plentiful organic substance on Earth, they cannot digest it. It does, however, create bulk, which helps to move food and wastes along the gastrointestinal tract. In humans and animals, carbohydrates function mainly as a source of chemical energy for generating ATP needed to drive metabolic reactions. Only a few carbohydrates are used for building structural units. One example is deoxyribose, a type of sugar that is a building block of deoxyribonucleic acid (DNA), the molecule that carries inherited genetic information.

Carbon, hydrogen, and oxygen are the elements found in carbohydrates. The ratio of hydrogen to oxygen atoms is usually 2 : 1, the same as in water. Although there are exceptions, carbohydrates generally contain one water molecule for each carbon atom. This is the reason they are called carbohydrates, which means "watered carbon." The three major groups of carbohydrates, based on their sizes, are monosaccharides, disaccharides, and polysaccharides (Table 2.6).

Monosaccharides and Disaccharides: The Simple Sugars

Monosaccharides and disaccharides are known as **simple sugars.** The monomers of carbohydrates, **monosaccharides** (mon'-ō-SAK-a-rīds; *sacchar-* = sugar), contain from three to seven carbon atoms. They are designated by names ending in "-ose" with a prefix that indicates the number of carbon atoms. For example, monosaccharides with three carbons are called *trioses* (*tri-* = three). There are also *tetroses* (four-carbon sugars), *pentoses* (five-carbon sugars), *hexoses* (six-carbon sugars), and *heptoses* (seven-carbon sugars). Cells throughout the body break down the hexose glucose to produce ATP.

Two monosaccharide molecules can combine by dehydration synthesis to form one **disaccharide** (dī-SAK-a-rīd; *di-* = two) molecule and a molecule of water. For example, molecules of the monosaccharides glucose and fructose combine to form a molecule of the disaccharide sucrose (table sugar), as shown in

Figure 2.14 **Alternate ways to write the structural formula for glucose.**

🔑 In standard shorthand, carbon atoms are understood to be at locations where two bond lines intersect, and single hydrogen atoms are not indicated.

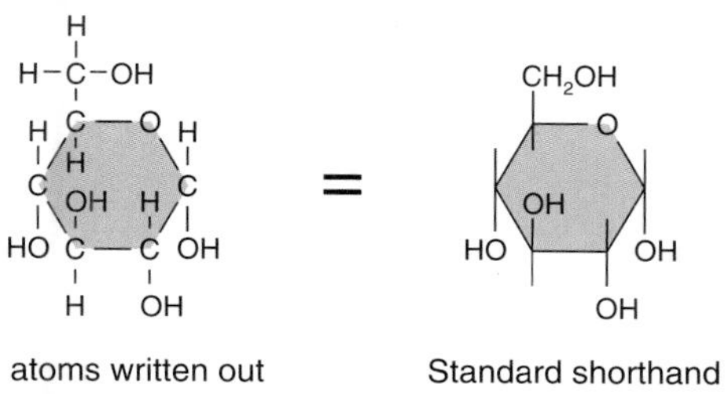

All atoms written out   Standard shorthand

❓ **How many hydroxyl groups does a molecule of glucose have? How many carbon atoms are part of glucose's carbon skeleton?**

Table 2.6	Major Carbohydrate Groups
Type of Carbohydrate	**Examples**
Monosaccharides	Glucose (the main blood sugar) Fructose (found in fruits) Galactose (in milk sugar) Deoxyribose (in DNA) Ribose (in RNA)
Disaccharides	Sucrose (table sugar) = glucose + fructose Lactose (milk sugar) = glucose + galactose Maltose = glucose + glucose
Polysaccharides	Glycogen, the stored form of carbohydrate in animals Starch, the stored form of carbohydrate in plants and main carbohydrate in food Cellulose, part of cell walls in plants; not digested by humans but aids movement of food through intestines

Figure 2.15 **Structural and molecular formulas for the monosaccharides glucose and fructose and the disaccharide sucrose.** In dehydration synthesis (read from left to right), two smaller molecules, glucose and fructose, are joined to form a larger molecule of sucrose. Note the loss of a water molecule. In hydrolysis (read from right to left), the addition of a water molecule to the larger sucrose molecule breaks the disaccharide into two smaller molecules, glucose and fructose.

🔑 **Monosaccharides are the monomers used to build carbohydrates.**

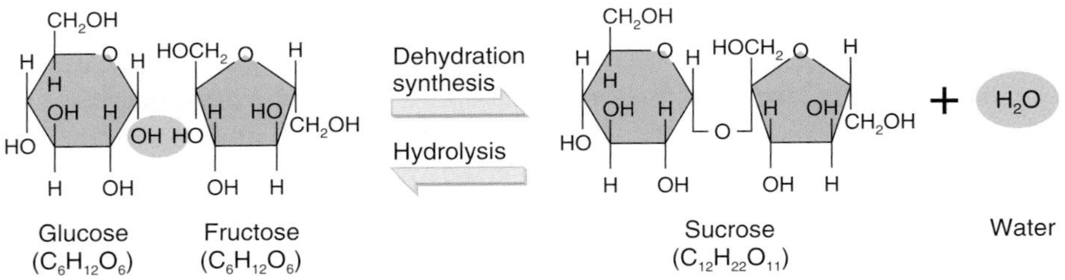

Dehydration synthesis

Hydrolysis

H_2O

Glucose ($C_6H_{12}O_6$)   Fructose ($C_6H_{12}O_6$)   Sucrose ($C_{12}H_{22}O_{11}$)   Water

❓ **How many carbon atoms can you count in fructose? In sucrose?**

Figure 2.15. Glucose and fructose are isomers. Although they have the same molecular formula, they are different monosaccharides because the relative positions of the oxygen and carbon atoms are different. Notice that the formula for sucrose is $C_{12}H_{22}O_{11}$, not $C_{12}H_{24}O_{12}$, because a molecule of water is removed as the two monosaccharides are joined.

Disaccharides can also be split into smaller, simpler molecules by hydrolysis. A molecule of sucrose, for example, may be hydrolyzed into its components, glucose and fructose, by the addition of water. Figure 2.15 also illustrates this reaction.

Polysaccharides

The third major group of carbohydrates are the **polysaccharides** (pol′-ē-SAK-a-rīds). Each polysaccharide molecule contains tens or hundreds of monosaccharides joined through dehydration synthesis reactions. Unlike simple sugars, polysaccharides usually are not soluble in water and do not taste sweet. The main polysaccharide in the human body is glycogen, which is made entirely of glucose monomers linked to one another in branching chains (Figure 2.16). A small amount of carbohydrates are stored as glycogen in the liver and skeletal muscles. Like disaccharides, polysaccharides can be broken down into monosaccharides through hydrolysis reactions. For example, when the blood glucose level falls, liver cells have the ability to break down glycogen into glucose and release it into the blood. In this way, glucose is provided for body cells, which break it down to synthesize ATP.

Lipids

A second important group of organic compounds is **lipids** (*lip-* = fat). Lipids make up 18–25% of body mass in lean adults. Like carbohydrates, lipids contain carbon, hydrogen, and oxygen. Unlike carbohydrates, they do not have a 2:1 ratio of hydrogen to oxygen. The proportion of electronegative oxygen atoms in lipids is usually smaller than in carbohydrates, so there are fewer polar covalent bonds. As a result, most lipids are insoluble in polar solvents such as water; they are *hydrophobic*. Non-

Figure 2.16 **Part of a glycogen molecule, the main polysaccharide in the human body.**

🔑 **Glycogen is made up of glucose monomers and is the storage form of carbohydrate in the human body.**

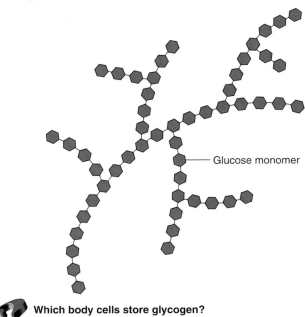

Glucose monomer

❓ **Which body cells store glycogen?**

polar solvents such as chloroform and ether, however, readily dissolve lipids. Because they are hydrophobic, only the smallest lipids (some fatty acids) can dissolve in watery blood plasma. To become more soluble in blood plasma, other lipid molecules form complexes with hydrophilic protein molecules. The resulting lipid and protein particles are termed **lipoproteins.**

The diverse lipid family includes triglycerides (fats and oils), phospholipids (lipids that contain phosphorus), steroids (lipids that contain rings of carbon atoms), eicosanoids (20-carbon lipids), and a variety of other lipids, including fatty acids, fat-soluble vitamins (vitamins A, D, E, and K), and lipoproteins. Table 2.7 introduces the various types of lipids and highlights their roles in the human body.

Triglycerides

The most plentiful lipids in your body and in your diet are the **triglycerides** (trī-GLI-cer-īdes; *tri-* = three), also known as **triacylglycerols.** At room temperature, triglycerides may be either solids (fats) or liquids (oils), and they are the body's most highly concentrated form of chemical energy. Triglycerides provide more than twice as much energy per gram as do carbohydrates and proteins. Our capacity to store triglycerides in adipose (fat) tissue is unlimited for all practical purposes. Excess dietary carbohydrates, proteins, fats, and oils all have the same fate: They are deposited in adipose tissue as triglycerides.

A triglyceride consists of two types of building blocks, a single glycerol molecule and three fatty acid molecules. A three-carbon **glycerol** molecule forms the backbone of a triglyceride (Figure 2.17). Three **fatty acids** are attached by dehydration synthesis reactions, one to each carbon of the glycerol backbone. The chemical bond formed where each water molecule is removed is an *ester linkage* (see Table 2.5). The reverse reaction, hydrolysis, breaks down a single molecule of a triglyceride into three fatty acids and glycerol.

Saturated fats are triglycerides that contain only *single covalent bonds* between fatty acid carbon atoms. Because they lack double bonds, each carbon atom is *saturated with hydrogen atoms* (see, for example, palmitic acid and stearic acid in Figure 2.17c). Triglycerides with mainly saturated fatty acids usually are solid at room temperature. Although saturated fats occur mostly in animal tissues, they also are found in a few plant products, such as cocoa butter, palm oil, and coconut oil.

Monounsaturated fats contain fatty acids with *one double covalent bond* between two fatty acid carbon atoms. Thus, they are not completely saturated with hydrogen atoms (see, for example, oleic acid in Figure 2.17c). Olive oil and peanut oil are rich in triglycerides with monounsaturated fatty acids.

Polyunsaturated fats contain *more than one double covalent bond* between fatty acid carbon atoms. An example is linoleic acid. Canola oil, corn oil, safflower oil, sunflower oil, and soybean oil contain a high percentage of polyunsaturated fatty acids.

Table 2.7	Types of Lipids in the Body			
Type of Lipid	**Functions**		**Type of Lipid**	**Functions**
Triglycerides *(fats and oils)*	Protection, insulation, energy storage.		**Other Lipids**	
Phospholipids	Major lipid component of cell membranes.		*Fatty acids*	Catabolized to generate adenosine triphosphate (ATP) or used to synthesize triglycerides and phospholipids.
Steroids				
Cholesterol	Minor component of all animal cell membranes; precursor of bile salts, vitamin D, and steroid hormones.		*Carotenes*	Needed for synthesis of vitamin A, which is used to make visual pigments in the eyes.
Bile salts	Needed for absorption of dietary lipids.		*Vitamin E*	Promotes wound healing, prevents tissue scarring, contributes to the normal structure and function of the nervous system, and functions as an antioxidant.
Vitamin D	Helps regulate calcium level in the body; needed for bone growth and repair.			
Adrenocortical hormones	Help regulate metabolism, resistance to stress, and salt and water balance.		*Vitamin K*	Required for synthesis of blood-clotting proteins.
Sex hormones	Stimulate reproductive functions and sexual characteristics.		*Lipoproteins*	Transport lipids in the blood, carry triglycerides and cholesterol to tissues, and remove excess cholesterol from the blood.
Eicosanoids	Have diverse effects on blood clotting, inflammation, immunity, stomach acid secretion, airway diameter, lipid breakdown, and smooth muscle contraction.			

Figure 2.17   The formation of a triglyceride (triacylglycerol) from a glycerol and three fatty acid molecules.   Each time a glycerol (a) and a fatty acid (b) are joined in dehydration synthesis, a molecule of water is removed. An ester linkage joins the glycerol to each of the three molecules of fatty acids, which vary in length and the number and location of double bonds between carbon atoms (C = C). Shown here (c) is a triglyceride molecule that contains two saturated fatty acids and a monounsaturated fatty acid. The kink (bend) in the oleic acid occurs at the double bond.

One glycerol and three fatty acids are the building blocks of triglycerides.

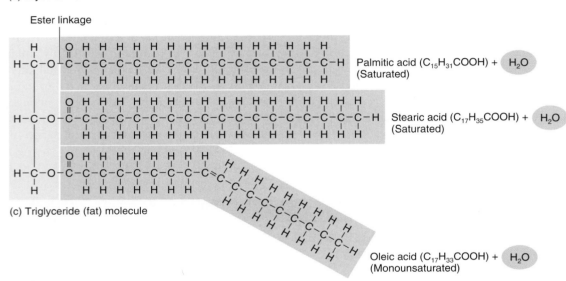

(a) Glycerol molecule

(b) Fatty acid molecule

Palmitic acid ($C_{15}H_{31}COOH$)

Ester linkage

Palmitic acid ($C_{15}H_{31}COOH$) + H_2O
(Saturated)

Stearic acid ($C_{17}H_{35}COOH$) + H_2O
(Saturated)

Oleic acid ($C_{17}H_{33}COOH$) + H_2O
(Monounsaturated)

(c) Triglyceride (fat) molecule

Does the oxygen in the water molecule removed during dehydration synthesis come from the glycerol or from a fatty acid?

Phospholipids

Like triglycerides, **phospholipids** have a glycerol backbone and two fatty acid chains attached to the first two carbons. In the third position, however, a phosphate group (PO_4^{3-}) links a small charged group that usually contains nitrogen (N) to the backbone (Figure 2.18). This portion of the molecule (the "head") is polar and can form hydrogen bonds with water molecules. The two fatty acids (the "tails"), by contrast, are nonpolar and can interact only with other lipids. Molecules that have both polar and nonpolar parts are said to be **amphipathic** (am-fi-PATH-ic; *amphi-* = on both sides; *-pathic* = feeling). Amphipathic phos-

pholipids line up tails-to-tails in a double row to make up much of the membrane that surrounds each cell (Figure 2.18c).

Steroids

The structure of **steroids** differs considerably from that of the triglycerides. Steroids have four rings of carbon atoms (colored gold in Figure 2.19). Body cells synthesize other steroids from cholesterol (Figure 2.19a), which has a large nonpolar region consisting of the four rings and a hydrocarbon tail. In the body, the commonly encountered steroids, such as cholesterol, estrogens, testosterone, cortisol, bile salts, and vitamin D, are further

Figure 2.18 Phospholipids. (a) In the synthesis of phospholipids, two fatty acids attach to the first two carbons of the glycerol backbone. A phosphate group links a small charged group to the third carbon in glycerol. In (b), the circle represents the polar head region, and the two wavy lines represent the two nonpolar tails. Double bonds in the fatty acid hydrocarbon chain often form kinks in the tail.

 Phospholipids are amphipathic molecules, having both polar and nonpolar regions.

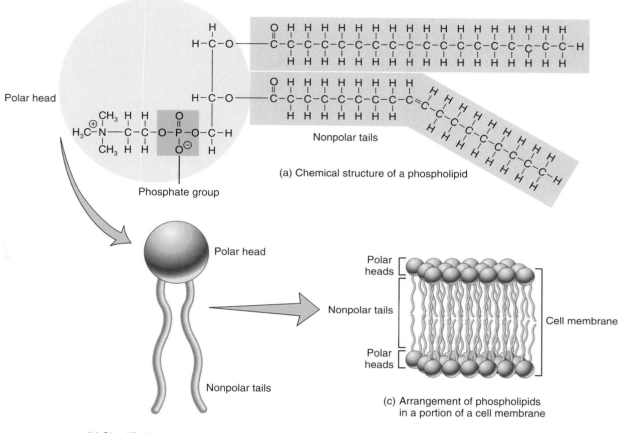

(a) Chemical structure of a phospholipid

(b) Simplified way to draw a phospholipid

(c) Arrangement of phospholipids in a portion of a cell membrane

 **Which portion of a phospholipid is hydrophilic, and which portion is hydrophobic?**

classified as **sterols** because they also have at least one hydroxyl (alcohol) group (−OH). The polar hydroxyl groups make sterols weakly amphipathic.

Saturated Fats, Cholesterol, and Atherosclerosis

Atherosclerosis is a progressively worsening disorder in which fatty plaques (deposits) form in the walls of arteries. Plaque buildup narrows the passageway and severely limits blood flow in advanced cases. Although many factors contribute, people who eat a diet high in saturated fats and cholesterol run a greater risk of developing atherosclerosis than do people who eat a diet lower in these lipids. To reduce dietary intake of saturated fats and cholesterol, eat less red meat (beef, pork, and lamb) and fat-laden dairy products (whole milk, butter, and cheese). ■

Other Lipids

Eicosanoids (ī-KŌ-sa-noids; *eicosan-* = twenty) are lipids derived from a 20-carbon fatty acid called arachidonic acid. The two principal subclasses of eicosanoids are the **prostaglandins** (pros′-ta-GLAN-dins) and the **leukotrienes** (loo′-kō-TRĪ-ēnz).

Figure 2.19 Steroids. All steroids have four rings of carbon atoms.

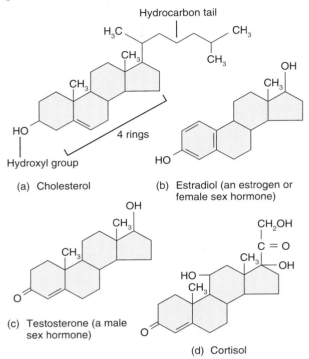

🔑 **Cholesterol, which is synthesized in the liver, is the starting material for synthesis of other steroids in the body.**

(a) Cholesterol

(b) Estradiol (an estrogen or female sex hormone)

(c) Testosterone (a male sex hormone)

(d) Cortisol

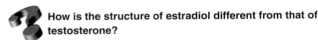

How is the structure of estradiol different from that of testosterone?

Prostaglandins have a wide variety of functions. They modify responses to hormones, contribute to the inflammatory response (Chapter 22), prevent stomach ulcers, dilate (enlarge) airways to the lungs, regulate body temperature, and influence formation of blood clots, to name just a few effects. Leukotrienes participate in allergic and inflammatory responses.

Body lipids also include fatty acids (which can undergo either hydrolysis to provide ATP or dehydration synthesis to build triglycerides and phospholipids); fat-soluble vitamins such as beta-carotenes (the yellow-orange pigments in egg yolk, carrots, and tomatoes that are converted to vitamin A); vitamins D, E, and K; and lipoproteins.

Proteins

Proteins are large molecules that contain carbon, hydrogen, oxygen, and nitrogen. Some proteins also contain sulfur. A normal, lean adult body is 12 – 18% protein. Much more complex in structure than carbohydrates or lipids, proteins have many roles in the body and are largely responsible for the structure of body tissues. Enzymes are proteins that speed up most biochemical reactions. Other proteins work as "motors" to drive muscle contraction. Antibodies are proteins that defend against invading microbes. Some hormones that regulate homeostasis also are proteins. Table 2.8 describes several important functions of proteins.

Amino Acids and Polypeptides

The monomers of proteins are **amino acids** (a-MĒ-nō). Each of the 20 different amino acids has three important functional

Table 2.8	Functions of Proteins		
Type of Protein	**Functions**	**Type of Protein**	**Functions**
Structural	Form structural framework of various parts of the body. *Examples:* collagen in bone and other connective tissues, and keratin in skin, hair, and fingernails.	**Immunological**	Aid responses that protect body against foreign substances and invading pathogens. *Examples:* antibodies and interleukins.
Regulatory	Function as hormones that regulate various physiological processes; control growth and development; as neurotransmitters, mediate responses of the nervous system. *Examples:* the hormone insulin, which regulates blood glucose level, and a neurotransmitter known as substance P, which mediates sensation of pain in the nervous system.	**Transport**	Carry vital substances throughout body. *Example:* hemoglobin, which transports most oxygen and some carbon dioxide in the blood.
		Catalytic	Act as enzymes that regulate biochemical reactions. *Examples:* salivary amylase, sucrase, and ATPase.
Contractile	Allow shortening of muscle cells, which produces movement. *Examples:* myosin and actin.		

groups attached to a central carbon atom (Figure 2.20a): (1) an amino group (−NH₂), (2) an acidic carboxyl group (−COOH), and (3) a side chain (R group). At the normal pH of body fluids, both the amino group and the carboxyl group are ionized (Figure 2.20b). The distinctive side chain gives each amino acid its individual chemical identity (Figure 2.20c).

Synthesis of a protein takes place in stepwise fashion—one amino acid is joined to a second, a third is then added to the first two, and so on. The covalent bond joining each pair of amino acids is a **peptide bond.** It always forms between the carbon of the carboxyl group (−COOH) of one amino acid and the nitrogen of the amino group (−NH₂) of another. At the site of formation of a peptide bond, a molecule of water is removed (Figure 2.21). Thus, this is a dehydration synthesis reaction. Breaking a peptide bond, as occurs during digestion of dietary proteins, is a hydrolysis reaction (Figure 2.21).

When two amino acids combine, a **dipeptide** results. Adding another amino acid to a dipeptide produces a **tripeptide.** Further additions of amino acids result in the formation of a chainlike **peptide** (4–9 amino acids) or **polypeptide** (10–2000 or more amino acids). Small proteins may consist of a single polypeptide chain with as few as 50 amino acids. Other proteins have hundreds or thousands of amino acids and may consist of two or more polypeptide chains folded together.

A great variety of proteins is possible because each variation in the number or sequence of amino acids can produce a different protein. The situation is similar to using an alphabet of 20 letters to form words. Each different amino acid is like a letter, and their various combinations (peptides, polypeptides, or proteins) are like the words.

Levels of Structural Organization in Proteins

Proteins exhibit four levels of structural organization. The **primary structure** is the unique sequence of amino acids that are linked by covalent peptide bonds to form a polypeptide chain

Figure 2.20 **Amino acids.** (a) In keeping with their name, amino acids have an amino group (shaded blue) and a carboxyl (acid) group (shaded red). The side chain (R group) is different in each amino acid. (b) At pH close to 7, both the amino group and the carboxyl group are ionized. (c) Glycine is the simplest amino acid; the side chain is a single H atom. Cysteine is one of two amino acids that contain sulfur (S). The side chain in tyrosine contains a six-carbon ring. Lysine has a second amino group at the end of its side chain.

Body proteins contain 20 different amino acids, each of which has a unique side chain.

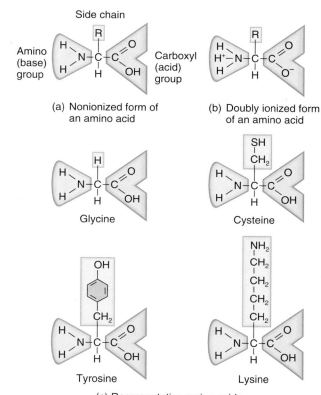

(a) Nonionized form of an amino acid

(b) Doubly ionized form of an amino acid

Glycine

Cysteine

Tyrosine

Lysine

(c) Representative amino acids

In an amino acid, what is the minimum number of carbon atoms? Of nitrogen atoms?

Figure 2.21 **Formation of a peptide bond between two amino acids during dehydration synthesis.** In this example, glycine is joined to alanine, forming a dipeptide (read from left to right). Breaking a peptide bond occurs via hydrolysis (read from right to left).

Amino acids are the monomers used to build proteins.

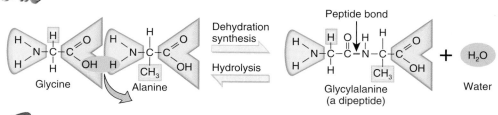

Glycine Alanine Dehydration synthesis Hydrolysis Peptide bond Glycylalanine (a dipeptide) Water

What type of reaction takes place during catabolism of proteins?

(Figure 2.22a). A protein's primary structure is genetically determined. Any changes in a protein's amino acid sequence can have serious consequences for body cells. In sickle-cell disease, for example, a nonpolar amino acid (valine) replaces a polar amino acid (glutamate) at just two locations in the oxygen-carrying protein hemoglobin. This change of amino acids diminishes hemoglobin's water solubility. As a result, the altered hemoglobin tends to form crystals inside red blood cells, producing deformed, sickle-shaped cells that cannot properly squeeze through narrow blood vessels.

The **secondary structure** of a protein is the repeated twisting or folding of neighboring amino acids in the polypeptide chain (Figure 2.22b). Two common secondary structures are *alpha helixes* (clockwise spirals) and *pleated sheets.* The secondary structure of a protein is stabilized by hydrogen bonds, which form at regular intervals along the polypeptide backbone.

The **tertiary structure** (TUR-shē-er′-ē) refers to the three-dimensional shape of a polypeptide chain. Each protein has a unique tertiary structure that determines how it will function. The tertiary folding pattern may allow amino acids at opposite ends of the chain to be close neighbors (Figure 2.22c). Several types of bonds can contribute to a protein's tertiary structure. The strongest but least common bonds are S−S covalent bonds, called *disulfide bridges.* These bonds form between the sulfhydryl groups of two monomers of the amino acid cysteine. Many weak bonds—hydrogen bonds, ionic bonds, and hydrophobic interactions—also help determine the folding pattern. Some parts of a polypeptide are attracted to water (hydrophilic), and other parts are repelled by it (hydrophobic). Because most proteins in our body exist in watery surroundings, the folding process places most amino acids that have hydrophobic side chains in the central core of the protein, away from the protein's surface. Often, helper molecules known as *chaperones* aid the folding process.

When a protein contains more than one polypeptide chain, the arrangement of the individual polypeptide chains relative to one another is the **quaternary structure** (KWA-ter-ner′-ē; Figure 2.22d). The bonds that hold polypeptide chains together are similar to those that maintain the tertiary structure.

Proteins vary tremendously in structure. Different proteins have different architectures and different three-dimensional shapes. This variation in structure and shape is directly related to their diverse functions. In practically every case, the function of a protein depends on its ability to recognize and bind to some other molecule, much as a key fits in a lock. Thus, a hormone binds to a specific protein on a cell whose function it will alter, and an antibody protein binds to a foreign substance (antigen) that has invaded the body. A protein's unique shape permits it to interact with other molecules to carry out a specific function.

Homeostatic mechanisms maintain the temperature and chemical composition of body fluids, which allow body proteins to keep their proper three-dimensional shapes. If a protein encounters an altered environment, it may unravel and lose its characteristic shape (secondary, tertiary, and quaternary structure). This process is called **denaturation.** Denatured proteins are no longer functional. A common example of permanent denaturation is seen in frying an egg. In a raw egg the soluble egg-white protein (albumin) is a clear, viscous fluid. When heat is applied to the egg, however, the protein denatures, becomes insoluble, and turns white.

Enzymes

In living cells, most catalysts are protein molecules called **enzymes** (EN-zīms). Some enzymes consist of two parts—a protein portion, called the **apoenzyme** (ā′-pō-EN-zīm), and a nonprotein portion, called a **cofactor.** The cofactor may be a metal ion (such as iron, magnesium, zinc, or calcium) or an organic molecule called a *coenzyme.* Coenzymes often are derived from vitamins. The names of enzymes usually end in the suffix *-ase.* All enzymes can be grouped according to the types of chemical reactions they catalyze. For example, *oxidases* add oxygen, *kinases* add phosphate, *dehydrogenases* remove hydrogen, *ATPases* split ATP, *anhydrases* remove water, *proteases* break down proteins, and *lipases* break down triglycerides.

Enzymes catalyze specific reactions. They do so with great efficiency and with many built-in controls. Three important properties of enzymes are as follows

1. *Enzymes are highly specific.* Each particular enzyme binds only to specific **substrates**—the reactant molecules on which the enzyme acts. Of the more than 1000 known enzymes in your body, each has a characteristic three-dimensional shape with a specific surface configuration, which allows it to recognize and bind to certain substrates. In some cases, the part of the enzyme that catalyzes the reaction, called the **active site,** is thought to "fit" the substrate like a key fits in a lock. In other cases the active site changes its shape to fit snugly around the substrate once the substrate enters the active site. This change in shape is known as an *induced fit.*

Not only is an enzyme matched to a particular substrate, it also catalyzes a specific reaction. From among the large number of diverse molecules in a cell, an enzyme must recognize the correct substrate and then take it apart or merge it with another substrate to form one or more specific products.

2. *Enzymes are very efficient.* Under optimal conditions, enzymes can catalyze reactions at rates that are from 100 million to 10 billion times more rapid than those of similar reactions occurring without enzymes. The number of substrate molecules that a single enzyme molecule can convert to product molecules in one second is generally between 1 and 10,000 and can be as high as 600,000.

3. *Enzymes are subject to a variety of cellular controls.* Their rate of synthesis and their concentration at any given time are under the control of a cell's genes. Substances within the cell may either enhance or inhibit the activity of a given enzyme. Many enzymes have both active and inactive forms in cells. The rate at which the inactive form becomes active or vice versa is determined by the chemical environment inside the cell.

Figure 2.22   Levels of structural organization in proteins.   (a) The primary structure is the sequence of amino acids in the polypeptide. (b) Common secondary structures include alpha helixes and pleated sheets. For simplicity, the amino acid side groups are not shown here. (c) The tertiary structure is the overall folding pattern that produces a distinctive, three-dimensional shape. (d) The quaternary structure in a protein is the arrangement of two or more polypeptide chains relative to one another.

🔑 The unique shape of each protein permits it to carry out specific functions.

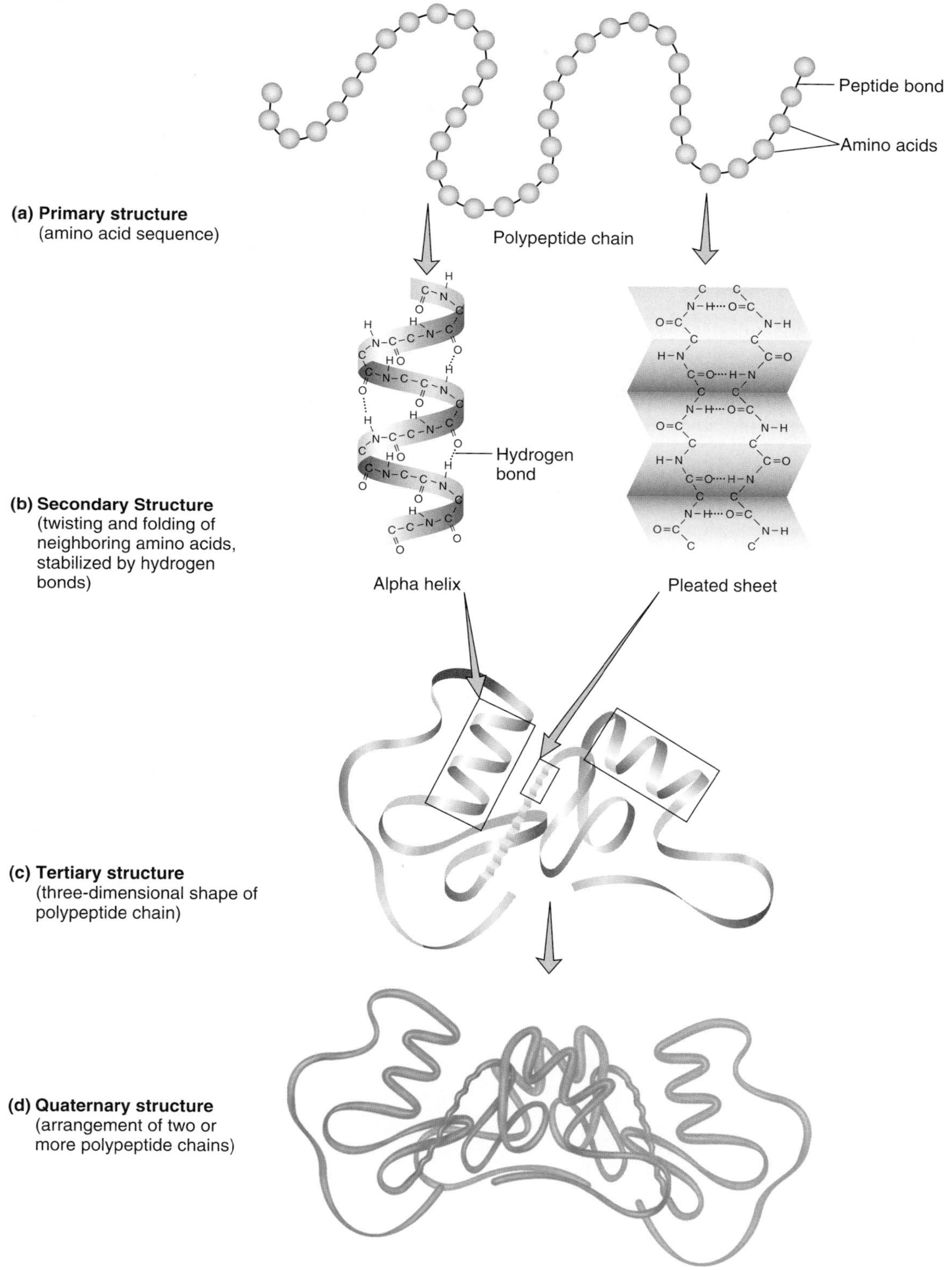

(a) Primary structure
(amino acid sequence)

Peptide bond

Amino acids

Polypeptide chain

(b) Secondary Structure
(twisting and folding of neighboring amino acids, stabilized by hydrogen bonds)

Hydrogen bond

Alpha helix

Pleated sheet

(c) Tertiary structure
(three-dimensional shape of polypeptide chain)

(d) Quaternary structure
(arrangement of two or more polypeptide chains)

❓ Do all proteins have a quaternary structure?

Enzymes lower the activation energy of a chemical reaction by decreasing the "randomness" of the collisions between molecules. They also help bring the substrates together in the proper orientation so that the reaction can occur. Figure 2.23 depicts how an enzyme works:

1 The substrates make contact with the active site on the surface of the enzyme molecule, forming a temporary intermediate compound called the **enzyme-substrate complex.** In this reaction the two substrate molecules are sucrose (a disaccharide) and water.

2 The substrate molecules are transformed by either the rearrangement of existing atoms, the breakdown of the substrate molecule, or the combination of several substrate molecules into the products of the reaction. Here the products are two monosaccharides: glucose and fructose.

3 After the reaction is completed and the reaction products move away from the enzyme, the unchanged enzyme is free to attach to other substrate molecules.

Sometimes a single enzyme may catalyze a reversible reaction in either direction, depending on the relative amounts of the substrates and products. For example, the enzyme *carbonic anhydrase* catalyzes the following reversible reaction:

$$CO_2 + H_2O \underset{\text{Carbonic anhydrase}}{\rightleftharpoons} H_2CO_3$$

Carbon dioxide   Water   Carbonic acid

During exercise, when more CO_2 is produced and released into the blood, the reaction flows to the right, increasing the amount of carbonic acid in the blood. Then, as you exhale CO_2, its level in the blood falls and the reaction flows to the left, converting carbonic acid to CO_2 and H_2O.

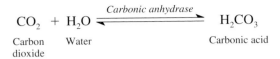

Galactosemia

Galactosemia (ga-lak-tō-SĒ-mē-a; *galactos-* = milk; *-emia-* = in the blood) is an inherited disorder in which the monosaccharide galactose builds up in the blood. Galactose level increases because the enzyme that converts galactose to glucose (galactose-1-phosphate uridylyl transferase) is absent. An infant with the disorder will fail to thrive within a week after birth due to anorexia (loss of appetite), vomiting, and diarrhea. The main source of galactose in the infant's diet is the disaccharide lactose (milk sugar), which undergoes hydrolysis into glucose and galactose. Hence, treatment of galactosemia consists of eliminating milk from the diet. If treatment is delayed, the infant will remain physically small and become mentally retarded. ■

Nucleic Acids: Deoxyribonucleic Acid (DNA) and Ribonucleic Acid (RNA)

Nucleic acids (nū-KLĒ-ic), so named because they were first discovered in the nuclei of cells, are huge organic molecules that contain carbon, hydrogen, oxygen, nitrogen, and phosphorus. Nucleic acids are of two varieties. The first, **deoxyribonucleic acid (DNA)** (dē-ok′-sē-rī-bō-nū-KLĒ-ik), forms the inherited genetic material inside each cell. In humans, each **gene** is a segment of a DNA molecule. Our genes determine the traits we inherit, and by controlling protein synthesis they regulate most of the activities that take place in body cells throughout a lifetime. When a cell divides, its hereditary information passes on to the next generation of cells. **Ribonucleic acid (RNA),** the second type of nucleic acid, relays instructions from the genes to guide each cell's synthesis of proteins from amino acids.

The monomers of nucleic acids are **nucleotides.** A nucleic acid is a chain of repeating nucleotides. Each nucleotide of DNA consists of three parts (Figure 2.24a):

1. *Nitrogenous base.* DNA contains four different nitrogenous bases, which contain atoms of C, H, O, and N. In DNA the four **nitrogenous bases** are adenine (A), thymine (T), cytosine (C), and guanine (G). Whereas adenine and guanine are larger, double-ring bases called **purines** (PŪR-ēnz), thymine and cytosine are smaller, single-ring bases called **pyrimidines** (pī-RIM-i-dēnz). The nucleotides are named according to the base that is present. For instance, a nucleotide containing thymine is called a thymine nucleotide, one containing adenine is called an adenine nucleotide, and so on.

Figure 2.23 How an enzyme works.

An enzyme speeds up a chemical reaction without being altered or consumed.

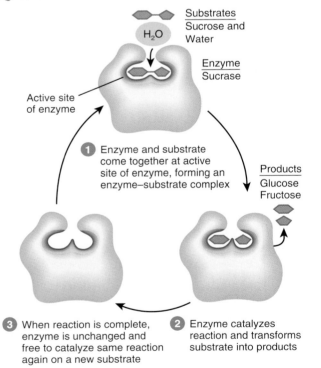

Substrates Sucrose and Water

H_2O

Enzyme Sucrase

Active site of enzyme

1 Enzyme and substrate come together at active site of enzyme, forming an enzyme–substrate complex

Products Glucose Fructose

3 When reaction is complete, enzyme is unchanged and free to catalyze same reaction again on a new substrate

2 Enzyme catalyzes reaction and transforms substrate into products

Why is it that sucrase cannot catalyze the formation of sucrose from glucose and fructose?

Figure 2.24 DNA molecule. (a) A nucleotide consists of a base, a pentose sugar, and a phosphate group. (b) The paired bases project toward the center of the double helix. The structure is stabilized by hydrogen bonds (dotted lines) between each base pair. There are two hydrogen bonds between adenine and thymine and three between cytosine and guanine.

Nucleotides are the monomers of nucleic acids.

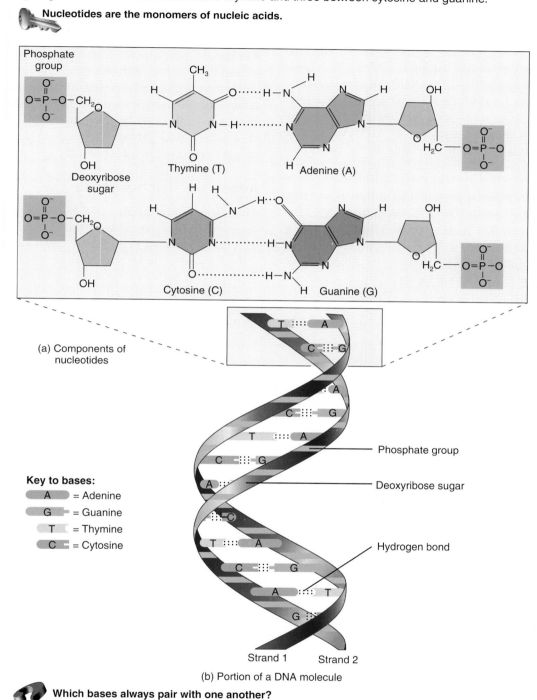

(a) Components of nucleotides

Key to bases:
A = Adenine
G = Guanine
T = Thymine
C = Cytosine

Phosphate group
Deoxyribose sugar
Hydrogen bond

Strand 1   Strand 2

(b) Portion of a DNA molecule

Which bases always pair with one another?

2. *Pentose sugar.* A five-carbon sugar called **deoxyribose** attaches to each base in DNA.

3. *Phosphate group.* Alternating phosphate groups (PO_4^{3-}) and pentoses form the "backbone" of a DNA strand; the bases project inward from the backbone chain (Figure 2.24b).

In 1953, F.H.C. Crick of Great Britain and J.D. Watson, a young American scientist, published a brief paper describing how these three components might be arranged in DNA. Their insights into data gathered by others led them to construct a model so elegant and simple that the scientific world immediately knew it was correct! In the Watson–Crick **double helix** model, DNA resembles a spiral ladder (Figure 2.24b). Two strands of alternating phosphate groups and deoxyribose sugars form the uprights of the ladder. Paired bases, held together by hydrogen bonds, form the rungs. Adenine always pairs with thymine, and cytosine always pairs with guanine. If you know the sequence of bases in one strand of DNA, you can predict the sequence on the complementary (second) strand. Each time DNA is copied, as when living cells divide to increase their number, the two strands unwind. Each strand serves as the template or mold on which to construct a new second strand. Any change that occurs in the base sequence of a DNA strand is called a *mutation.* Some mutations can result in the death of a cell, cause cancer, or produce genetic defects in future generations.

RNA, the second variety of nucleic acid, differs from DNA in several respects. In humans, RNA is single-stranded. The sugar in the RNA nucleotide is the pentose **ribose,** and RNA contains the pyrimidine base uracil (U) instead of thymine. Cells contain three different kinds of RNA: messenger RNA, ribosomal RNA, and transfer RNA. Each has a specific role to perform in carrying out the instructions coded in DNA (described on page 87).

DNA Fingerprinting

A technique called DNA fingerprinting is used in research and in courts of law to ascertain whether a person's DNA matches the DNA obtained from samples or pieces of legal evidence such as blood stains or hairs. In each person, certain DNA segments contain base sequences that are repeated several times. Both the number of repeat copies in one region and the number of regions subject to repeat are different from one person to another. DNA fingerprinting can be done with minute quantities of DNA—for example, from a single strand of hair, a drop of semen, or a spot of blood. It also can be used to identify a crime victim or a child's biological parents and even to determine whether two people have a common ancestor. ■

Adenosine Triphosphate

Adenosine triphosphate (a-DEN-ō-sēn) or **ATP** is the "energy currency" of living systems (Figure 2.25). It functions to transfer the energy liberated in exergonic catabolic reactions to cellular activities that require energy (endergonic reactions). Among these cellular activities are muscular contractions, movement of chromosomes during cell division, movement of structures within cells, transport of substances across cell membranes, and

Figure 2.25 Structures of ATP and ADP. "Squiggles" (~) indicate the two phosphate bonds that can be used to transfer energy. Energy transfer typically involves hydrolysis of the last phosphate bond of ATP.

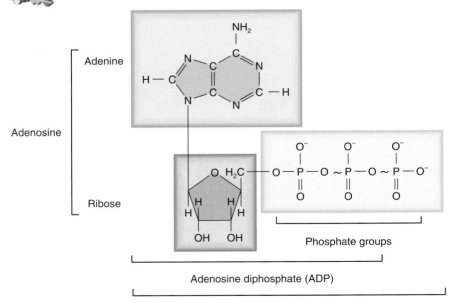

ATP transfers chemical energy to power cellular activities.

What are some cellular activities that depend on energy supplied by ATP?

synthesis of larger molecules from smaller ones. Structurally, ATP consists of three phosphate groups attached to an adenosine unit composed of adenine and the five-carbon sugar ribose.

When the third phosphate group (PO_4^{3-}), symbolized by Ⓟ in the following discussion, is hydrolyzed by the addition of a water molecule, the overall reaction liberates energy. This energy is used by the cell to power its activities. The enzyme that catalyzes the hydrolysis of ATP is called *ATPase*. Removal of the third phosphate group leaves a molecule called **adenosine diphosphate (ADP)**. This reaction is as follows:

$$\text{ATP} + \text{H}_2\text{O} \xrightarrow{\textit{ATPase}} \text{ADP} + \text{Ⓟ} + \text{E}$$

ATP	H₂O		ADP	Ⓟ	E
Adenosine triphosphate	Water		Adenosine diphosphate	Phosphate group	Energy

The energy supplied by the catabolism of ATP into ADP is constantly being used by the cell. As the supply of ATP at any given time is limited, a mechanism exists to replenish it: The enzyme *ATP synthase* catalyzes the addition of a phosphate group to ADP. The reaction is as follows:

$$\text{ADP} + \text{Ⓟ} + \text{E} \xrightarrow{\textit{ATP synthase}} \text{ATP} + \text{H}_2\text{O}$$

ADP	Ⓟ	E		ATP	H₂O
Adenosine diphosphate	Phosphate group	Energy		Adenosine triphosphate	Water

As you can see from this reaction, energy is required to produce ATP. The energy needed to attach a phosphate group to

ADP is supplied mainly by the catabolism of glucose in a process called cellular respiration. Cellular respiration has two phases, anaerobic and aerobic:

1. ***Anaerobic phase.*** In the absence of oxygen, glucose is partially broken down by a series of catabolic reactions into pyruvic acid. Each glucose molecule that is converted into a pyruvic acid molecule yields two molecules of ATP.

2. ***Aerobic phase.*** In the presence of oxygen, glucose is completely broken down into carbon dioxide and water. These reactions generate heat and 36 to 38 ATP molecules.

Chapters 10 and 25 cover the details of cellular respiration.

▶ C H E C K P O I N T

16. How are carbohydrates classified?

17. How are dehydration synthesis and hydrolysis reactions related?

18. What is the importance to the body of triglycerides, phospholipids, steroids, lipoproteins, and eicosanoids?

19. Distinguish among saturated, monounsaturated, and polyunsaturated fats.

20. Define a protein. What is a peptide bond?

21. What are the levels of structural organization in proteins?

22. How do DNA and RNA differ?

23. In the reaction catalyzed by ATP synthase, what are the substrates and products? Is this an exergonic or endergonic reaction?

STUDY OUTLINE

HOW MATTER IS ORGANIZED (p. 27)

1. All forms of matter are composed of chemical elements.
2. Oxygen, carbon, hydrogen, and nitrogen make up about 96% of body mass.
3. Each element is made up of small units called atoms.
4. Atoms consist of a nucleus, which contains protons and neutrons, plus electrons that move about the nucleus in regions called electron shells.
5. The number of protons (the atomic number) distinguishes the atoms of one element from those of another element.
6. The mass number of an atom is the sum of its protons and neutrons.
7. Different atoms of an element that have the same number of protons but different numbers of neutrons are called isotopes. Radioactive isotopes are unstable and decay.
8. The atomic mass of an element is the average mass of all naturally occurring isotopes of that element.
9. An atom that *gives up* or *gains* electrons becomes an ion—an atom that has a positive or negative charge because it has unequal numbers of protons and electrons. Positively charged ions are cations; negatively charged ions are anions.

10. If two atoms share electrons, a molecule is formed. Compounds contain atoms of two or more elements.
11. A free radical is an electrically charged atom or group of atoms with an unpaired electron in its outermost shell. A common example is superoxide, which is formed by the addition of an electron to an oxygen molecule.

CHEMICAL BONDS (p. 31)

1. Forces of attraction called chemical bonds hold atoms together. These bonds result from gaining, losing, or sharing electrons in the valence shell.
2. Most atoms become stable when they have an octet of eight electrons in their valence (outermost) electron shell.
3. When the force of attraction between ions of opposite charge holds them together, an ionic bond has formed.
4. In a covalent bond, atoms share pairs of valence electrons. Covalent bonds may be single, double, or triple and either nonpolar or polar.
5. An atom of hydrogen that forms a polar covalent bond with an oxygen atom or a nitrogen atom may also form a weaker bond,

called a hydrogen bond, with an electronegative atom. The polar covalent bond causes the hydrogen atom to have a partial positive charge (δ^+) that attracts the partial negative charge (δ^-) of neighboring electronegative atoms, often oxygen or nitrogen.

CHEMICAL REACTIONS (p. 34)

1. When atoms combine with or break apart from other atoms, a chemical reaction occurs. The starting substances are the reactants, and the ending ones are the products.
2. Energy, the capacity to do work, is of two principal kinds: potential (stored) energy and kinetic energy (energy of motion).
3. Endergonic reactions require energy, whereas exergonic reactions release energy. ATP couples endergonic and exergonic reactions.
4. The initial energy investment needed to start a reaction is the activation energy. Reactions are more likely when the concentrations and the temperatures of the reacting particles are higher.
5. Catalysts accelerate chemical reactions by lowering the activation energy. Most catalysts in living organisms are protein molecules called enzymes.
6. Synthesis reactions involve the combination of reactants to produce larger molecules. The reactions are anabolic and usually endergonic.
7. In decomposition reactions, a substance is broken down into smaller molecules. The reactions are catabolic and usually exergonic.
8. Exchange reactions involve the replacement of one atom or atoms by another atom or atoms.
9. In reversible reactions, end products can revert to the original reactants.

INORGANIC COMPOUNDS AND SOLUTIONS (p. 37)

1. Inorganic compounds usually are small and usually lack carbon. Organic substances always contain carbon, usually contain hydrogen, and always have covalent bonds.
2. Water is the most abundant substance in the body. It is an excellent solvent and suspending medium, participates in hydrolysis and dehydration synthesis reactions, and serves as a lubricant. Because of its many hydrogen bonds, water molecules are cohesive, which causes a high surface tension. Water also has a high capacity for absorbing heat and a high heat of vaporization.
3. Inorganic acids, bases, and salts dissociate into ions in water. An acid ionizes into anions and hydrogen ions (H^+); a base ionizes into cations and hydroxide ions (OH^-). A salt ionizes into neither H^+ nor OH^-.
4. Mixtures are combinations of elements or compounds that are physically blended together but are not bound by chemical bonds. Solutions, colloids, and suspensions are mixtures with different properties.

5. Two ways to express the concentration of a solution are percentage (mass per volume) and moles per liter. A mole (abbreviated mol) is the amount in grams of any substance that has a mass equal to the combined atomic mass of all its atoms.
6. The pH of body fluids must remain fairly constant for the body to maintain homeostasis. On the pH scale, 7 represents neutrality. Values below 7 indicate acidic solutions, and values above 7 indicate alkaline solutions.
7. Buffer systems usually consist of a weak acid and a weak base. They react with excess H^+ and excess OH^-, which helps maintain pH homeostasis.
8. One important buffer system is the carbonic acid – bicarbonate buffer system. The bicarbonate ion (HCO_3^-) acts as a weak base, and carbonic acid (H_2CO_3) acts as a weak acid.

ORGANIC COMPOUNDS (p. 42)

1. Carbon, with its four valence electrons, bonds covalently with other carbon atoms to form large molecules of many different shapes. Attached to the carbon skeletons of organic molecules are functional groups that confer distinctive chemical properties.
2. Small organic molecules are joined together to form larger molecules by dehydration synthesis reactions in which a molecule of water is removed. In the reverse process, called hydrolysis, large molecules are broken down into smaller ones by adding water.
3. Carbohydrates provide most of the chemical energy needed to generate ATP. They may be monosaccharides, disaccharides, or polysaccharides.
4. Lipids are a diverse group of compounds that include triglycerides (fats and oils), phospholipids, steroids, and eicosanoids. Triglycerides protect, insulate, provide energy, and are stored. Phospholipids are important cell membrane components. Eicosanoids (prostaglandins and leukotrienes) modify hormone responses, contribute to inflammation, dilate airways, and regulate body temperature.
5. Proteins are constructed from amino acids. They give structure to the body, regulate processes, provide protection, help muscles contract, transport substances, and serve as enzymes. Levels of structural organization among proteins are primary, secondary, tertiary, and quaternary.
6. Deoxyribonucleic acid (DNA) and ribonucleic acid (RNA) are nucleic acids consisting of nitrogenous bases, five-carbon (pentose) sugars, and phosphate groups. DNA is a double helix and is the primary chemical in genes. RNA takes part in protein synthesis.
7. Adenosine triphosphate (ATP) is the principal energy-transferring molecule in living systems. When it transfers energy to an endergonic reaction, it is decomposed to adenosine diphosphate (ADP) and Ⓟ. ATP is synthesized from ADP and Ⓟ using the energy supplied by various decomposition reactions, particularly those of glucose.

❓ SELF-QUIZ QUESTIONS

Fill in the blanks in the following statements.

1. Atoms consist of _____, _____, and _____.
2. To be chemically stable, the atoms of all elements except hydrogen and helium must hold _____ electrons in the valence shell.
3. The number of protons located in an atom's nucleus is known as its _____ .
4. The factors that influence the chance that a collision will occur and cause a chemical reaction are _____ and _____.

Indicate whether the following statements are true or false.

5. The four elements which compose most of the body's mass are carbon, hydrogen, oxygen, and nitrogen.
6. Ionic bonds occur when atoms lose, gain, and share electrons in their valence shell.

Choose the one best answer to the following questions.

7. Which of the following statements are *true* of covalent bonds? (1) The atoms form a molecule by sharing one, two, or three pairs of their valence electrons. (2) Covalent bonds can form only between atoms of different elements. (3) The greater the number of electron pairs shared, the stronger the covalent bond. (4) Covalent bonds are the least common chemical bonds in the body. (5) Covalent bonds can be either polar or nonpolar. (a) 1, 2, and 3, (b) 1, 3, and 5, (c) 2, 4, and 5, (d) 1, 2, and 5, (e) 1, 3, and 4.
8. Organic compounds that function primarily to provide a readily available source of chemical energy to generate the ATP that drives metabolic reactions are (a) lipids, (b) nucleic acids, (c) water, (d) carbohydrates, (e) proteins.
9. The body's most highly concentrated form of chemical energy is (a) a triglyceride, (b) a lipoprotein, (c) glycogen, (d) a steroid, (e) an eicosanoid.
10. The organic compounds that serve in structural, regulatory, contractile, immunological, transport, and catalytic capacities are (a) lipids, (b) carbohydrates, (c) proteins, (d) nucleic acids, (e) buffers.
11. Which of the following statements is *not true* concerning DNA? (a) It contains the five-carbon sugar deoxyribose. (b) It composed of a single strand of nucelotides. (c) DNA has a double helix construction. (d) DNA contains the nitrogenous bases adenine, guanine, cytosine and thymine. (e) It is the cell's genetic material.
12. Match the following reactions with the term that describes them.
    (1) synthesis reaction (2) exchange reaction
    (3) decomposition reaction (4) reversible reaction
    _____ (a) $H_2 + Cl_2 \longrightarrow 2\ HCl$
    _____ (b) $3\ NaOH + H_3PO_4 \longrightarrow Na_3PO_4 + 3\ H_2O$
    _____ (c) $CaCO_3 + CO_2 + H_2O \longrightarrow Ca(HCO_3)_2$
    _____ (d) $NH_3 + H_2O \rightleftharpoons NH_4^+ + OH^-$

13. Which of the following statements regarding ATP are *true?* (1) ATP is the energy currency for the cell. (2) The energy supplied by the hydrolysis of ATP is constantly being used by the cell. (3) Energy is required to produce ATP. (4) The production of ATP involves both aerobic and anaerobic phases. (5) The process of producing energy in the form of ATP is termed the law of conservation of energy. (a) 1, 2, 3, and 4, (b) 1, 2, 3, and 5, (c) 2, 4, and 5, (d) 1, 2, and 4, (e) 3, 4, and 5.
14. Match the following.
    (1) acid (2) base (3) buffer (4) enzyme
    (5) pH (6) ATP (7) salt (8) water
    _____ (a) a polar covalent molecule that serves as a solvent, has a high heat capacity, creates a high surface tension, and serves as a lubricant
    _____ (b) a substance that dissociates into one or more hydrogen ions and one or more anions
    _____ (c) a substance that dissociates into cations and anions, neither of which is a hydrogen ion or a hydroxyl ion
    _____ (d) a proton acceptor
    _____ (e) a measure of hydrogen ion concentration
    _____ (f) a chemical compound that can convert strong acids and bases into weak ones
    _____ (g) a catalyst for chemical reactions that is specific, efficient, and under cellular control
    _____ (h) a compound that functions to temporarily store and then transfer energy liberated in exergonic reactions to cellular activities that require energy
15. Match the following.
    (1) chemical element (2) matter (3) atom
    (4) compound (5) molecule (6) isotope
    (7) free radical (8) ion
    _____ (a) an electrically charged atom or group of atoms with an unpaired electron in its outermost shell
    _____ (b) a combination resulting when two or more atoms share electrons
    _____ (c) a building block of matter that cannot be split into a simpler chemical substance by ordinary chemical reactions
    _____ (d) anything that occupies space and has mass
    _____ (e) an atom of an element that differs in the number of neutrons, and therefore mass number, from other atoms of the same element
    _____ (f) the smallest unit of matter that retains the properties and characteristics of an element
    _____ (g) an atom that has given up or gained electrons
    _____ (h) a substance that can be broken down into two or more different elements by ordinary chemical means

CRITICAL THINKING QUESTIONS

1. When Selma joined the "Be Kind to Invertebrates Society" (their slogan is "Have you hugged a bug today?"), she decided to eat a diet free of sugars and fatty acids. How likely is it that Selma will stick to her diet?

   HINT   *Selma's diet makes even bread and water look good.*

2. Ming read an article in the *Daily News* describing a DNA fingerprinting technique being used as evidence in a criminal trial. The article explained that DNA is formed from 20 unique com-binations of amino acids that are different in every individual. What should Ming's letter to the editor say?

   HINT   *What do the initials "DNA" stand for?*

3. When asked the pH of blood, math major Sean guessed "the same as water." Kareem told him that he was way off! Explain in terms that Sean will understand.

   HINT   *Express the pH of blood mathematically.*

ANSWERS TO FIGURE QUESTIONS

2.1 In carbon, the first shell contains two electrons and the second shell contains four electrons.

2.2 The four most plentiful elements in living organisms are oxygen, carbon, hydrogen, and nitrogen.

2.3 Antioxidants such as vitamins C and E can inactivate oxygen free radicals.

2.4 A cation is a positively charged ion; an anion is a negatively charged ion.

2.5 An ionic bond involves the *loss* and *gain* of electrons; a covalent bond involves the *sharing* of pairs of electrons.

2.6 The oxygen atom in a water molecule has greater electronegativity than the hydrogen atoms.

2.7 The N atom in ammonia is electronegative. Because it attracts electrons more strongly than do the H atoms, the nitrogen end of ammonia acquires a slight negative charge. H atoms in water molecules (or in other ammonia molecules) can form hydrogen bonds with the nitrogen. Likewise, O atoms in water molecules can form hydrogen bonds with H atoms in ammonia molecules.

2.8 The law of conservation of mass stipulates that the number of hydrogen atoms in the reactants equals the number in the products—in this case, four hydrogen atoms total. Put another way, two molecules of H_2 are needed so that the number of H atoms and O atoms in the reactants is the same as the number of H atoms and O atoms in the products.

2.9 This reaction is exergonic because the reactants have more potential energy than the products.

2.10 No. A catalyst does not change the potential energies of the products and reactants; it only lowers the activation energy needed to get the reaction going.

2.11 Because sugar easily dissolves in a polar solvent (water), you can correctly predict that it has several polar covalent bonds.

2.12 $CaCO_3$ is a salt, and H_2SO_4 is an acid.

2.13 At pH = 6, $[H^+] = 10^{-6}$ mol/liter and $[OH^-] = 10^{-8}$ mol/liter. A pH of 6.82 is more acidic than a pH of 6.91. Both pH = 8.41 and pH = 5.59 are 1.41 pH units from neutral (pH = 7).

2.14 Glucose has five $-OH$ groups and 6 carbon atoms.

2.15 There are 6 carbons in fructose and 12 in sucrose.

2.16 Cells in the liver and in skeletal muscle store glycogen.

2.17 The oxygen in the water molecule comes from a fatty acid.

2.18 The polar head is hydrophilic, and the nonpolar tails are hydrophobic.

2.19 The only differences are the number of double bonds in and the types of functional groups attached to ring A.

2.20 An amino acid has a minimum of two carbon atoms and one nitrogen atom. See the structure of glycine in Figure 2.20c.

2.21 Hydrolysis occurs during catabolism of proteins.

2.22 Proteins consisting of a single polypeptide chain do not have a quaternary structure.

2.23 Sucrase has specificity for the sucrose molecule and thus would not "recognize" glucose and fructose.

2.24 Thymine always pairs with adenine, and cytosine always pairs with guanine.

2.25 A few cellular activities that depend on energy supplied by ATP are muscular contractions, movement of chromosomes, transport of substances across cell membranes, and synthesis (anabolic) reactions.

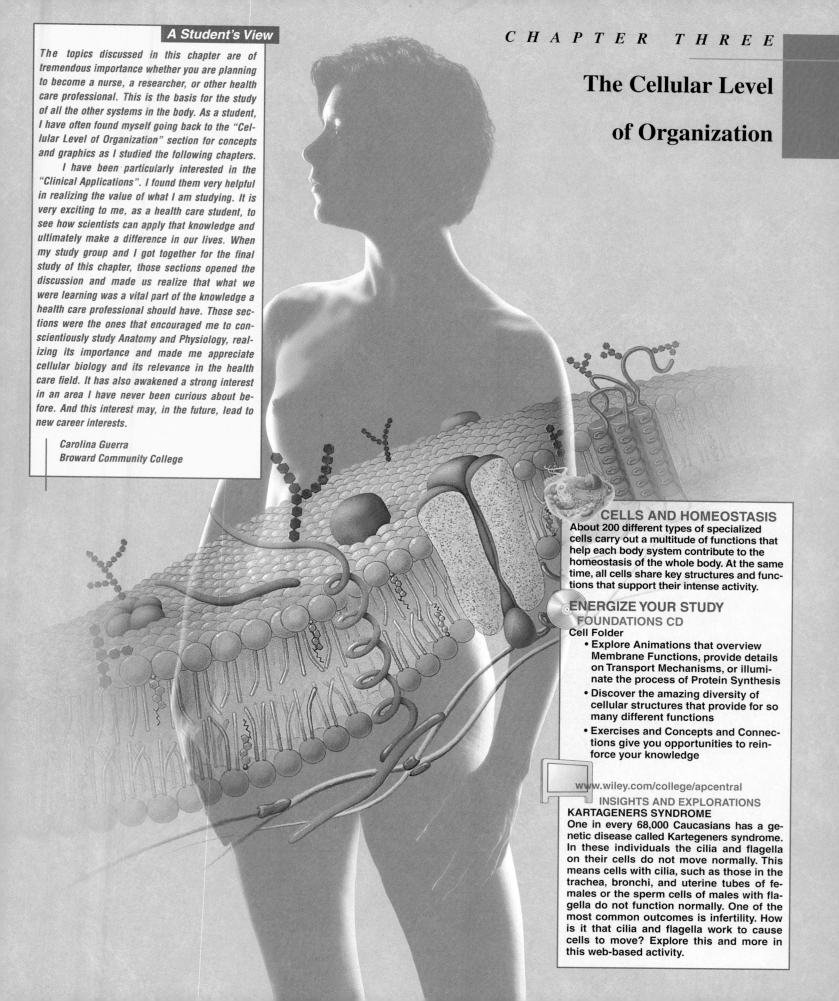

The Cellular Level of Organization

CELLS AND HOMEOSTASIS

About 200 different types of specialized cells carry out a multitude of functions that help each body system contribute to the homeostasis of the whole body. At the same time, all cells share key structures and functions that support their intense activity.

ENERGIZE YOUR STUDY
FOUNDATIONS CD
Cell Folder

- Explore Animations that overview Membrane Functions, provide details on Transport Mechanisms, or illuminate the process of Protein Synthesis
- Discover the amazing diversity of cellular structures that provide for so many different functions
- Exercises and Concepts and Connections give you opportunities to reinforce your knowledge

www.wiley.com/college/apcentral

INSIGHTS AND EXPLORATIONS
KARTAGENERS SYNDROME

One in every 68,000 Caucasians has a genetic disease called Kartegeners syndrome. In these individuals the cilia and flagella on their cells do not move normally. This means cells with cilia, such as those in the trachea, bronchi, and uterine tubes of females or the sperm cells of males with flagella do not function normally. One of the most common outcomes is infertility. How is it that cilia and flagella work to cause cells to move? Explore this and more in this web-based activity.

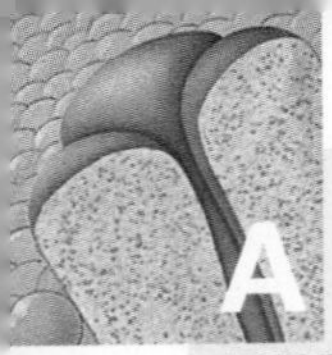

About 200 different types of cells are the basic units that compose your body. Each **cell** is a living structural and functional unit that is enclosed by a membrane. All cells arise from existing cells by the process of **cell division,** in which one "parent" cell divides into two new "daughter" cells. In your body, different types of cells fulfill unique roles that support homeostasis and contribute to the many functional capabilities of the human organism. **Cell biology** is the study of cellular structure and function. As you study the various parts of a cell and their relationships to each other, you will learn that cell structure and function are intimately related. Cells carry out a dazzling array of chemical reactions to create and maintain life processes. They do so, in part, by isolating specific types of chemical reactions within specialized structures inside the cell. In this chapter, you will learn about cellular structures and how they carry out their functions.

PARTS OF A CELL

OBJECTIVE

• **Name and describe the three main parts of a cell.**

Figure 3.1 provides an overview of the typical structures found in body cells. Whereas most cells have many of the structures shown in this diagram, no one cell has all of them. For ease of study, we can divide a cell into three main parts: plasma membrane, cytoplasm, and nucleus.

- The **plasma membrane** forms the cell's flexible outer surface, separating the cell's internal environment from the external environment outside the cell. It is a selective barrier that regulates the flow of materials into and out of a cell. This selectivity helps establish and maintain the appropriate environment for normal cellular activities. The plasma membrane also plays a key role in communication, both among cells and between cells and their external environment.

- The **cytoplasm** (SĪ-tō-plasm; *-plasm* = formed or molded) consists of all the cellular contents between the plasma

Figure 3.1 **Typical structures found in body cells.**

The cell is the basic living, structural and functional unit of the body.

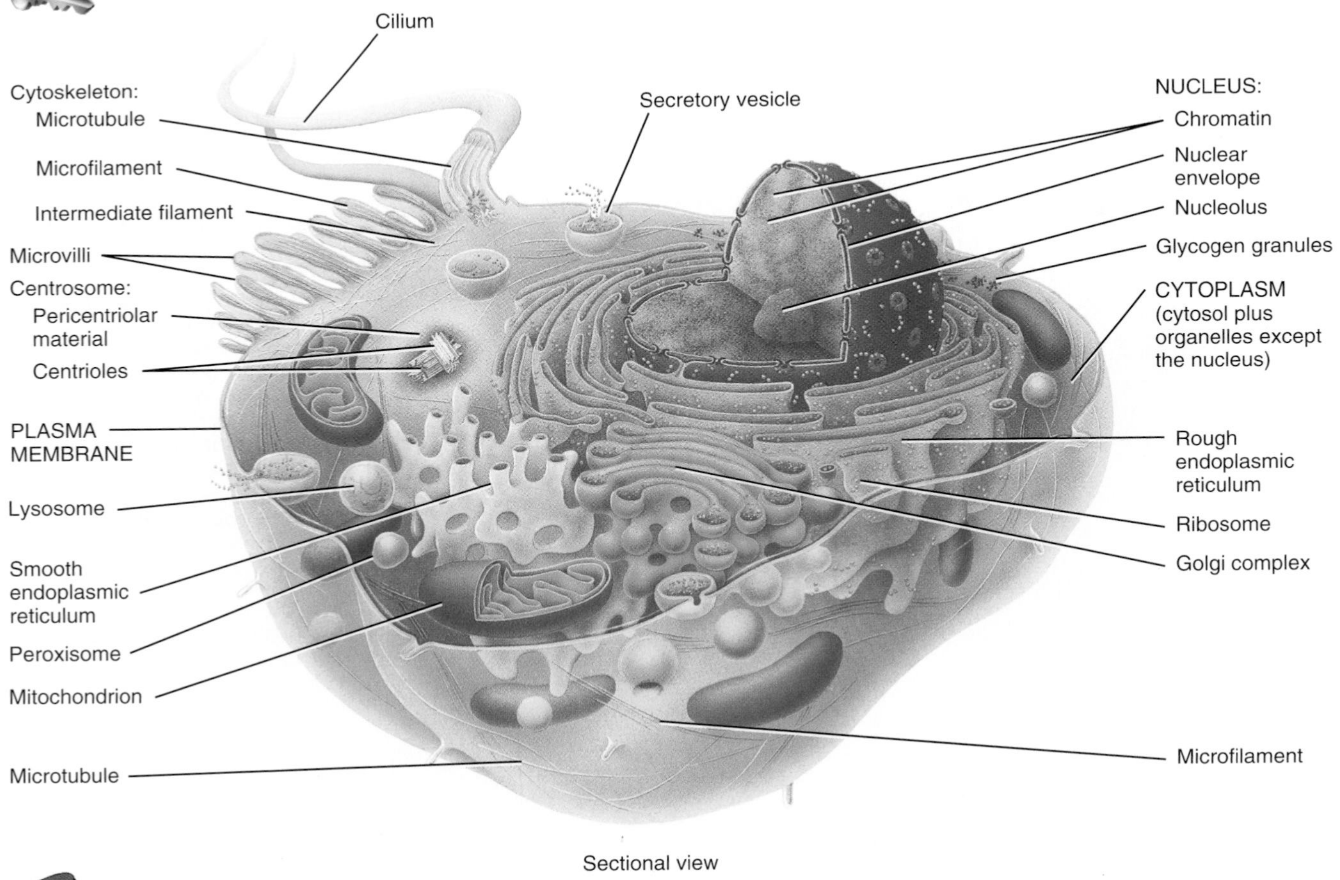

Cilium

Cytoskeleton:
  Microtubule
  Microfilament
Intermediate filament
Microvilli
Centrosome:
  Pericentriolar material
  Centrioles
PLASMA MEMBRANE
Lysosome
Smooth endoplasmic reticulum
Peroxisome
Mitochondrion
Microtubule

Secretory vesicle

NUCLEUS:
  Chromatin
  Nuclear envelope
  Nucleolus
Glycogen granules
CYTOPLASM (cytosol plus organelles except the nucleus)
Rough endoplasmic reticulum
Ribosome
Golgi complex
Microfilament

Sectional view

 What are the three principal parts of a cell?

membrane and the nucleus. This compartment has two components: cytosol and organelles. **Cytosol** (SĪ-tō-sol) is the fluid portion of cytoplasm. It contains water, dissolved solutes, and suspended particles. Surrounded by cytosol are several different types of **organelles** (or-ga-NELZ = little organs). Each type of organelle has a characteristic shape and specific functions. Examples are the cytoskeleton, ribosomes, endoplasmic reticulum, Golgi complex, lysosomes, peroxisomes, and mitochondria.

- The **nucleus** (NOO-klē-us = nut kernel) is a large organelle that houses most of a cell's DNA. Within the nucleus are the **chromosomes** (*chromo-* = colored), each of which consists of a single molecule of DNA associated with several proteins. A chromosome contains thousands of hereditary units called **genes** that control most aspects of cellular structure and function.

THE PLASMA MEMBRANE

OBJECTIVE

- **Describe the structure and functions of the plasma membrane.**

The **plasma membrane** is a flexible yet sturdy barrier that surrounds and contains the cytoplasm of a cell. The *fluid mosaic model* describes its structure. According to this model, the molecular arrangement of the plasma membrane resembles an ever-moving sea of fluid lipids that contains a mosaic of many different proteins (Figure 3.2). Some proteins float freely like icebergs in the lipid sea whereas others are anchored at specific locations like boats at a dock. The membrane lipids allow passage of several types of lipid-soluble molecules but act as a barrier to the entry or exit of charged or polar substances. By contrast, some of the proteins in the plasma membrane allow movement of polar molecules and ions into and out of the cell.

The Lipid Bilayer

The basic structural framework of the plasma membrane is the **lipid bilayer.** It consists of two back-to-back layers made up of three types of lipid molecules—phospholipids, cholesterol, and glycolipids (Figure 3.2). About 75% of the membrane lipids are **phospholipids,** lipids that contain phosphate groups. Present in smaller amounts are **cholesterol,** a steroid with an attached −OH (hydroxyl) group, and various **glycolipids,** lipids with attached carbohydrate groups.

The bilayer arrangement occurs because the lipids are **amphipathic** (am-fē-PATH-ik) molecules, which means that they have both polar and nonpolar parts. In phospholipids (see Figure 2.18 on page 47), the polar part is the phosphate-containing

Figure 3.2   The fluid mosaic arrangement of lipids and proteins in the plasma membrane.

Membranes are fluid structures because the lipids and many of the proteins are free to rotate and move sideways in their own half of the bilayer.

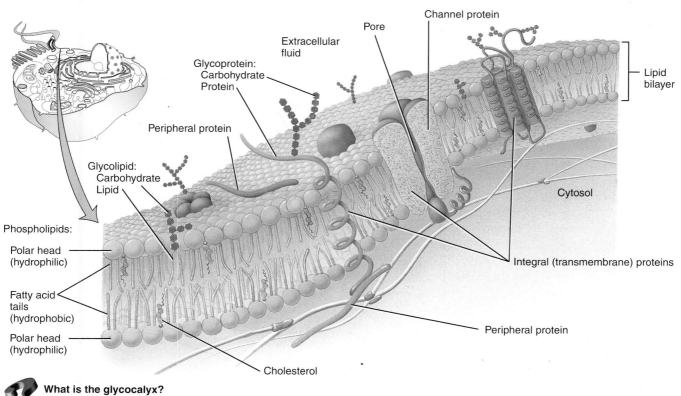

What is the glycocalyx?

"head," which is *hydrophilic* (*hydro-* = water; *-philic* = loving). The nonpolar parts are the two long fatty acid "tails," which are *hydrophobic* (*-phobic* = fearing) hydrocarbon chains. Because "like seeks like," the phospholipid molecules orient themselves in the bilayer with their hydrophilic heads facing outward. In this way, the heads face a watery fluid on either side—cytosol on the inside and extracellular fluid on the outside. The hydrophobic fatty acid tails in each half of the bilayer point toward the center of the membrane, forming a nonpolar, hydrophobic region in the membrane's interior.

Glycolipids account for about 5% of membrane lipids. Their carbohydrate groups form a polar "head," whereas their fatty acid "tails" are nonpolar. Glycolipids appear only in the membrane layer that faces the extracellular fluid, which is one reason the bilayer is asymmetric, or different, on its two sides. The remaining 20% of plasma membrane lipids are weakly amphipathic cholesterol molecules (see Figure 2.19a on page 48), which are interspersed among the other lipids in both layers of the membrane. The tiny $-OH$ group is the only polar region of cholesterol, and it forms hydrogen bonds with the polar heads of phospholipids and glycolipids. The stiff steroid rings and hydrocarbon tail of cholesterol are nonpolar; they fit among the fatty acid tails of the phospholipids and glycolipids.

Arrangement of Membrane Proteins

Membrane proteins are divided into two categories—integral and peripheral—according to whether or not they are firmly embedded in the membrane (Figure 3.2). **Integral proteins** extend into or through the lipid bilayer among the fatty acid tails and can be removed only by methods that disrupt membrane structure. Most integral proteins are **transmembrane proteins,** which span the entire lipid bilayer and protrude into both the cytosol and extracellular fluid. A few integral proteins are tightly attached to one side of the bilayer by covalent bonding to fatty acids in the lipid bilayer. **Peripheral proteins,** by contrast, associate more loosely with the polar heads of membrane lipids or with integral proteins at the inner or outer surface of the membrane. In addition, peripheral proteins can be removed without disrupting membrane integrity. Like membrane lipids, integral membrane proteins are amphipathic. Their hydrophilic regions protrude into either the watery extracellular fluid or the cytosol, and their hydrophobic regions extend among the fatty acid tails.

Many membrane proteins are **glycoproteins,** proteins with carbohydrate groups attached to the ends that protrude into the extracellular fluid. The carbohydrates are *oligosaccharides* (*oligo-* = few; *saccharides* = sugars), chains of 2 to 60 monosaccharides that may be straight or branched. The carbohydrate portions of glycolipids and glycoproteins form an extensive sugary coat called the **glycocalyx** (glī-kō-KĀL-iks). The composition of the glycocalyx acts like a molecular "signature" that enables cells to recognize one another. For example, a white blood cell's ability to detect a "foreign" glycocalyx is one basis

of the immune response that helps us destroy invading organisms. In addition, the glycocalyx enables cells to adhere to one another in some tissues, and it protects cells from being digested by enzymes in the extracellular fluid. The hydrophilic properties of the glycocalyx attract a film of fluid to the surface of many cells. This action makes red blood cells slippery as they flow through narrow blood vessels and protects cells that line the airways and the gastrointestinal tract from drying out.

Functions of Membrane Proteins

Generally, the types of lipids in cellular membranes vary only slightly from one membrane to another. In contrast, the membranes of different cells and various intracellular organelles have remarkably different assortments of proteins. Their proteins determine many of the functions that membranes can perform (Figure 3.3). Some integral membrane proteins are **ion channels** that have a *pore* or hole through which specific ions, such as potassium ions (K^+), can flow into or out of the cell. Most ion channels are *selective;* they allow only a single type of ion to pass through. Other membrane proteins act as **transporters,** which selectively move a polar substance from one side of the membrane to the other. Integral proteins called **receptors** serve as cellular recognition sites. Each type of receptor recognizes and binds a specific type of molecule. For instance, insulin receptors bind the hormone insulin. A specific molecule that binds to a receptor is called a **ligand** (LĪ-gand; *liga* = tied) of that receptor. Some integral and peripheral proteins are **enzymes** that catalyze specific chemical reactions at the inside or outside surface of the cell. Membrane glycoproteins and glycolipids often are **cell-identity markers.** They may enable a cell to recognize other cells of the same kind during tissue formation or to recognize and respond to potentially dangerous foreign cells. The ABO blood type markers are one example of cell identity markers. When you receive a blood transfusion, the blood type must be compatible with your own. Integral and peripheral proteins may serve as **linkers,** which anchor proteins in the plasma membranes of neighboring cells to one another or to protein filaments inside and outside the cell.

Membrane Fluidity

Membranes are fluid structures, rather like cooking oil, because most of the membrane lipids and many of the membrane proteins easily rotate and move sideways in their own half of the bilayer. Neighboring lipid molecules exchange places about 10 million times per second and may wander completely around a cell in only a few minutes! Membrane fluidity depends both on the number of double bonds in the fatty acid tails of the lipids that make up the bilayer and on the amount of cholesterol present. Each double bond puts a "kink" in the fatty acid tail (see Figure 2.18 on page 47), which increases membrane fluidity because it prevents lipid molecules from packing tightly in the

Figure 3.3 Functions of membrane proteins.

 Membrane proteins largely reflect the functions a cell can perform.

☐ Extracellular fluid ◼ Plasma membrane ☐ Cytosol

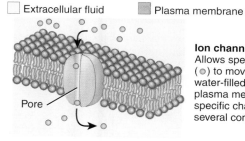

Ion channel
Allows specific ion
(⊙) to move through
water-filled pore. Most
plasma membranes include
specific channels for
several common ions.

Pore

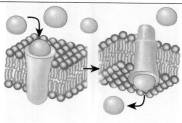

Transporter
Transports specific
substances (⊙) across
membrane by changing
shape. For example, amino
acids, needed to synthesize
new proteins, enter body
cells via transporters.

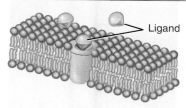

Ligand

Receptor
Recognizes specific ligand
(▽) and alters cell's
function in some way.
For example, antidiuretic
hormone binds to receptors
in the kidneys and changes
the water permeability of
certain plasma membranes.

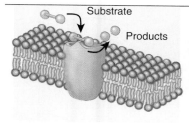

Substrate

Products

Enzyme
Catalyzes reaction inside or
outside cell (depending on
which direction the active
site faces). For example,
lactase protruding from
epithelial cells lining your
small intestine splits the
disaccharide lactose in the
milk you drink.

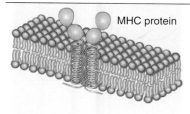

MHC protein

Cell Identity Marker
Distinguishes your cells
from anyone else's (unless
you are an identical twin).
An important class of such
markers are the major
histocompatability
(MHC) proteins.

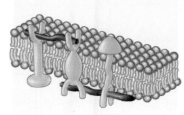

Linker
Anchors filaments inside
and outside to the plasma
membrane, providing
structural stability and shape
for the cell. May also
participate in movement
of the cell or link
two cells together.

**When stimulating a cell, the hormone insulin first binds to a
protein in the plasma membrane. This action best represents
which membrane protein function?**

membrane. Due to the fluidity of membrane lipids, the lipid bilayer self-seals if it is torn or punctured. When a needle is pushed through a plasma membrane and pulled out, the puncture site seals spontaneously, and the cell does not burst. This property of lipid bilayers allows scientists to fertilize an oocyte by injecting a sperm cell through a tiny syringe. It also permits removal and replacement of a cell's nucleus in cloning experiments, such as was done to create the cloned sheep named Dolly.

Despite the great mobility of membrane lipids and proteins in their own half of the bilayer, they seldom flip-flop from one half of the bilayer to the other. Flip-flopping is rare because it is difficult for hydrophilic parts of membrane molecules to pass through the hydrophobic core of the membrane. Thus, the halves of the membrane bilayer remain asymmetric. The outside half contains some lipids and proteins that are different from those in the inside half.

Because of the way it forms hydrogen bonds with neighboring phospholipid and glycolipid heads and fills the space between bent fatty acid tails, cholesterol makes the lipid bilayer stronger but less fluid at normal body temperature. At low temperatures, cholesterol has the opposite effect—it increases membrane fluidity. This effect of cholesterol is significant in cold-blooded animals but is not important in warm-blooded humans. In atherosclerosis (one form of arteriosclerosis), commonly called "hardening of the arteries," smooth muscle cells and cholesterol build up to form atherosclerotic plaques inside blood vessels. The added cholesterol stiffens the plaque, thus contributing to the decreased flexibility of blood vessels that characterizes this disease.

Membrane Permeability

A membrane is said to be *permeable* to substances that can pass through it and *impermeable* to those that cannot. Although plasma membranes are not completely permeable to any substance, they do permit some substances to pass more readily than others. This property of membranes is termed **selective permeability.**

The lipid bilayer portion of the membrane is permeable to nonpolar, uncharged molecules, such as oxygen, carbon dioxide, and steroids, but is impermeable to ions and charged or polar molecules, such as glucose. It is also slightly permeable to water and urea, a waste product from the breakdown of amino acids. The slight permeability to water and urea is an unexpected property given that water and urea are polar molecules. These two small molecules are thought to pass through the lipid bilayer in the following way. As the fatty acid tails of membrane phospholipids and glycolipids randomly move about, small gaps briefly appear in the hydrophobic environment of the membrane's interior. Water and urea molecules are small enough to move from one gap to another until they have crossed the membrane.

Transmembrane proteins that act as channels and transporters increase the plasma membrane's permeability to a variety of small- and medium-sized polar and charged substances (including ions) that cannot cross the lipid bilayer. Channels and transporters are very selective. Each one helps only a specific molecule or ion to cross the membrane. Macromolecules, such as proteins, are unable to pass across the plasma membrane except by endocytosis and exocytosis (discussed later in this chapter).

Gradients Across the Plasma Membrane

Because the plasma membrane of a cell is selectively permeable, it admits some substances into the cytosol while excluding others. This property allows a living cell to maintain different concentrations of selected substances on either side of the plasma membrane. A **concentration gradient** is a difference in the concentration of a chemical from one place to another, such as from the inside to the outside of the plasma membrane. Many ions and molecules are more concentrated in either the cytosol or the extracellular fluid. For instance, oxygen molecules and sodium ions (Na^+) are more concentrated in the extracellular fluid than in the cytosol, whereas the opposite is true for carbon dioxide molecules and potassium ions (K^+) (Figure 3.4a).

The plasma membrane also creates a difference in the distribution of positively and negatively charged ions between one side of the plasma membrane and the other. Typically, the inner surface of the plasma membrane is more negatively charged and the outer surface is more positively charged (Figure 3.4b). A difference in electrical charges between two regions constitutes an *electrical potential*. Because it occurs across the plasma membrane, this charge difference is termed the *membrane potential.*

As you will see shortly, the concentration gradient and membrane potential are important because they help move substances across the plasma membrane. In many cases a substance will move across a plasma membrane *down its concentration gradient.* That is to say, a substance will move "downhill," from where it is more concentrated to where it is less concentrated. Similarly, a positively charged substance will tend to move toward a negatively charged area, and a negatively charged substance will tend to move toward a positively charged area. Because the concentration gradient and the membrane potential both affect ion movements, the combined influence is termed an ion's **electrochemical gradient.**

▶ CHECKPOINT

1. How do hydrophobic and hydrophilic regions govern the arrangement of membrane lipids in a bilayer?

2. What substances can and cannot diffuse through the lipid bilayer?

3. Defend this statement: The proteins present in a plasma membrane determine the functions that a membrane can perform.

4. How does cholesterol affect membrane fluidity?

5. Why are membranes said to have selective permeability?

6. What factors contribute to an electrochemical gradient?

Figure 3.4   Gradients across the plasma membrane. (a) Sodium ions and oxygen molecules are more concentrated in the extracellular fluid, whereas potassium ions and carbon dioxide are more concentrated in cytosol. (b) Because the inner surface of the plasma membrane of most cells is negative relative to the outer surface, an electrical gradient exists across the membrane.

🔑 **Each substance moves across the membrane down its own electrochemical gradient.**

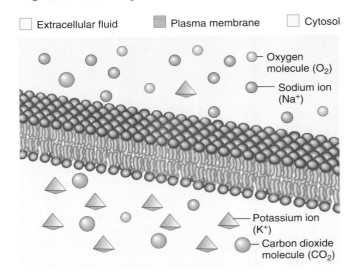

▢ Extracellular fluid   ▨ Plasma membrane   ▢ Cytosol

Oxygen molecule (O_2)
Sodium ion (Na^+)
Potassium ion (K^+)
Carbon dioxide molecule (CO_2)

(a) Concentration gradients

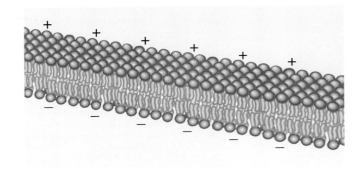

(b) Electrical gradient

 Will the electrochemical gradient favor flow of Na^+ into or out of cells?

TRANSPORT ACROSS THE PLASMA MEMBRANE

▶ O B J E C T I V E

• **Describe the processes that transport substances across the plasma membrane.**

Transport of materials across the plasma membrane is essential to the life of a cell. Certain substances must move into the cell to support metabolic reactions. Others that have been produced by the cell for export or cellular waste products must move out. As you have seen, some substances can cross the lipid bilayer part of the plasma membrane, whereas others use ion channels or transporters formed by membrane proteins.

Substances move across cellular membranes via transport processes that are classified as active or passive depending on whether they require cellular energy (Figure 3.5). In *passive transport,* a substance moves down its concentration or electrochemical gradient across the membrane using only its own kinetic energy. The three types of passive transport are diffusion through the lipid bilayer, diffusion through channels, and facilitated diffusion. In *active transport,* cellular energy is used to drive the substance "uphill" against its concentration gradient.

Besides passive and active transport, materials can enter and leave cells by endocytosis and exocytosis. In these processes, the material becomes enclosed within a tiny membrane-surrounded sac called a vesicle. In *endocytosis,* vesicles detach from the plasma membrane while bringing materials into the cell. In *exocytosis,* vesicles merge with the plasma membrane to release materials from the cell. Even large particles, such as whole bacteria and red blood cells, and macromolecules, such as polysaccharides and proteins, may enter cells via endocytosis and leave cells via exocytosis.

Principles of Diffusion

Learning why materials diffuse across membranes requires an understanding of how diffusion occurs in a solution. **Diffusion** is the random mixing of particles that occurs in a solution because of the particles' kinetic energy. Both the *solutes,* the dissolved substances, and the *solvent,* the liquid that does the dissolving, undergo diffusion. If a particular solute is present in high concentration in one area of a solution and in low concentration in another area, solute molecules will diffuse toward the area where they are less concentrated. They move *down their concentration gradient.* After some time, an equilibrium is reached and the particles are evenly distributed throughout the solution. At equilibrium, the particles continue to move about randomly, due to their kinetic energy, without further change in their concentrations.

An example of diffusion occurs if you place a crystal of dye in a water-filled container (Figure 3.6). Just next to the dye, the color is intense because the dye concentration is higher there. At increasing distances, the color is lighter and lighter because the dye concentration is lower. Some time later, at equilibrium, the solution of water and dye has a uniform color. The dye molecules and water molecules have diffused down their concentration gradients until they are evenly mixed in solution.

In the example of dye diffusion, no membrane was involved. Substances may also diffuse through a membrane, if the membrane is permeable to them. Several factors influence the diffusion rate of substances across plasma membranes:

1. *Steepness of the concentration gradient.* The greater the difference in concentration between the two sides of the membrane, the higher the rate of diffusion. When charged particles are diffusing, it is the steepness of the electrochemical gradient that determines diffusion rate across the membrane.

Figure 3.5 Processes for transport of materials across the plasma membrane.

 In passive transport, a substance moves down its electrochemical gradient, whereas in active transport, cellular energy is used to drive the substance "uphill," against its electrochemical gradient.

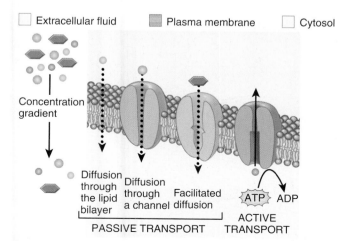

Figure 3.6 Diffusion. A crystal of dye placed in a cylinder of water dissolves (beginning) and then diffuses from the region of higher dye concentration to regions of lower dye concentration (intermediate). At equilibrium, dye concentration is uniform throughout although random movement continues.

 In diffusion, a substance moves down its concentration gradient.

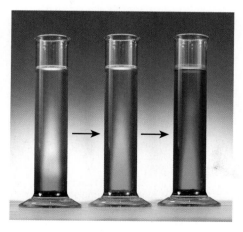

Beginning   Intermediate   Equilibrium

 In which transport processes do transmembrane proteins take part?

 How would having a fever affect body processes that involve diffusion?

2. Temperature. The higher the temperature, the faster the rate of diffusion. In a person who has a fever, all of the body's diffusion processes occur more rapidly.

3. Mass of the diffusing substance. The larger the mass of the diffusing particle, the slower its diffusion rate. Molecules with smaller molecular weights diffuse more rapidly than larger ones.

4. Surface area. The larger the membrane surface area available for diffusion, the faster the diffusion rate. For example, the air sacs of the lungs have a large surface area available for diffusion of oxygen from the air into the blood. Some lung diseases, such as emphysema, reduce the surface area. This slows the rate of diffusion and makes breathing more difficult.

5. Diffusion distance. The greater the distance over which diffusion must occur, the longer it takes. Diffusion across a plasma membrane takes only a fraction of a second because the membrane is so thin. In pneumonia, fluid collects in the lungs and increases diffusion distance, thus slowing diffusion of oxygen into the blood.

We will begin to explore membrane transport by seeing how the solvent water crosses membranes and then continue by looking at the various categories of solutes.

Osmosis

Osmosis (oz-MŌ-sis) is the net movement of a solvent through a selectively permeable membrane. In living systems, the solvent is water, which moves by osmosis across plasma membranes from an area of *higher water concentration* to an area of *lower water concentration*. Another way to understand this idea is to consider the solute concentration: In osmosis, water moves through a selectively permeable membrane from an area of *lower solute concentration* to an area of *higher solute concentration*. During osmosis, water molecules pass through a plasma membrane in two ways: (1) by moving through the lipid bilayer, as previously described, and (2) by moving through **aquaporins** (*aqua-* = water), integral membrane proteins that function as water channels.

Osmosis occurs only when a membrane is permeable to water but is not permeable to certain solutes. A simple experiment can demonstrate osmosis. Consider a U-shaped tube in which a selectively permeable membrane separates the left and right arms of the tube. A volume of pure water is poured into the left arm, and the same volume of a solution containing an impermeable solute is poured into the right arm (Figure 3.7a). Because

Figure 3.7 Principle of osmosis. Water molecules move through the selectively permeable membrane; the solute molecules in the right arm cannot pass through the membrane. (a) As the experiment starts, water molecules move from the left arm into the right arm, down the water concentration gradient. (b) After some time, the volume of water in the left arm has decreased and the volume of solution in the right arm has increased. At equilibrium, net osmosis has stopped. Hydrostatic pressure forces just as many water molecules to move from right to left as osmotic pressure forces water molecules to move from left to right. (c) If pressure is applied to the solution in the right arm, the starting conditions can be restored. This pressure, which stops osmosis, is equal to the osmotic pressure.

🔑 **Osmosis is the movement of water molecules through a selectively permeable membrane.**

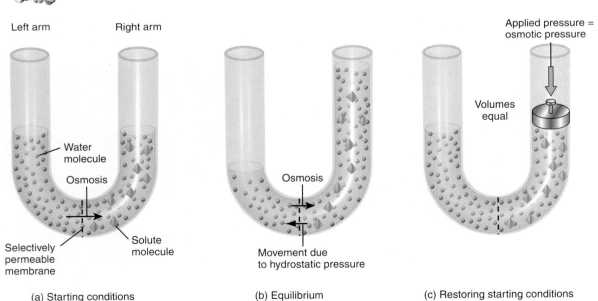

(a) Starting conditions   (b) Equilibrium   (c) Restoring starting conditions

❓ **Will the fluid level in the right arm rise until the water concentrations are the same in both arms?**

the *water* concentration is higher on the left and lower on the right, net movement of water molecules—osmosis—occurs from left to right. Note that water is moving down its concentration gradient. At the same time, the membrane prevents diffusion of the solute from the right arm into the left arm. As a result, the volume of water in the left arm decreases, and the volume of solution in the right arm increases (Figure 3.7b).

You might think that osmosis would continue until no water remained on the left side, but this is *not* what happens. In this experiment, the higher the column of solution in the right arm becomes, the more pressure it exerts on its side of the membrane. Pressure exerted in this way by a liquid is known as *hydrostatic pressure* and it forces water molecules to move back into the left arm. Equilibrium is reached when just as many water molecules move from right to left due to the hydrostatic pressure as move from left to right due to osmosis (Figure 3.7b).

A solution containing solute particles that cannot cross the membrane exerts a force, called the **osmotic pressure.** The osmotic pressure of a solution is proportional to the concentration of the solute particles that cannot cross the membrane—the higher the solute concentration, the higher the solution's osmotic pressure. Consider what would happen if a piston were used to apply more pressure to the fluid in the right arm. With enough pressure, the volume of fluid in each arm could be restored to the starting volume, and the concentration of solute in the right arm would be the same as it was in the beginning (Figure 3.7c). The amount of pressure needed to restore the starting condition equals the osmotic pressure. So, osmotic pressure is the pressure needed to stop the movement of water into a solution containing solutes when a membrane permeable only to the water separates the two solutions. Notice that the osmotic pressure of a solution does not produce the movement of water during osmosis. Rather it is the pressure that would *prevent* such water movement.

Normally, the osmotic pressure of the cytosol is the same as the osmotic pressure of the interstitial fluid outside cells. Because the osmotic pressure on both sides of the plasma membrane (which is selectively permeable) is the same, cell volume remains relatively constant. When body cells are placed in a solution having a different osmotic pressure than their cytosol, however, the shape and volume of the cells changes. As water moves by osmosis into or out of the cells, their volume increases or decreases. A solution's **tonicity** (*tonic* = tension) is a measure of the solution's ability to change the volume of cells by altering their water content.

Any solution in which a cell—for example, a red blood cell (RBC)—maintains its normal shape and volume is an **isotonic solution** (*iso-* = same) (Figure 3.8a). The concentrations of solutes that cannot cross the plasma membrane are the same on both sides of the membrane in this solution. For instance, a 0.9% NaCl solution (0.9 grams of sodium chloride in 100 mL of solution), called a *normal (physiological) saline solution,* is isotonic for RBCs. Whereas the RBC plasma membrane permits the water to move back and forth, it behaves as though it is impermeable to Na$^+$ and Cl$^-$, the solutes. (Any Na$^+$ or Cl$^-$ ions that

Figure 3.8 Tonicity and its effects on red blood cells (RBCs). One example of an isotonic solution for RBCs is 0.9% NaCl.

 Cells placed in an isotonic solution maintain their shape because there is no net water movement into or out of the cell.

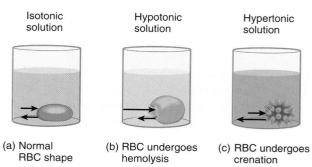

(a) Normal RBC shape (b) RBC undergoes hemolysis (c) RBC undergoes crenation

 **Will a 2% solution of NaCl cause hemolysis or crenation of RBCs? Why?**

enter the cell through channels or transporters are immediately moved back out by active transport or other means. These ions thus behave as though they cannot penetrate the membrane.) When RBCs are bathed in 0.9% NaCl, water molecules enter and exit at the same rate, allowing the RBCs to keep their normal shape and volume.

A different situation results if RBCs are placed in a **hypotonic solution** (*hypo-* = less than), a solution that has a *lower* concentration of solutes than the cytosol inside the RBCs (Figure 3.8b). In this case, water molecules enter the cells faster than they leave, causing the RBCs to swell and eventually to burst. The rupture of RBCs in this manner is called **hemolysis** (hē-MOL-i-sis; *hemo-* = blood; *-lysis* = to loosen or split apart). Pure water is very hypotonic and causes rapid hemolysis.

A **hypertonic solution** (*hyper-* = greater than) has a *higher* concentration of solutes than does the cytosol inside RBCs (Figure 3.8c). One example of a hypertonic solution is a 2% NaCl solution. In such a solution, water molecules move out of the cells faster than they enter, causing the cells to shrink. The shrinkage of RBCs in this manner is called **crenation** (kre-NĀ-shun).

Medical Uses of Isotonic, Hypertonic, and Hypotonic Solutions

RBCs and other body cells may be damaged or destroyed if exposed to hypertonic or hypotonic solutions. For this reason, most **intravenous (IV) solutions,** liquids infused into the blood of a vein, are isotonic. Examples are isotonic saline (0.9% NaCl) and D5W, which stands for dextrose 5% in water. Sometimes infusion of a hypertonic solution is useful to treat patients who have *cerebral edema,* excess interstitial fluid in the brain. Infusion of a hypertonic solution, such as mannitol, relieves such fluid overload by causing osmosis of water from interstitial fluid into the blood. Then, the kidneys excrete excess water from the blood into the urine. Hypotonic solutions, either IV or oral, can be used to treat people who are dehydrated. The water in the hypotonic solution moves from blood into interstitial fluid and then into body cells to rehydrate them. The sports drink or water that you consume to "rehydrate" after a workout is hypotonic. ■

Diffusion Through the Lipid Bilayer

Nonpolar, hydrophobic molecules diffuse through the lipid bilayer of the plasma membrane, into and out of cells. Such molecules include oxygen, carbon dioxide, and nitrogen gases; fatty acids, steroids, and fat-soluble vitamins (A, E, D, and K); small alcohols; and ammonia. Two polar molecules—water and urea—also can diffuse through the lipid bilayer. Because the plasma membrane is somewhat permeable to all these substances, they do not contribute to the tonicity of body fluids. Diffusion through the lipid bilayer is important in the movement of oxygen and carbon dioxide between blood and body cells, and between blood and air within the lungs during breathing. It also is the route for absorption of some nutrients and excretion of some wastes by body cells.

Diffusion Through Membrane Channels

Most membrane channels are *ion channels;* they allow passage of small, inorganic ions, which are too hydrophilic to penetrate the nonpolar interior of the lipid bilayer. Each ion can diffuse across the membrane only at sites containing channels that allow that specific ion to pass. In typical plasma membranes, the most numerous ion channels are selective for K^+ (potassium ions) or Cl^- (chloride ions); fewer channels are available for Na^+ (sodium ions) or Ca^{2+} (calcium ions). Diffusion of solutes through channels is generally slower than diffusion through the lipid bilayer because channels occupy a smaller fraction of the membrane's total surface area. Still, diffusion through channels is a very fast process: More than a million potassium ions can flow through a K^+ channel in one second!

A channel is said to be "gated" when part of the channel protein acts as a "plug" or "gate," changing shape in one way to open the pore and in another way to close it (Figure 3.9). Some gated channels randomly alternate between the open and closed positions, whereas others are regulated by chemical or electrical changes inside and outside the cell. When the gates of a channel are open, ions diffuse into or out of cells, down their electrochemical gradients. The plasma membranes of different types of cells may have fewer or more ion channels and thus display different permeabilities to various ions.

Facilitated Diffusion

Several solutes that are too polar or highly charged to diffuse through the lipid bilayer and are too big to diffuse through membrane channels can cross the plasma membrane by **facilitated**

Figure 3.9   Diffusion of K^+ through a gated membrane channel. Most membranes have channels that are selective for potassium ions (K^+), sodium ions (Na^+), calcium ions (Ca^{2+}), and chloride ions (Cl^-).

An ion channel is an integral transmembrane protein that permits specific small, inorganic ions to diffuse through its pore.

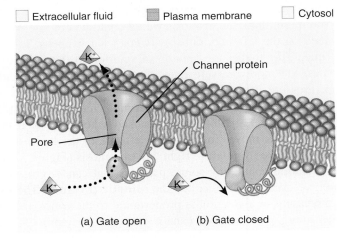

☐ Extracellular fluid   ▨ Plasma membrane   ☐ Cytosol

Channel protein

Pore

(a) Gate open   (b) Gate closed

How does a gated channel work?

diffusion. This process originally was envisioned as one in which a "carrier" protein bound to a solute and diffused with it across the membrane. The name is unfortunate, however, because the process is not truly diffusion. Instead, a solute binds to a specific transporter on one side of the membrane and is released on the other side after the transporter undergoes a change in shape. For this reason, some experts advocate changing the name to *facilitated transport.* Also, the proteins that mediate facilitated diffusion are no longer termed "carriers." We now know that they don't move across the membrane like a ferry boat carrying passengers across a lake.

Like diffusion, nonetheless, the net result of facilitated diffusion is movement down a concentration gradient. This happens because the solute binds more often to the transporter on the side of the membrane where the solute is more concentrated before being moved to the side where it is less concentrated. Once the concentration is the same on both sides of the membrane, solute molecules bind to the transporter on the cytosolic side and move out to the extracellular fluid as rapidly as they bind to the transporter on the extracellular side and move into the cytosol. Thus, the rate of facilitated diffusion is determined by the steepness of the concentration gradient between the two sides of the membrane.

The number of facilitated diffusion transporters available in a plasma membrane places an upper limit, called the *transport maximum,* on the rate at which facilitated diffusion can occur. Once all the transporters are occupied, the transport maximum is reached, and a further increase in the concentration gradient does not produce faster facilitated diffusion. Thus, the process of facilitated diffusion exhibits *saturation.*

Solutes that can move across plasma membranes by facilitated diffusion include glucose, fructose, galactose, and some vitamins. Glucose enters many body cells by facilitated diffusion, as follows (Figure 3.10):

1 Glucose binds to a glucose transporter protein called GluT at the extracellular surface of the membrane.

2 As the GluT transporter changes shape, glucose passes through the membrane.

3 GluT releases glucose into the cytosol.

After glucose enters a cell by facilitated diffusion, an enzyme called *hexokinase* attaches a phosphate group to produce a different molecule, namely glucose 6-phosphate. This reaction keeps the intracellular concentration of glucose very low, so that the glucose concentration gradient always favors facilitated diffusion of glucose into, not out of, cells.

The selective permeability of the plasma membrane is often regulated to achieve homeostasis. For instance, the hormone insulin promotes the insertion of many copies of a specific type of glucose transporter into the plasma membranes of certain cells. Thus, the effect of insulin is to increase the transport maximum for facilitated diffusion of glucose into cells. With more transporters available, body cells can pick up glucose from the blood more rapidly.

Active Transport

Some polar or charged solutes that must enter or leave body cells cannot cross the plasma membrane through any form of passive transport because they would need to move "uphill," *against* their concentration gradient. Such solutes may be able to cross the membrane by **active transport.** Active transport is an energy-requiring process in which transporter proteins move solutes across the membrane against a concentration gradient. Two sources of cellular energy can be used to drive active transport: (1) Energy obtained from hydrolysis of ATP is the source in *primary active transport;* (2) Energy stored in an ionic concentration gradient is the source in *secondary active transport.* Like facilitated diffusion, active transport processes exhibit a transport maximum and saturation. Solutes actively transported across the plasma membrane include several ions, such as Na^+, K^+, H^+ (hydrogen ions), Ca^{2+}, I^- (iodide ions), and Cl^-; amino acids; and monosaccharides. (Note that some of these substances also cross the membrane via channels or facilitated diffusion when the proper channel proteins or transporters are present.)

Primary Active Transport

In **primary active transport,** energy derived from hydrolysis of ATP changes the shape of a transporter protein, which "pumps"

Figure 3.10 Facilitated diffusion of glucose across a plasma membrane. The transporter (GluT) binds to glucose in the extracellular fluid, changes its shape, and releases glucose into the cytosol.

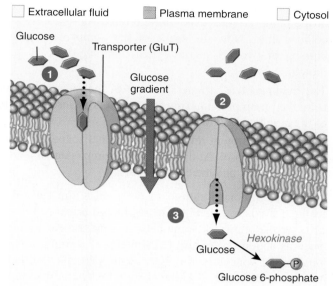

Facilitated diffusion requires a transporter but does not use ATP.

| Extracellular fluid | Plasma membrane | Cytosol |

What factors affect the rate of facilitated diffusion?

a substance across a plasma membrane against its concentration gradient. Indeed, transporter proteins that carry out primary active transport are often called **pumps.** A typical body cell expends about 40% of the ATP it generates on primary active transport. Chemicals that turn off ATP production—for example, the poison cyanide—are lethal because they shut down active transport in cells throughout the body.

The most prevalent primary active transport mechanism expels sodium ions (Na^+) from cells and brings potassium ions (K^+) in. Because of the specific ions it moves, this transporter is called the **sodium-potassium pump.** Because a part of the sodium-potassium pump acts as an *ATPase,* an enzyme that hydrolyzes ATP, another name for this pump is **Na^+/K^+ ATPase.** All cells have thousands of sodium-potassium pumps in their plasma membranes. These sodium-potassium pumps maintain a low concentration of Na^+ in the cytosol by pumping them into the extracellular fluid against the Na^+ concentration gradient. At the same time, the pumps move K^+ into cells against the K^+ concentration gradient. Because K^+ and Na^+ slowly leak back across the plasma membrane down their electrochemical gradients—through passive transport or secondary active transport—the sodium-potassium pumps must work nonstop to maintain a low concentration of Na^+ and a high concentration of K^+ in the cytosol.

Figure 3.11 depicts the operation of the sodium-potassium pump.

1 Three Na$^+$ in the cytosol bind to the pump protein.

2 Binding of Na$^+$ triggers the hydrolysis of ATP into ADP, a reaction that also attaches a phosphate group (P) to the pump protein. This chemical reaction changes the shape of the pump protein, expelling the three Na$^+$ into the extracellular fluid. Now the shape of the pump protein favors binding of two K$^+$ in the extracellular fluid to the pump protein.

3 The binding of K$^+$ triggers release of the phosphate group from the pump protein. This reaction again causes the shape of the pump protein to change.

4 As the pump protein reverts to its original shape, it releases K$^+$ into the cytosol. At this point, the pump is again ready to bind three Na$^+$, and the cycle repeats.

The different concentrations of Na$^+$ and K$^+$ in cytosol and extracellular fluid are crucial for maintaining normal cell volume and also for the ability of some cells to generate electrical signals such as action potentials. The tonicity of a solution is proportional to the concentration of its solute particles that cannot penetrate the membrane. Because sodium ions that diffuse into a cell or enter through secondary active transport are immediately pumped out, it is as if they never entered. In effect, sodium ions behave as if they cannot penetrate the membrane. Thus, sodium ions are an important contributor to the tonicity of the extracellular fluid. A similar condition holds for K$^+$ in the cytosol. By helping to maintain normal tonicity on each side of the plasma membrane, the sodium-potassium pumps ensure that cells neither shrink nor swell due to osmotic movement of water out of or into cells by osmosis.

Secondary Active Transport

In **secondary active transport,** the energy stored in a Na$^+$ or H$^+$ concentration gradient is used to drive other substances across the membrane against their own concentration gradients. Because a Na$^+$ or H$^+$ gradient is established by primary active transport, secondary active transport *indirectly* uses energy obtained from the hydrolysis of ATP.

The sodium-potassium pump maintains a steep concentration gradient of Na$^+$ across the plasma membrane. As a result, the sodium ions have stored or potential energy, just as water behind a dam has potential energy. Accordingly, if there is a route for Na$^+$ to leak back in, some of the stored energy can be converted to kinetic energy (energy of motion) and used to transport other substances *against their concentration gradients.* In essence, secondary active transport proteins harness the energy in the Na$^+$ concentration gradient by providing routes for Na$^+$ to leak into cells. In secondary active transport, a transporter protein simultaneously binds to Na$^+$ and another substance and then changes its shape so that both substances cross the membrane at the same time. These transporters are either *symporters,* which move two substances in the same direction, or *antiporters,* which move two substances in opposite directions across the membrane.

Plasma membranes contain several antiporters and symporters that are powered by the Na$^+$ gradient (Figure 3.12a). For instance, the concentration of calcium ions (Ca^{2+}) is low in the cytosol because Na$^+$/Ca^{2+} antiporters eject calcium ions. Like-

Figure 3.11 **The sodium-potassium pump (Na$^+$/K$^+$ ATPase) expels sodium ions (Na$^+$) and brings potassium ions (K$^+$) into the cell.**

 Sodium-potassium pumps maintain a low intracellular concentration of sodium ions.

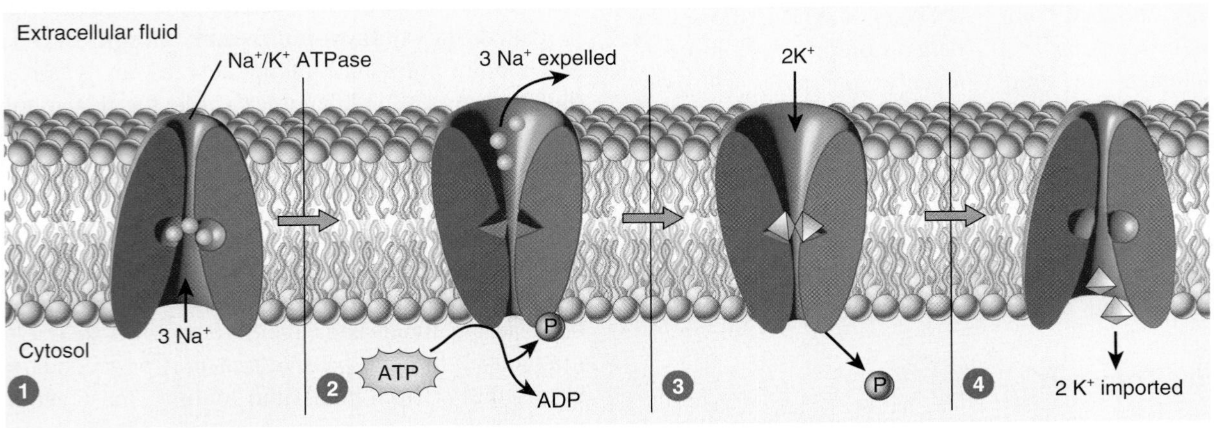

Extracellular fluid

Na$^+$/K$^+$ ATPase   3 Na$^+$ expelled   2K$^+$

Cytosol   3 Na$^+$

1   **2** ATP → ADP, P   **3** P   **4** 2 K$^+$ imported

What is the role of ATP in the operation of this pump?

Figure 3.12 Secondary active transport mechanisms. (a) Antiporters carry two substances across the membrane in opposite directions. (b) Symporters carry two substances across the membrane in the same direction.

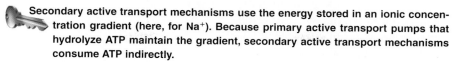

Secondary active transport mechanisms use the energy stored in an ionic concentration gradient (here, for Na^+). Because primary active transport pumps that hydrolyze ATP maintain the gradient, secondary active transport mechanisms consume ATP indirectly.

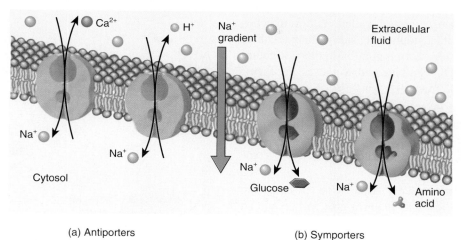

(a) Antiporters   (b) Symporters

What is the main difference between primary and secondary active transport mechanisms?

wise, Na^+/H^+ antiporters help regulate the cytosol's pH (H^+ concentration) by expelling excess H^+. By contrast, dietary glucose and amino acids are absorbed into cells that line the small intestine by Na^+-glucose and Na^+-amino acid symporters (Figure 3.12b). In each case, sodium ions are moving down their concentration gradient while the other solutes move "uphill," against their concentration gradients. Keep in mind that all these symporters and antiporters can do their job because the sodium-potassium pumps maintain a low concentration of Na^+ in the cytosol.

Digitalis Increases Ca^{2+} in Heart Muscle Cells

Because it strengthens the heartbeat, digitalis often is given to patients with *heart failure,* a condition of weakened pumping action by the heart. Digitalis exerts its effect by slowing the sodium-potassium pumps, which lets more Na^+ accumulate inside heart muscle cells. The result is a smaller Na^+ concentration gradient across the plasma membrane, which causes the Na^+/Ca^{2+} antiporters to slow down. As a result, more Ca^{2+} remains inside heart muscle cells. The slight increase in the level of Ca^{2+} in the cytosol of heart muscle cells increases the force of their contractions and thus strengthens the heartbeat. ∎

Transport in Vesicles

A **vesicle,** as noted earlier, is a small, spherical sac that has budded off from an existing membrane. Vesicles transport a variety of substances from one structure to another within cells, as you will see later in this chapter. Vesicles also import materials from and release materials into extracellular fluid. During **endocytosis** (*endo-* = within), materials move into a cell in a vesicle formed from the plasma membrane. In **exocytosis** (*exo-* = out), materials move out of a cell by the fusion of vesicles formed inside the cell with the plasma membrane. Both endocytosis and exocytosis require energy supplied by ATP.

Endocytosis

Here we consider three types of endocytosis: receptor-mediated endocytosis, phagocytosis, and pinocytosis.

Receptor-mediated endocytosis is a highly selective type of endocytosis by which cells take up specific ligands. (Recall that ligands are molecules that bind to specific receptors.) A vesicle forms after a receptor protein in the plasma membrane recognizes and binds to a particular particle in the extracellular fluid. For instance, cells take up cholesterol contained in low-density lipoproteins (LDLs), transferrin (an iron-transporting protein in the blood), some vitamins, antibodies, and certain hormones by receptor-mediated endocytosis. Receptor-mediated endocytosis of LDLs (and other ligands) occurs as follows (Figure 3.13):

1 *Binding.* On the extracellular side of the plasma membrane, an LDL particle that contains cholesterol binds to a specific receptor in the plasma membrane to form a receptor-LDL complex. The receptors are integral membrane proteins that are concentrated in regions of the plasma membrane, called *clathrin-coated pits.* Here, a protein called clathrin attaches to the membrane on its cytoplasmic side. Many clathrin molecules come together, forming a basketlike structure around the receptor-LDL complexes that causes the membrane to invaginate (fold inward).

2 *Vesicle formation.* The invaginated edges of the membrane around the clathrin-coated pit fuse and a small piece of the membrane pinches off. The resulting vesicle, known as a *clathrin-coated vesicle,* contains the receptor-LDL complexes.

3 *Uncoating.* Almost immediately after it is formed, the clathrin-coated vesicle loses its clathrin coat to become an *uncoated vesicle.* Clathrin molecules either return to the inner surface of the plasma membrane or help form coats on other vesicles inside the cell.

4 *Fusion with endosome.* The uncoated vesicle quickly fuses with a vesicle known as an *endosome.* Within an endosome, the LDL particles separate from their receptors.

5 *Recycling of receptors to plasma membrane.* Most of the receptors accumulate in elongated protrusions of the endosome. These pinch off, forming a *transport vesicle* that contains the receptors and returns them to the plasma membrane. An LDL receptor is returned to the plasma membrane about 10 minutes after it enters a cell.

6 *Degradation in lysosome.* Other transport vesicles, which contain the LDL particles, bud off the endosome and soon fuse with a *lysosome.* Within lysosomes are many digestive enzymes. Certain enzymes break down the large protein and lipid molecules of the LDL particle into amino acids, fatty acids, and cholesterol. These smaller molecules then leave the lysosome. The cell uses cholesterol for rebuilding its membranes and for synthesis of steroids, such as estrogen. Fatty acids and amino acids can be used for ATP production or to build other molecules needed by the cell.

Viruses and Receptor-mediated Endocytosis

Although receptor-mediated endocytosis normally imports needed materials, some viruses are able to use this mechanism to enter and infect body cells. For example, the human immunodeficiency virus (HIV), which causes acquired immunodeficiency syndrome (AIDS), can attach to a receptor called CD4. This receptor is present in the plasma membrane of white blood cells called helper T cells. After binding to CD4, HIV enters the helper T cell via receptor-mediated endocytosis. ■

Figure 3.13 Receptor-mediated endocytosis of a low-density lipoprotein (LDL) particle.

Receptor-mediated endocytosis imports materials that are needed by cells.

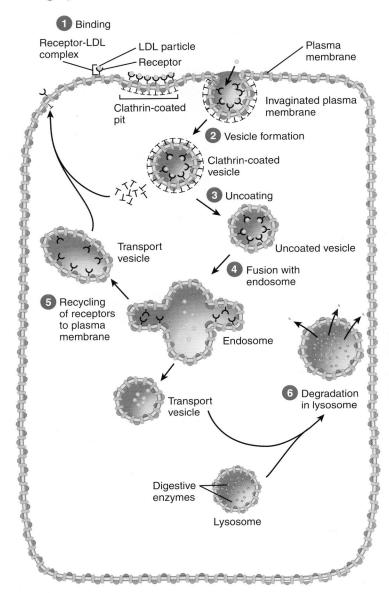

Besides LDL particles, what are several examples of ligands that can undergo receptor-mediated endocytosis?

Phagocytosis (fag′-ō-sī-TŌ-sis; *phago-* = to eat) is a form of endocytosis in which the cell engulfs large solid particles, such as worn-out cells, whole bacteria, or viruses (Figure 3.14). Only a few body cells, termed **phagocytes,** are able to carry out phagocytosis. Two main types of phagocytes are *macrophages,* located in many body tissues, and *neutrophils,* a type of white blood cell. Phagocytosis begins when the particle binds to a plasma membrane receptor, causing the cell to extend **pseudopods** (SOO-dō-pods; *pseudo-* = false; *-pods* = feet), which are

Figure 3.14 Phagocytosis. Pseudopods surround a particle and the membranes fuse to form a phagosome.

Phagocytosis is a vital defense mechanism that helps protect the body from disease.

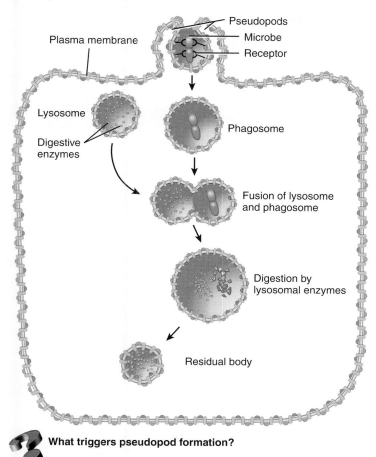

What triggers pseudopod formation?

Figure 3.15 Pinocytosis. The plasma membrane folds inward, forming a pinocytic vesicle.

Most body cells carry out pinocytosis, the nonselective uptake of tiny droplets of extracellular fluid.

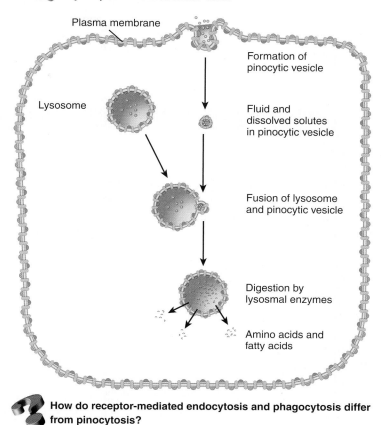

How do receptor-mediated endocytosis and phagocytosis differ from pinocytosis?

projections of its plasma membrane and cytoplasm. Pseudopods surround the particle outside the cell, and the membranes fuse to form a vesicle called a *phagosome,* which enters the cytoplasm. The phagosome fuses with one or more lysosomes, and lysosomal enzymes break down the ingested material. In most cases, any undigested materials in the phagosome remain indefinitely in a vesicle called a *residual body.* The process of phagocytosis is a vital defense mechanism that helps protect the body from disease. Through phagocytosis, macrophages dispose of billions of aged, worn-out red blood cells every day and neutrophils help rid the body of invading microorganisms.

Most body cells carry out **pinocytosis** (pi-nō-sī-TŌ-sis; *pino-* = to drink), a form of endocytosis in which tiny droplets of extracellular fluid are taken up (Figure 3.15). No receptor proteins are involved; all solutes dissolved in the extracellular fluid are brought into the cell. During pinocytosis, the plasma membrane folds inward and forms a *pinocytic vesicle* containing a droplet of extracellular fluid. The pinocytic vesicle detaches or "pinches off" from the plasma membrane and enters the cytosol. Within the cell, the pinocytic vesicle fuses with a lysosome,

where enzymes degrade the engulfed solutes. The resulting smaller molecules, such as amino acids and fatty acids, leave the lysosome to be used elsewhere in the cell.

Exocytosis

In contrast with endocytosis, which brings materials into a cell, **exocytosis** results in *secretion,* the liberation of materials from a cell. All cells carry out exocytosis, but it is especially important in two types of cells: (1) Secretory cells liberate digestive enzymes, hormones, mucus, or other secretions; (2) Nerve cells release substances called *neurotransmitters* via exocytosis (see Figure 12.14 on page 405). During exocytosis, membrane-enclosed vesicles called *secretory vesicles* form inside the cell, fuse with the plasma membrane, and release their contents into the extracellular fluid.

Segments of the plasma membrane lost through endocytosis are recovered or recycled by exocytosis. The balance between endocytosis and exocytosis keeps the surface area of a cell's plasma membrane relatively constant. Membrane exchange is quite extensive in certain cells. In your pancreas, for example, the cells that secrete digestive enzymes can recycle an amount of plasma membrane equal to the cell's entire surface area in 90 minutes.

Transcytosis

In **transcytosis,** vesicles undergo endocytosis on one side of a cell, move across the cell, and then undergo exocytosis on the opposite side. As the vesicles fuse with the plasma membrane, the vesicular contents are released into the extracellular fluid. Transcytosis occurs most often across the endothelial cells that line blood vessels and is a means for materials to move between blood plasma and interstitial fluid. For instance, when a woman is pregnant, some of her antibodies cross the placenta into the fetal circulation via transcytosis.

Table 3.1 summarizes the processes by which materials are transported into and out of cells.

Table 3.1	Transport of Materials Into and Out of Cells	
Transport Process	**Description**	**Substances Transported**
Osmosis	Movement of water molecules across a selectively permeable membrane from an area of higher water concentration to an area of lower water concentration.	Solvent: water in living systems.
Diffusion	Random mixing of molecules or ions due to their kinetic energy. A substance diffuses down a concentration gradient until it reaches equilibrium.	
Diffusion through the lipid bilayer	Passive diffusion of a substance through the lipid bilayer of the plasma membrane.	Nonpolar, hydrophobic solutes: oxygen, carbon dioxide, and nitrogen; fatty acids, steroids, and fat-soluble vitamins; glycerol, small alcohols; ammonia. Polar molecules: water and urea.
Diffusion through membrane channels	Passive diffusion of a substance down its electrochemical gradient through channels that span a lipid bilayer; some channels are gated.	Small inorganic solutes, mainly ions: K^+, Cl^-, Na^+, and Ca^{2+}. Water.
Facilitated Diffusion	Passive movement of a substance down its concentration gradient via transmembrane proteins that act as transporters; maximum diffusion rate is limited by number of available transporters.	Polar or charged solutes: glucose, fructose, galactose, and some vitamins.
Active Transport	Transport in which cell expends energy to move a substance across the membrane against its concentration gradient through transmembrane proteins that act as transporters; maximum transport rate is limited by number of available transporters.	Polar or charged solutes.
Primary active transport	Transport of a substance across the membrane against its concentration gradient by pumps; transmembrane proteins that use energy supplied by hydrolysis of ATP.	Na^+, K^+, Ca^{2+}, H^+, I^-, Cl^-, and other ions.
Secondary active transport	Coupled transport of two substances across the membrane using energy supplied by a Na^+ or H^+ concentration gradient maintained by primary active transport pumps. Antiporters move Na^+ (or H^+) and another substance in opposite directions across the membrane; symporters move Na^+ (or H^+) and another substance in the same direction across the membrane.	Antiport: Ca^{2+}, H^+ out of cells. Symport: glucose, amino acids into cells.
Transport In Vesicles	Movement of substances into or out of a cell in vesicles that bud from the plasma membrane; requires energy supplied by ATP.	
Endocytosis	Movement of substances into a cell in vesicles.	
Receptor-mediated endocytosis	Ligand-receptor complexes trigger infolding of a clathrin-coated pit that forms a vesicle containing ligands.	Ligands: transferrin, low-density lipoproteins (LDLs), some vitamins, certain hormones, and antibodies.
Phagocytosis	"Cell eating"; movement of a solid particle into a cell after pseudopods engulf it to form a phagosome.	Bacteria, viruses, and aged or dead cells.
Pinocytosis	"Cell drinking"; movement of extracellular fluid into a cell by infolding of plasma membrane to form a pinocytic vesicle.	Solutes in extracellular fluid.
Exocytosis	Movement of substances out of a cell in secretory vesicles that fuse with the plasma membrane and release their contents into the extracellular fluid.	Neurotransmitters, hormones, and digestive enzymes.

7. What is the key difference between passive and active transport?

8. How do symporters and antiporters carry out their functions?

9. What factors can increase the rate of diffusion?

10. What is osmotic pressure?

11. Which substances can diffuse directly through the lipid bilayer?

12. How does diffusion through membrane channels compare to facilitated diffusion?

13. What is the difference between primary and secondary active transport?

14. In what ways are endocytosis and exocytosis similar and different?

CYTOPLASM

► O B J E C T I V E

- **Describe the structure and function of cytoplasm, cytosol, and organelles.**

Cytoplasm has two components: (1) the cytosol and (2) a variety of organelles, tiny structures that perform different functions in the cell.

Cytosol

The **cytosol** is the fluid portion of the cytoplasm that surrounds organelles (see Figure 3.1) and constitutes about 55% of total cell volume. Although it varies in composition and consistency from one part of a cell to another, cytosol is 75–90% water plus various dissolved and suspended components. Among these are different types of ions, glucose, amino acids, fatty acids, proteins, lipids, ATP, and waste products. Also present in some cells are various organic molecules that aggregate into masses for storage. These aggregations may appear and disappear at different times in the life of a cell. Examples include *lipid droplets* that contain triglycerides and clusters of glycogen molecules called *glycogen granules* (see Figure 3.1).

The cytosol is the site of many chemical reactions required for a cell's existence. For example, enzymes in cytosol catalyze *glycolysis,* a series of 10 chemical reactions that produces two molecules of ATP from one molecule of glucose. Other types of cytosolic reactions provide the building blocks for maintaining cell structures and for growth.

Organelles

As noted earlier, **organelles** are specialized structures that have characteristic shapes and that perform specific functions in cellular growth, maintenance, and reproduction. Despite the many chemical reactions occurring in a cell at the same time, there is little interference between one type of reaction and another be-cause they occur in different organelles. Each type of organelle has its own set of enzymes that carry out specific reactions, and each is a functional compartment where specific biochemical processes take place. The numbers and types of organelles vary in different cells, depending on the cell's function. Moreover, organelles often cooperate with each other to maintain homeostasis. Although the nucleus is a large organelle, it is discussed separately because of its special importance in directing the life of a cell.

The Cytoskeleton

The **cytoskeleton** is a network of several different kinds of protein filaments that extend throughout the cytosol (see Figure 3.1). The cytoskeleton provides a structural framework for the cell, serving as a scaffold that helps to determine a cell's shape and to organize the cellular contents. The cytoskeleton also aids movement of organelles within the cell, of chromosomes during cell division, and of whole cells such as phagocytes. Although its name implies rigidity, the cytoskeleton is continually reorganizing as a cell moves and changes shape, as, for example, during cell division. Three types of filamentous proteins contribute to the cytoskeleton. In the order of their increasing diameter, these structures are microfilaments, intermediate filaments, and microtubules.

Microfilaments, the thinnest elements of the cytoskeleton, are composed of the protein *actin* and are most prevalent at the periphery of a cell (Figure 3.16a). Microfilaments have two general functions: helping generate movement and providing mechanical support. With respect to movement, microfilaments are involved in muscle contraction, cell division, and cell locomotion, such as occurs during the migration of embryonic cells during development, the invasion of tissues by white blood cells to fight infection, or the migration of skin cells during wound healing.

Microfilaments provide much of the mechanical support that is responsible for the basic strength and shapes of cells. They anchor the cytoskeleton to integral proteins in the plasma membrane. Microfilaments also provide mechanical support for cell extensions called **microvilli** (*micro-* = small; *-villi* = tufts of hair), which are nonmotile, microscopic fingerlike projections of the plasma membrane. Within each microvillus is a core of parallel microfilaments. Because they greatly increase the surface area of the cell, microvilli are abundant on cells involved in absorption, such as the epithelial cells that line the small intestine.

As their name suggests, **intermediate filaments** are thicker than microfilaments but thinner than microtubules (Figure 3.16b). Several different proteins can compose intermediate filaments, which are exceptionally strong. They are found in parts of cells subject to mechanical stress and also help stabilize the position of organelles such as the nucleus.

The largest of the cytoskeletal components, **microtubules** are long, unbranched hollow tubes composed mainly of the protein *tubulin.* An organelle called the centrosome (discussed

Figure 3.16   Cytoskeleton.

 The cytoskeleton is a network of three types of protein filaments that extend throughout the cytoplasm: microfilaments, intermediate filaments, and microtubules.

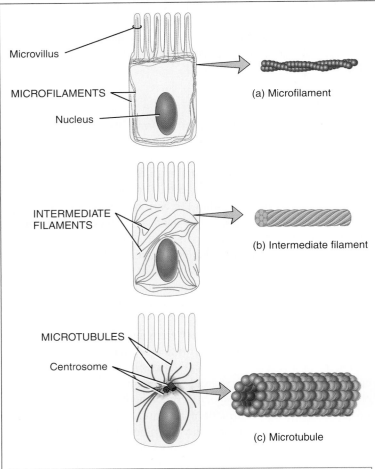

Microvillus

MICROFILAMENTS

Nucleus

(a) Microfilament

INTERMEDIATE FILAMENTS

(b) Intermediate filament

MICROTUBULES

Centrosome

(c) Microtubule

Functions

1. The cytoskeleton serves as a scaffold that helps to determine a cell's shape and to organize the cellular contents.
2. The cytoskeleton aids movement of organelles within the cell, of chromosomes during cell division, and of whole cells such as phagocytes.

Which cytoskeletal component helps form the structure of centrioles, cilia, and flagella?

shortly) serves as the site where assembly of microtubules begins. The microtubules grow outward from the centrosome toward the periphery of the cell (Figure 3.16c). Microtubules help determine cell shape and function in the movement of organelles, such as secretory vesicles. They also participate in the movement of chromosomes during cell division and in the movement of specialized cell projections, such as cilia and flagella. *Motor proteins* called *kinesins* and *dyneins* are responsible for powering the movements in which microtubules participate.

These proteins act like miniature engines that propel substances and organelles along a microtubule, much as a train engine pushes or pulls cars along a railroad track.

Centrosome

The **centrosome,** located near the nucleus, consists of two components: a pair of centrioles and pericentriolar material (Figure 3.17a). The two **centrioles** are cylindrical structures, each composed of nine clusters of three microtubules (triplets) arranged in a circular pattern (Figure 3.17b). The long axis of one centriole is at a right angle to the long axis of the other (Figure 3.17c). Surrounding the centrioles is **pericentriolar material,** which contains hundreds of ring-shaped complexes composed of the protein *tubulin.* These tubulin complexes are the organizing centers for growth of the mitotic spindle, which plays a critical role in cell division, and for microtubule formation in nondividing cells.

Cilia and Flagella

Microtubules are the dominant components of cilia and flagella, both of which are motile projections of the cell surface. **Cilia** (SIL-ē-a = eyelashes; singular is *cilium*) are numerous, short, hairlike projections that extend from the surface of the cell (see Figure 3.1). Each cilium contains a core of 20 microtubules surrounded by plasma membrane (Figure 3.18a). The microtubules are arranged such that one pair in the center is surrounded by nine clusters of two fused microtubules (doublets). Each cilium is anchored to a *basal body* just below the surface of the plasma membrane. A basal body is similar in structure to a centriole and functions in initiating the assembly of cilia and flagella.

A cilium displays an oarlike pattern of beating in which it is relatively stiff during the power stroke, but relatively flexible during the recovery stroke (Figure 3.18b). The coordinated movement of many cilia on the surface of a cell causes the steady movement of fluid along the cell's surface. Many cells of the respiratory tract, for example, have hundreds of cilia that help sweep foreign particles trapped in mucus away from the lungs. Their movement is paralyzed by nicotine in cigarette smoke. For this reason, smokers cough often to remove foreign particles from their airways. Cells that line the uterine (Fallopian) tubes also have cilia that sweep oocytes (egg cells) toward the uterus.

Flagella (fla-JEL-a = whip; singular is *flagellum*) are similar in structure to cilia but are typically much longer. Flagella usually move an entire cell. A flagellum generates forward motion along its axis by rapidly wiggling in a wavelike pattern (Figure 3.18c). The only example of a flagellum in the human body is a sperm cell's tail, which propels the sperm toward its rendezvous with an oocyte.

Ribosomes

Ribosomes (RĪ-bō-sōms) are the sites of protein synthesis. The name of these tiny organelles reflects their high content of one

Figure 3.17 Centrosome.

Located near the nucleus, the centrosome consists of a pair of centrioles and pericentriolar material.

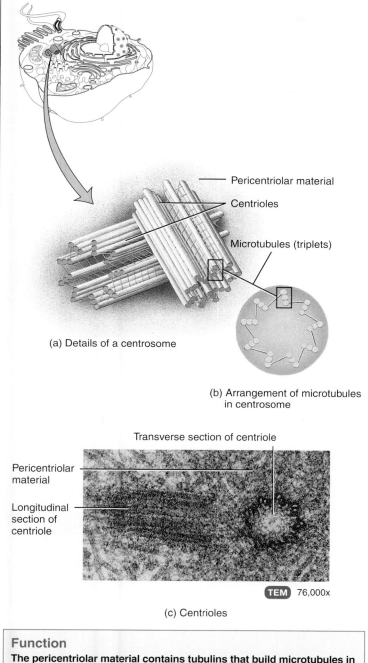

(a) Details of a centrosome

— Pericentriolar material

— Centrioles

Microtubules (triplets)

(b) Arrangement of microtubules in centrosome

Transverse section of centriole

Pericentriolar material

Longitudinal section of centriole

TEM 76,000x

(c) Centrioles

Function
The pericentriolar material contains tubulins that build microtubules in nondividing cells and form the mitotic spindle during cell division.

 If you observed that a cell did not have a centrosome, what could you predict about its capacity for cell division?

Figure 3.18 Cilia and flagella.

A cilium contains a core of microtubules with one pair in the center surrounded by nine clusters of doublet microtubules.

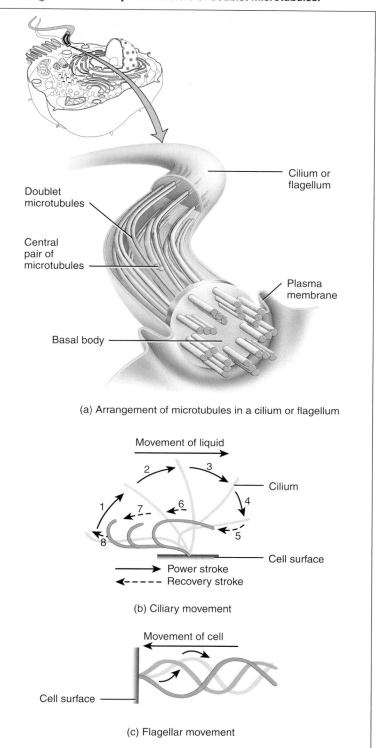

Cilium or flagellum

Doublet microtubules

Central pair of microtubules

Plasma membrane

Basal body

(a) Arrangement of microtubules in a cilium or flagellum

Movement of liquid

Cilium

Cell surface

→ Power stroke
← - - - Recovery stroke

(b) Ciliary movement

Movement of cell

Cell surface

(c) Flagellar movement

Functions
1. Cilia move fluids along a cell's surface.
2. A flagellum moves an entire cell.

 What is the functional difference between cilia and flagella?

type of ribonucleic acid, **ribosomal RNA (rRNA),** but each one also includes more than 50 proteins. Structurally, a ribosome consists of two subunits, one about half the size of the other (Figure 3.19). The large and small subunits are made separately in the nucleolus, a spherical body inside the nucleus. Once produced, the large and small subunits exit the nucleus and then come together in the cytoplasm.

Some ribosomes are attached to the outer surface of the nuclear membrane and to an extensively folded membrane called the endoplasmic reticulum. These ribosomes synthesize proteins destined for specific organelles, for insertion in the plasma membrane, or for export from the cell. Other ribosomes are "free" or unattached to other cytoplasmic structures. Free ribosomes synthesize proteins used in the cytosol. Ribosomes are also located within mitochondria, where they synthesize mitochondrial proteins.

Endoplasmic Reticulum

The **endoplasmic reticulum** (en′-dō-PLAS-mik re-TIK-ū-lum; *-plasmic* = cytoplasm; *reticulum* = network) or **ER** is an extensive network of folded membranes that extends from the nuclear

envelope (the membrane around the nucleus), to which it is connected, throughout the cytoplasm (Figure 3.20). ER membranes are shaped like flattened sacs or tubules. The ER is so extensive that it constitutes more than half of the membranous surfaces within most cells. Two distinct, but interrelated, forms of ER exist in cells. Ribosomes adhere to the surface of *rough* ER whereas *smooth* ER looks "smooth" because it lacks ribosomes.

Figure 3.20   Endoplasmic reticulum.

 The endoplasmic reticulum is a network of membrane-enclosed sacs or tubules that extend throughout the cytoplasm and connect to the nuclear envelope.

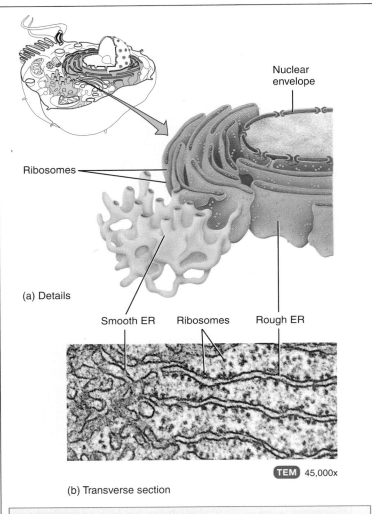

(a) Details

(b) Transverse section

TEM 45,000x

Functions
1. **Rough ER synthesizes glycoproteins and phospholipids that are transferred into cellular organelles, inserted into the plasma membrane, or secreted during exocytosis.**
2. **Smooth ER synthesizes fatty acids and steroids, such as estrogens and testosterone; inactivates or detoxifies drugs and other potentially harmful substances; removes the phosphate group from glucose 6-phosphate; and stores and releases calcium ions that trigger contraction in muscle cells.**

Figure 3.19   Ribosomes.

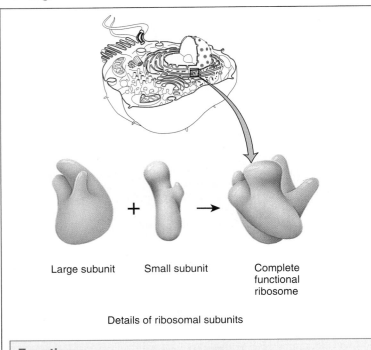 Ribosomes are the sites of protein synthesis.

Details of ribosomal subunits

Functions
1. **Ribosomes associated with endoplasmic reticulum synthesize proteins destined for insertion in the plasma membrane or secretion from the cell.**
2. **Free ribosomes synthesize proteins used in the cytosol.**

 Where are subunits of ribosomes synthesized and assembled?

What are the structural and functional differences between rough and smooth ER?

Proteins synthesized by ribosomes attached to **rough ER** enter the space within the ER for processing and sorting. In some cases, enzymes attach carbohydrates to the proteins to form glycoproteins. In other cases, enzymes attach the proteins to phospholipids, also synthesized by the rough ER. These molecules may be incorporated into organelle membranes, inserted into the plasma membrane, or secreted via exocytosis.

Smooth ER extends from the rough ER to form a network of membranous tubules (Figure 3.20). Synthesis of fatty acids and steroids, such as estrogens and testosterone, occurs in the smooth ER. In liver cells, enzymes located in the smooth ER inactivate or detoxify drugs and other potentially harmful substances, for example, alcohol. In liver, kidney, and intestinal cells a smooth ER enzyme removes the phosphate group from glucose 6-phosphate, which allows the "free" glucose to enter the bloodstream. In muscle cells, calcium ions that trigger con-

traction are released from the sarcoplasmic reticulum, a form of smooth ER.

Golgi Complex

Most of the proteins synthesized by ribosomes attached to rough ER are ultimately transported to other regions of the cell. The first step in the transport pathway is through an organelle called the **Golgi complex** (GOL-jē). It consists of 3 to 20 **Golgi cisternae** (sis-TER-nē = cavities; singular is *cisterna*). These small, flattened membranous sacs with bulging edges resemble a stack of pita bread (Figure 3.21). The cisternae are often curved, giving the Golgi complex a cuplike shape. Most cells have several Golgi complexes, and Golgi complexes are more extensive in cells that secrete proteins. As you will see, this fact is a clue to the organelle's role in the cell.

Figure 3.21 Golgi complex.

The opposite faces of a Golgi complex differ in size, shape, content, and enzymatic activities.

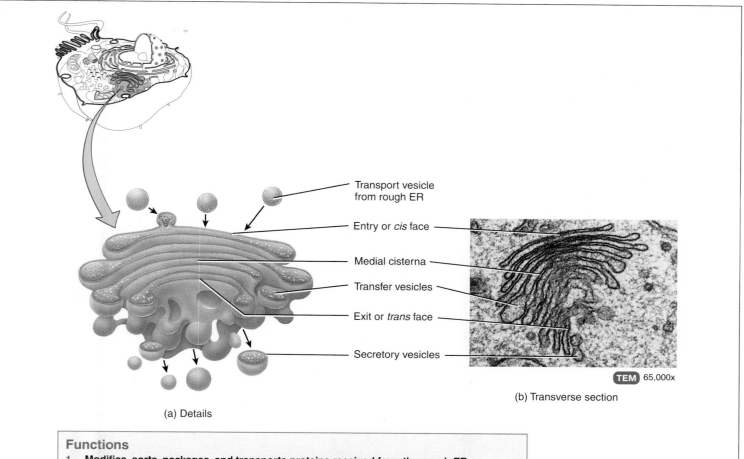

Transport vesicle from rough ER

Entry or *cis* face

Medial cisterna

Transfer vesicles

Exit or *trans* face

Secretory vesicles

TEM 65,000x

(b) Transverse section

(a) Details

Functions
1. **Modifies, sorts, packages, and transports proteins received from the rough ER.**
2. **Forms secretory vesicles that discharge processed proteins via exocytosis into extracellular fluid; forms membrane vesicles that ferry new molecules to the plasma membrane; forms transport vesicles that carry molecules to other organelles, such as lysosomes.**

 How do the entry and exit faces differ in function?

The cisternae at the opposite ends of a Golgi complex differ from each other in size, shape, and enzymatic activity. The convex **entry** or *cis* **face** is a cisterna that faces the rough ER. The concave **exit** or *trans* **face** is a cisterna that faces the plasma membrane. Sacs between the entry and exit faces are called **medial cisternae.** Transport vesicles from the ER merge to form the entry face. From the entry face, the cisterna are thought to mature, in turn becoming medial and then exit cisternae.

The entry, medial, and exit regions of the Golgi complex contain different enzymes that permit each to modify, sort, and package proteins for transport to different destinations. The entry face receives and modifies proteins produced by the rough ER. The medial cisternae add carbohydrates to proteins to form glycoproteins and lipids to proteins to form lipoproteins. The exit face modifies the molecules further and then sorts and packages them for transport to their destinations.

Proteins arriving at, passing through, and exiting the Golgi complex do so through maturation of the cisternae and exchanges that occur via transfer vesicles (Figure 3.22):

1 Proteins synthesized by ribosomes on the rough ER are surrounded by a piece of the ER membrane, which eventually buds from the membrane surface in the form of transport vesicles.

2 Transport vesicles move toward the entry face of the Golgi complex.

3 Fusion of several transport vesicles creates the entry face of the Golgi complex and releases proteins into its lumen.

4 The proteins move from the entry face into one or more medial cisternae. Enzymes in the medial cisternae modify the proteins to form glycoproteins, glycolipids, and lipoproteins. Transfer vesicles that bud from the edges of the cisternae move specific enzymes back toward the entry face and also move some partially modified proteins forward toward the exit face.

5 The products of the medial cisternae move into the lumen of the exit face.

6 Within the exit face cisterna, the products are further modified and are sorted and packaged.

7 Some of the processed proteins leave the exit face in **secretory vesicles.** These vesicles deliver the proteins to the plasma membrane, where they are discharged by exocytosis into the extracellular fluid. For example, certain pancreatic cells release the hormone insulin in this way.

8 Other processed proteins leave the exit face in **membrane vesicles** that deliver their contents to the plasma membrane

Figure 3.22   Processing and packaging of proteins by the Golgi complex.

All proteins exported from the cell are processed in the Golgi complex.

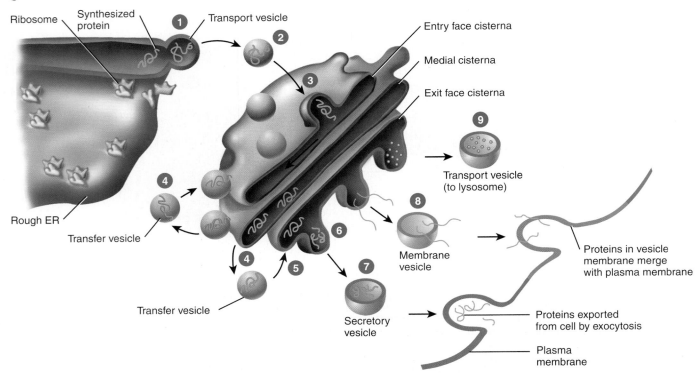

What are the three general destinations for proteins that leave the Golgi complex?

for incorporation into the membrane. In doing so, the Golgi complex adds new segments of plasma membrane as existing segments are lost and modifies the number and distribution of membrane molecules.

9 Finally, some processed proteins leave the exit face in **transport vesicles** that will carry the proteins to another cellular destination. For instance, transport vesicles ferry digestive enzymes to lysosomes, whose structure and functions are discussed next.

Lysosomes

Lysosomes (LĪ-sō-sōms; *lyso-* = dissolving; *-somes* = bodies) are membrane-enclosed vesicles that form from the Golgi complex (Figure 3.23). Inside are as many as 60 kinds of powerful digestive and hydrolytic enzymes that can break down a wide variety of molecules. Lysosomal enzymes work best at an acidic pH, and the lysosomal membrane includes active transport pumps that move hydrogen ions (H^+) into the lysosome. Thus, the lysosomal interior has a pH of 5, which is 100 times more acidic than the cytosolic pH of 7. The lysosomal membrane also includes transporters that move the final products of digestion, such as glucose, fatty acids, and amino acids, into the cytosol.

Lysosomal enzymes also help recycle the cell's own structures. A lysosome can engulf another organelle, digest it, and return the digested components to the cytosol for reuse. The process by which worn-out organelles are digested is known as **autophagy** (aw-TOF-a-jē; *auto-* = self; *-phagy* = eating). During autophagy, the organelle to be digested is enclosed by a membrane derived from the ER to create a vesicle called an *autophagosome,* which then fuses with a lysosome. In this way, a human liver cell, for example, recycles about half of its cytoplasmic contents every week. Lysosomal enzymes may also destroy their own cell, a process known as **autolysis** (aw-TOL-i-sis). Autolysis occurs in some pathological conditions and is also responsible for the tissue deterioration that occurs just after death.

Although most lysosomal enzymes act within a cell, there are some instances in which the enzymes operate in extracellular digestion. One example occurs during fertilization. The head of a sperm cell releases lysosomal enzymes that aid its penetration of the oocyte.

Tay-Sachs Disease

Some disorders are caused by faulty or absent lysosomal enzymes. For instance, **Tay-Sachs disease,** which most often affects children of Ashkenazi (eastern European Jewish) descent, is an inherited condition characterized by the absence of a single lysosomal enzyme called Hex A. This enzyme normally breaks down a membrane glycolipid called ganglioside G_{M2} that is especially prevalent in nerve cells. As ganglioside G_{M2} accumulates, because it is not broken down, the nerve cells function less efficiently. Children with Tay-Sachs disease typically experience

seizures and muscle rigidity. They gradually become blind, demented, and uncoordinated and usually die before the age of 5. Tests can now reveal whether an adult is a carrier of the defective gene. ■

Peroxisomes

Another group of organelles similar in structure to lysosomes, but smaller, are the **peroxisomes** (pe-ROKS-i-sōms; *peroxi-* = peroxide; *somes* = bodies; see Figure 3.1). Peroxisomes contain

Figure 3.23 Lysosomes.

Lysosomes contain several types of powerful digestive enzymes.

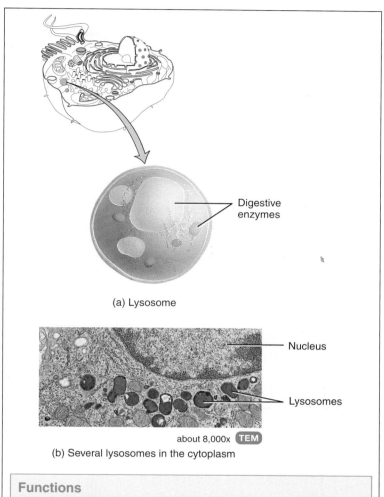

Digestive enzymes

(a) Lysosome

Nucleus

Lysosomes

about 8,000x **TEM**

(b) Several lysosomes in the cytoplasm

Functions

1. Digest substances that enter a cell via endocytosis and transport final products of digestion into cytosol.
2. Carry out autophagy, the digestion of worn-out organelles.
3. Carry out autolysis, the digestion of entire cell.
4. Carry out extracellular digestion.

 What is the name of the process by which worn-out organelles are digested by lysosomes?

several *oxidases,* enzymes that can oxidize (remove hydrogen atoms from) various organic substances. For instance, amino acids and fatty acids are oxidized in peroxisomes as part of normal metabolism. In addition, enzymes in peroxisomes oxidize toxic substances, such as alcohol. A byproduct of the oxidation reactions is hydrogen peroxide (H_2O_2), a potentially toxic compound. However, peroxisomes also contain the enzyme *catalase,* which decomposes H_2O_2. Because production and degradation of H_2O_2 occurs within the same organelle, peroxisomes protect other parts of the cell from the toxic effects of H_2O_2. New peroxisomes form by budding off from preexisting ones.

Proteasomes

Although lysosomes degrade proteins delivered to them in vesicles, cytosolic proteins also require disposal at certain times in the life of a cell. Continuous destruction of unneeded, damaged, or faulty proteins is the function of tiny structures called **proteasomes** (PRŌ-tē-a-sōmes = protein bodies). For example, proteins that are part of metabolic pathways are degraded after they have accomplished their function. Such protein destruction halts the pathway once the appropriate response has been achieved. A typical body cell contains many thousands of proteasomes, in both the cytosol and the nucleus. They were discovered only recently because they are far too small to discern under the light microscope and do not show up well in electron micrographs. Proteasomes were so-named because they contain myriad *proteases,* enzymes that cut proteins into small peptides. Once the enzymes of a proteasome have chopped up a protein into smaller chunks, other enzymes then break down the peptides into amino acids, which can be recycled into new proteins.

 **Proteasomes and Disease**

Proteasomes are thought to be a factor in several diseases. For instance, people who have **cystic fibrosis** produce a misshapen membrane transporter protein whose mission normally is to help pump chloride ions (Cl^-) out of certain cells. Proteosomes degrade the defective transporter before it can reach the plasma membrane. The result is an imbalance in the transport of fluid and ions across the plasma membrane that causes the buildup of thick mucus outside certain types of cells. The accumulated mucus clogs the airways in the lungs, causing breathing difficulty, and prevents proper secretion of digestive enzymes by the pancreas, causing digestive problems. Disease could also result from failure of proteasomes to degrade abnormal proteins. For example, clumps of misfolded proteins accumulate in brain cells of people with Parkinson disease and Alzheimer disease. Discovering why the proteasomes fail to clear these abnormal proteins is a goal of ongoing research. ■

Mitochondria

Because they generate most of the ATP, the "powerhouses" of the cell are its **mitochondria** (mī-tō-KON-drē-a; *mito-* = thread; *-chondria* = granules; singular is *mitochondrion*). A cell may have as few as a hundred or as many as several thousand mitochondria. Active cells, such as those found in the muscles, liver, and kidneys, have a large number of mitochondria because they use ATP at a high rate. Mitochondria are usually located in a cell where the energy need is greatest, such as in among the contractile proteins in muscle cells.

A mitochondrion consists of an **outer mitochondrial membrane** and an **inner mitochondrial membrane** with a small fluid-filled space between them (Figure 3.24). Both membranes are similar in structure to the plasma membrane. The inner mitochondrial membrane contains a series of folds called **cristae** (KRIS-tē = ridges). The large central fluid-filled cavity of a mitochondrion, enclosed by the inner mitochondrial membrane, is the **matrix.** The elaborate folds of the cristae provide an enormous surface area for the chemical reactions that are part of the aerobic phase of *cellular respiration.* These reactions produce most of a cell's ATP and the enzymes that catalyze them are located on the cristae and in the matrix.

Like peroxisomes, mitochondria self-replicate, a process that occurs during times of increased cellular energy demand or before cell division. Because ribosomes are also present in the mitochondrial matrix, synthesis of some proteins needed for mitochondrial functions occurs there. Mitochondria even have their own DNA, in the form of multiple copies of a circular DNA molecule that contains 37 genes. These mitochondrial genes control the synthesis of 2 ribosomal RNAs, 22 transfer RNAs, and 13 proteins that build mitochondrial components.

Although the nucleus of each somatic cell contains genes from both your mother and father, mitochondrial genes are inherited only from your mother. The head of a sperm (the part that penetrates and fertilizes an oocyte) normally lacks most organelles, such as mitochondria, ribosomes, endoplasmic reticulum, and the Golgi complex. Also, any sperm mitochondria that do enter the oocyte are soon destroyed.

▶ C H E C K P O I N T

15. What does cytoplasm have that cytosol lacks?

16. Which organelles are surrounded by a membrane and which are not?

17. Which organelles contribute to synthesizing protein hormones and packaging them into secretory vesicles?

18. What happens on the cristae and in the matrix of mitochondria?

NUCLEUS

▶ O B J E C T I V E

• **Describe the structure and function of the nucleus.**

The **nucleus** is a spherical or oval-shaped structure that usually is the most prominent feature of a cell (Figure 3.25). Most body

Figure 3.24 Mitochondria.

 Within mitochondria, chemical reactions called cellular respiration generate ATP.

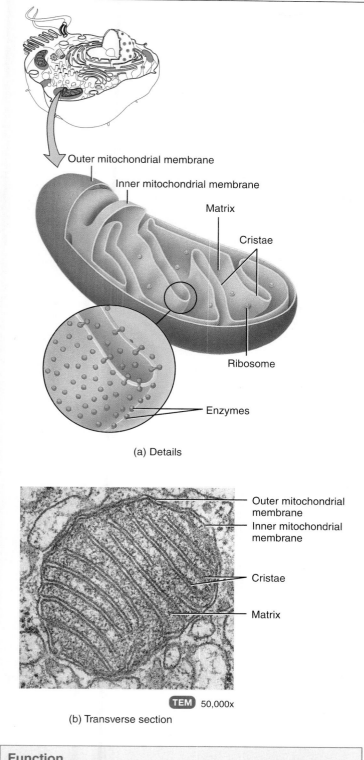

Outer mitochondrial membrane

Inner mitochondrial membrane

Matrix

Cristae

Ribosome

Enzymes

(a) Details

Outer mitochondrial membrane
Inner mitochondrial membrane

Cristae

Matrix

TEM 50,000x

(b) Transverse section

Function
Generate ATP through reactions of aerobic cellular respiration.

How do the cristae of a mitochondrion contribute to its ATP-producing function?

cells have a single nucleus, although some, such as mature red blood cells, have none. In contrast, skeletal muscle cells and a few other types of cells have more than one nucleus. A double membrane called the **nuclear envelope** separates the nucleus from the cytoplasm. Both layers of the nuclear envelope are lipid bilayers similar to the plasma membrane. The outer membrane of the nuclear envelope is continuous with rough ER and resembles it in structure. Many openings called **nuclear pores** extend through the nuclear envelope. Each nuclear pore consists of a circular arrangement of proteins that surrounds a large central opening that is about 10 times wider than the pore of a channel protein in the plasma membrane.

Nuclear pores control the movement of substances between the nucleus and the cytoplasm. Small molecules and ions diffuse passively through the pores. Most large molecules, however, such as RNAs and proteins, cannot pass through the nuclear pores by diffusion. Instead, their passage involves an active transport process in which the molecules are recognized and selectively transported through the nuclear pore into or out of the nucleus. For example, proteins needed for nuclear functions move from the cytosol into the nucleus whereas RNA molecules move from the nucleus into the cytosol.

Inside the nucleus are one or more spherical bodies called **nucleoli** (noo′-KLĒ-ō-lī; singular is *nucleolus*) that function in producing ribosomes. Each nucleolus is a cluster of protein, DNA, and RNA that is not enclosed by a membrane. Nucleoli are the sites of synthesis of rRNA and assembly of rRNA and proteins into ribosomal subunits. Nucleoli are quite prominent in cells that synthesize large amounts of protein, such as muscle and liver cells. Nucleoli disperse and disappear during cell division and reorganize once new cells are formed.

Within the nucleus are most of the cell's hereditary units, called **genes,** which control cellular structure and direct cellular activities. Genes are arranged in single file along **chromosomes** (*chromo-* = colored). Human somatic (body) cells have 46 chromosomes, 23 inherited from each parent. Each chromosome is a long molecule of DNA that is coiled together with several proteins (Figure 3.26 on page 85). This complex of DNA, proteins, and some RNA is called **chromatin.** The total genetic information carried in a cell or an organism is its **genome.**

In cells that are not dividing, the chromatin appears as a diffuse, granular mass. Electron micrographs reveal that chromatin has a "beads-on-a-string" structure. Each "bead" is a **nucleosome** and consists of double-stranded DNA wrapped twice around a core of eight proteins called **histones,** which help organize the coiling and folding of DNA. The "string" between the "beads" is **linker DNA,** which holds adjacent nucleosomes together. Another histone promotes coiling of nucleosomes into a larger diameter **chromatin fiber,** which then folds into large loops. In cells that are not dividing, this is how DNA is packed. Just before cell division takes place, however, the DNA replicates (duplicates) and the loops condense even more, forming a pair of **chromatids.** As you will see shortly, during cell division a pair of chromatids constitutes a chromosome.

Figure 3.25   Nucleus.

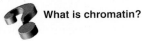

The nucleus contains most of the cell's genes, which are located on chromosomes.

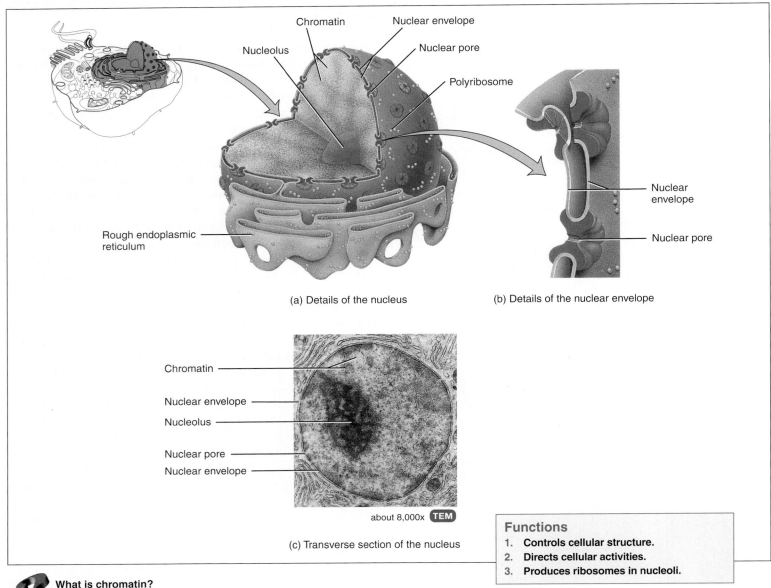

(a) Details of the nucleus

(b) Details of the nuclear envelope

about 8,000x **TEM**

(c) Transverse section of the nucleus

Functions
1. **Controls cellular structure.**
2. **Directs cellular activities.**
3. **Produces ribosomes in nucleoli.**

What is chromatin?

The main parts of a cell and their functions are summarized in Table 3.2 on page 86.

Genomics

In the last decade of the 20th century, the genomes of humans, mice, fruit flies, and more than 50 microbes were sequenced. As a result, research in the field of **genomics,** the study of the relationships between the genome and the biological functions of an organism, has flourished. The Human Genome Project began in 1990 as an effort to sequence all of the nearly 3.2 billion nu-

cleotides of our genome. By June 2000, scientists had produced a "working draft" of the human genome, having sequenced about 90% of the nucleotides. More than 99.9% of the nucleotide bases are identical in everyone. Less than 0.1% of our DNA (1 in each 1000 bases) accounts for inherited differences among humans. Surprisingly, at least half of the human genome consists of repeated sequences that do not code for proteins, so-called "junk" DNA. The average gene consists of 3,000 nucleotides, but sizes vary greatly. The largest known human gene codes for the protein dystrophin and has 2.4 million nucleotides. Scientists now estimate the total number of genes in

Figure 3.26 **Packing of DNA into a chromosome in a dividing cell.** When packing is complete, two identical DNA molecules and their histones form a pair of chromatids, which are held together by a centromere.

🔑 **A chromosome is a highly coiled and folded DNA molecule that is combined with protein molecules.**

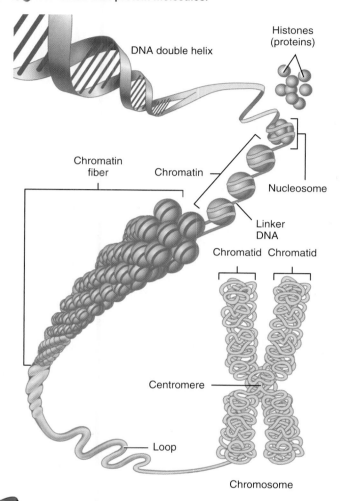

What are the components of a nucleosome?

the human genome at about 35,000 to 45,000, far fewer than the 100,000 previously predicted to exist. With an increasingly complete understanding of the human genome, genomic medicine aims to detect and treat genetic diseases at an early stage. Soon physicians may be able to provide counseling and treatment that is more effective for disorders that have significant genetic components, for instance, hypertension (high blood pressure), obesity, diabetes, and cancer. ∎

▶ CHECKPOINT
19. How is DNA packed in the nucleus?

PROTEIN SYNTHESIS

▶ OBJECTIVE

• **Describe the sequence of events that take place during protein synthesis.**

Although cells synthesize many chemicals to maintain homeostasis, much of the cellular machinery is devoted to synthesizing large numbers of diverse proteins. The proteins, in turn, determine the physical and chemical characteristics of cells and, therefore, of organisms. Just as genome means all of the genes in an organism, **proteome** means all of an organism's proteins. Some proteins are used to assemble cellular structures such as the plasma membrane, the cytoskeleton, and other organelles. Other proteins serve as hormones, antibodies, and contractile elements in muscle tissue. Still others are enzymes that regulate the rates of the numerous chemical reactions that occur in cells or transporters that carry various materials in the blood, such as transport of oxygen by hemoglobin.

Gene expression is the process whereby a gene's DNA is used to direct synthesis of a specific protein. First, the information encoded in a specific region of DNA is *transcribed* (copied) to produce a specific molecule of RNA (ribonucleic acid). Then, the RNA attaches to a ribosome, where the information contained in RNA is *translated* into a corresponding sequence of amino acids to form a new protein molecule (Figure 3.27).

Figure 3.27 **Overview of gene expression.** Synthesis of a specific protein requires transcription of a gene's DNA into RNA and translation of RNA into a corresponding sequence of amino acids.

🔑 **Whereas transcription occurs in the nucleus, translation occurs in the cytoplasm.**

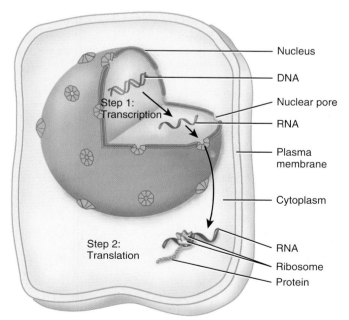

Why are proteins important in the life of a cell?

Table 3.2 Cell Parts and Their Functions

Part	Structure	Functions
Plasma Membrane	Fluid-mosaic lipid bilayer (phospholipids, cholesterol, and glycolipids) studded with proteins; surrounds cytoplasm.	Protects cellular contents; makes contact with other cells; contains channels, transporters, receptors, enzymes, cell-identity markers, and linker proteins; mediates the entry and exit of substances.
Cytoplasm	Cellular contents between the plasma membrane and nucleus—cytosol and organelles.	Site of all intracellular activities except those occurring in the nucleus.
Cytosol	Composed of water, solutes, suspended particles, lipid droplets, and glycogen granules.	Medium in which many of cell's metabolic reactions occur.
Organelles	Specialized structures with characteristic shapes.	Each organelle has specific functions.
Cytoskeleton	Network of three types of protein filaments: microfilaments, intermediate filaments, and microtubules.	Maintains shape and general organization of cellular contents; responsible for cellular movements.
Centrosome	A pair of centrioles plus pericentriolar material.	The pericentriolar material contains tubulins, which are used for growth of the mitotic spindle and microtubule formation.
Cilia and flagella	Motile cell surface projections that contain 20 microtubules and a basal body.	Cilia move fluids over a cell's surface; flagella move an entire cell.
Ribosome	Composed of two subunits containing ribosomal RNA and proteins; may be free in cytosol or attached to rough ER.	Protein synthesis.
Endoplasmic reticulum (ER)	Membranous network of flattened sacs or tubules. Rough ER is covered by ribosomes and is attached to nuclear envelope; smooth ER lacks ribosomes.	Rough ER synthesizes glycoproteins and phospholipids that are transferred to cellular organelles, inserted into the plasma membrane or secreted during exocytosis. Smooth ER synthesizes fatty acids and steroids; inactivates or detoxifies drugs; removes phosphate group from glucose 6-phosphate; and stores and releases calcium ions in muscle cells.
Golgi complex	Consists of 3–20 flattened membranous sacs called cisternae; structurally and functionally divided into entry (cis) face, medial cisternae, and exit (trans) face.	Entry (cis) face accepts proteins from rough ER; medial cisternae form glycoproteins, glycolipids, and lipoproteins; exit (trans) face modifies the molecules further, then sorts and packages them for transport to their destinations.
Lysosome	Vesicle formed from Golgi complex; contains digestive enzymes.	Fuses with and digests contents of endosomes, pinocytic vesicles, and phagosomes and transports final products of digestion into cytosol; digests worn-out organelles (autophagy), entire cells (autolysis), and extracellular materials.
Peroxisome	Vesicle containing oxidases (oxidative enzymes) and catalase (decomposes hydrogen peroxide); new peroxisomes bud from preexisting ones.	Oxidizes amino acids and fatty acids; detoxifies harmful substances, such as alcohol; produces hydrogen peroxide.
Proteasome	Tiny structure that contains proteases (proteolytic enzymes).	Degrades unneeded, damaged, or faulty proteins by cutting them into small peptides.
Mitochondrion	Consists of outer and inner mitochondrial membranes, cristae, and matrix; new mitochondria form from preexisting ones.	Site of aerobic cellular respiration reactions that produce most of a cell's ATP.
Nucleus	Consists of nuclear envelope with pores, nucleoli, and chromosomes, which exist as a tangled mass of chromatin in interphase cells.	Contains genes, which control cellular structure and direct cellular functions.

DNA and RNA store genetic information as sets of three nucleotides. A sequence of three such nucleotides in DNA is called a **base triplet.** Each DNA base triplet is transcribed as a complementary sequence of three RNA nucleotides, called a **codon.** A given codon specifies a particular amino acid. The **genetic code** is the set of rules that relate the base triplet sequence of DNA to the corresponding codons of RNA and the amino acids they specify.

The first step in protein synthesis is transcription, which occurs inside the nucleus.

Transcription

During **transcription,** the genetic information represented by the sequence of base triplets in DNA serves as a template for copying the information into a complementary sequence of codons in a strand of RNA. Three kinds of RNA are made from the DNA template:

- **Messenger RNA (mRNA)** directs the synthesis of a protein.
- **Ribosomal RNA (rRNA)** joins with ribosomal proteins to make ribosomes.
- **Transfer RNA (tRNA)** binds to an amino acid and holds it in place on a ribosome until it is incorporated into a protein during translation. One end of the tRNA carries a specific amino acid, and the opposite end consists of a triplet of nucleotides called an **anticodon.** By pairing between complementary bases, the tRNA anticodon attaches to the mRNA codon. Each of the more than 20 different types of tRNA binds to only one of the 20 different amino acids.

The enzyme **RNA polymerase** catalyzes transcription of DNA. However, the enzyme must be instructed where to start the transcription process and where to end it. Only one of the two DNA strands serves as a template for RNA synthesis. The segment of DNA where transcription begins is a special nucleotide sequence called a **promoter,** which is located near the beginning of a gene (Figure 3.28a). This is where RNA polymerase attaches to the DNA. During transcription, bases pair in a complementary manner: The bases cytosine (C), guanine (G), and thymine (T) in the DNA template pair with guanine, cytosine, and adenine (A), respectively, in the RNA strand (Figure 3.28b). However, adenine in the DNA template pairs with uracil (U), not thymine, in RNA:

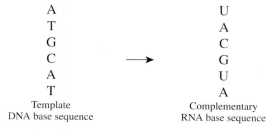

A	U
T	A
G	C
C	G
A	U
T	A
Template DNA base sequence	Complementary RNA base sequence

Transcription of the DNA strand ends at another special nucleotide sequence called a **terminator,** which specifies the end of the gene (Figure 3.28a). When RNA polymerase reaches the

Figure 3.28 Transcription. DNA transcription begins at a promoter and ends at a terminator.

 During transcription, the genetic information in DNA is copied to RNA.

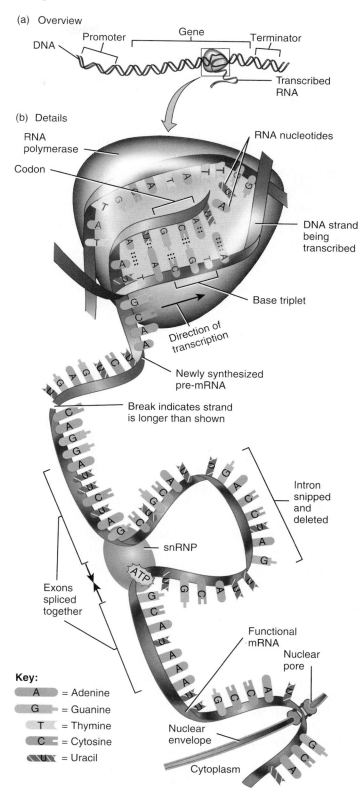

If the DNA template had the base sequence AGCT, what would be the mRNA base sequence? What enzyme catalyzes transcription of DNA?

terminator, the enzyme detaches from the transcribed RNA molecule and the DNA strand.

Not all parts of a gene actually code for parts of a protein. Within a gene are regions called **introns** that *do not* code for parts of proteins. They are located between regions called **exons** that *do* code for segments of a protein. Immediately after transcription, the transcript includes information from both introns and exons and is called **pre-mRNA.** The introns are removed from pre-mRNA by **small nuclear ribonucleoproteins** (snRNPs, pronounced "snurps"; Figure 3.28b). The snRNPs are enzymes that cut out the introns and splice together the exons. The resulting product is a functional mRNA molecule that passes through a pore in the nuclear envelope to reach the cytoplasm, where translation takes place.

Although the human genome contains 35,000 to 45,000 genes, there are probably 500,000 to 1 million human proteins. How can so many proteins be coded for by so few genes? Part of the answer lies in **alternative splicing** of mRNA, a process in which the pre-mRNA transcribed from a gene is spliced in different ways to produce several different mRNAs. The different mRNAs are then translated into different proteins. In this way, one gene may code for ten or more different proteins. In addition, chemical modifications are made to proteins after translation, for example, as proteins pass through the Golgi complex. Such chemical alterations can produce two or more different proteins from a single translation.

Antisense Therapy

In 1998 the U.S. Food and Drug Administration (FDA) approved Fomivirsen (Vitravene), the first drug designed to shut down gene expression by blocking the action of mRNA. The drug is an "antisense" DNA oligonucleotide (a few DNA nucleotides strung together) that is complementary to the "sense" of a specific mRNA. The DNA oligonucleotide base pairs with the mRNA and thereby blocks translation of the mRNA into a key protein needed for replication of the cytomegalovirus retinitis virus. The drug is injected into the eye to inhibit infection by this virus, which can rapidly cause blindness. Other antisense DNA oligonucleotides are being tested in clinical trials for treatment of certain cancers. ■

Translation

Translation is the process whereby the nucleotide sequence in an mRNA molecule specifies the amino acid sequence of a protein. Ribosomes in the cytoplasm carry out translation. The small subunit of a ribosome has a *binding site* for mRNA; the large subunit has two binding sites for tRNA molecules, a *P site* and an *A site* (Figure 3.29). The first tRNA molecule bearing its specific amino acid attaches to mRNA at the P site. The A site holds the next tRNA molecule bearing its amino acid. Translation occurs in the following way (Figure 3.30):

Figure 3.29   Translation. During translation, an mRNA molecule binds to a ribosome. Then, the mRNA nucleotide sequence specifies the amino acid sequence of a protein.

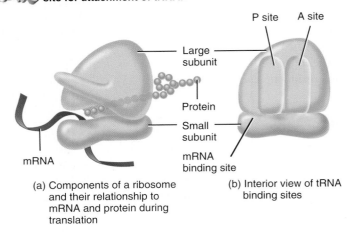

Ribosomes have a binding site for mRNA and a P site and an A site for attachment of tRNA.

(a) Components of a ribosome and their relationship to mRNA and protein during translation

(b) Interior view of tRNA binding sites

What roles do the P and A sites serve?

① An mRNA molecule binds to the small ribosomal subunit at the mRNA binding site. A special tRNA, called *initiator tRNA*, binds to the start codon (AUG) on mRNA, where translation begins. The tRNA anticodon (UAC) attaches to the mRNA codon (AUG) by pairing between the complementary bases. Besides being the start codon, AUG is also the codon for the amino acid methionine. Thus, methionine is always the first amino acid in a growing polypeptide.

② Next, the large ribosomal subunit attaches to the small ribosomal subunit–mRNA complex, creating a functional ribosome. The initiator tRNA, with its amino acid (methionine), fits into the P site of the ribosome.

③ The anticodon of another tRNA with its attached amino acid pairs with the second mRNA codon at the A site of the ribosome.

④ A component of the large ribosomal subunit catalyzes the formation of a peptide bond between methionine, which separates from its tRNA at the P site, and the amino acid carried by the tRNA at the A site.

⑤ After peptide bond formation, the tRNA at the P site detaches from the ribosome, and the ribosome shifts the mRNA strand by one codon. The tRNA in the A site bearing the dipeptide shifts into the P site, allowing another tRNA with its amino acid to bind to a newly exposed codon at the A site. Steps ③ through ⑤ occur repeatedly, and the polypeptide progressively lengthens.

⑥ Protein synthesis ends when the ribosome reaches a stop codon at the A site, at which time the completed protein detaches from the final tRNA. When the tRNA vacates the A site, the ribosome splits into its large and small subunits.

Figure 3.30 **Protein elongation and termination of protein synthesis during translation.**

 During protein synthesis the small and large ribosomal subunits join. When the process is complete, they separate.

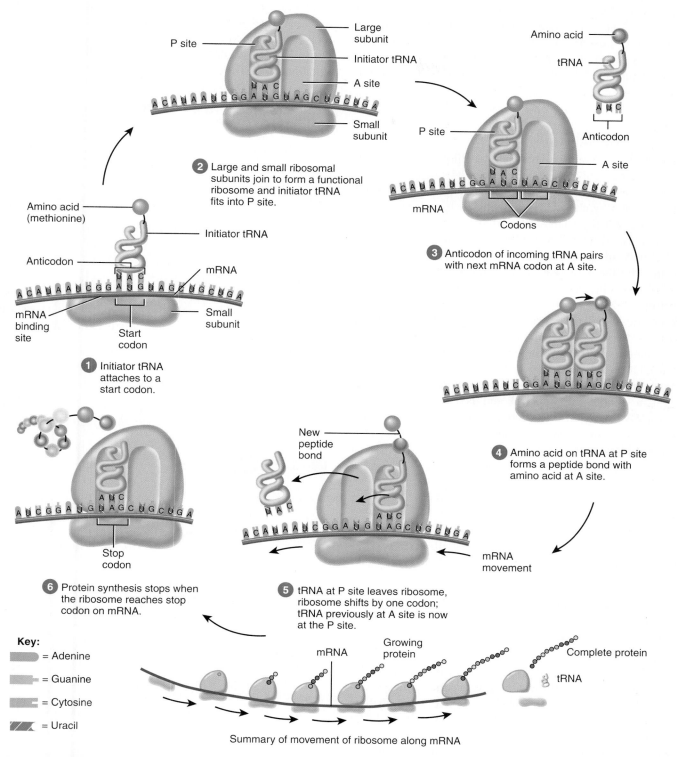

Key:

= Adenine

= Guanine

= Cytosine

= Uracil

Summary of movement of ribosome along mRNA

What is the function of a stop codon?

Protein synthesis progresses at a rate of about 15 peptide bonds per second. As the ribosome moves along the mRNA and before it completes synthesis of the whole protein, another ribosome may attach behind it and begin translation of the same mRNA strand. Several ribosomes attached to the same mRNA constitute a **polyribosome.** The simultaneous movement of several ribosomes along the same mRNA molecule permits the translation of one mRNA into several identical proteins at the same time.

Recombinant DNA

Beginning in 1973, scientists developed techniques for inserting genes from other organisms into a variety of host cells. This manipulation can cause the host organism to produce proteins it normally does not synthesize. Organisms so altered are called **recombinants,** and their DNA—a combination of DNA from different sources—is called **recombinant DNA.** When recombinant DNA functions properly, the host will synthesize the protein specified by the new gene it has acquired. The technology that has arisen from manipulating genetic material is **genetic engineering.**

The practical applications of recombinant DNA technology are enormous. Strains of recombinant bacteria are now producing large quantities of many important therapeutic substances. Just a few of these are *human growth hormone (hGH),* required for normal growth and metabolism; *insulin,* a hormone that helps regulate blood glucose level and is used by diabetics; *interferon (IFN),* an antiviral (and possibly anticancer) substance; and *erythropoietin (EPO),* a hormone that stimulates production of red blood cells. ■

▶ C H E C K P O I N T

20. What is the difference between transcription and translation?

CELL DIVISION

▶ O B J E C T I V E S

• **Discuss the stages, events, and significance of somatic cell division.**
• **Describe the signals that induce cell division.**

As cells become damaged, diseased, or worn out, they are replaced by cell division. Cell division also occurs during growth of tissues. The two types of cell division—somatic cell division and reproductive cell division—accomplish different goals for the organism.

In **somatic cell division,** a cell undergoes a nuclear division called **mitosis** and a cytoplasmic division called **cytokinesis** to produce two identical **daughter cells.** Each daughter cell has the same number and kind of chromosomes as the original cell. Somatic cell division replaces dead or injured cells and adds new ones for tissue growth.

Reproductive cell division is the mechanism that produces gametes—sperm and oocytes. This process consists of a special two-step division called **meiosis,** in which the number of chromosomes in the nucleus is reduced by half, followed by cytokinesis. Meiosis is described in Chapter 28; here we focus on somatic cell division.

The Cell Cycle in Somatic Cells

The **cell cycle** is an orderly sequence of events by which a somatic cell duplicates its contents and divides in two. Human cells, except for gametes, contain 23 pairs of chromosomes. The two chromosomes of each pair—one originally derived from the mother and the other from the father—are called **homologous chromosomes** or **homologues.** Each homologue carries one copy of each gene located on that chromosome. The two members of the same gene pair are located at identical sites on the two homologues. When a cell reproduces, it must replicate (duplicate) all its chromosomes so the genes may be passed on to the next generation of cells. The cell cycle consists of two major periods: interphase, when a cell is not dividing, and the mitotic phase, when a cell is dividing (Figure 3.31).

Figure 3.31 The cell cycle. Not illustrated is cytokinesis, division of the cytoplasm, which occurs during late anaphase or early telophase of the mitotic phase.

In a complete cell cycle, a cell duplicates its contents and divides into two daughter cells.

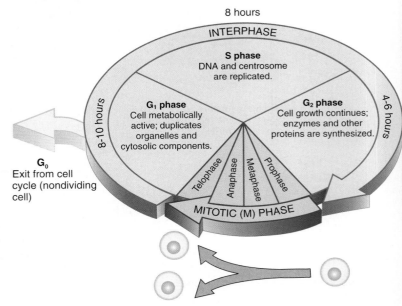

In which phase of the cell cycle does DNA replication occur?

Interphase

During **interphase** the cell replicates its DNA. It also produces additional organelles and cytosolic components in anticipation of cell division. Interphase is a state of high metabolic activity, and during this time the cell does most of its growing. Interphase consists of three phases: G_1, S, and G_2 (Figure 3.31). The S stands for synthesis of DNA. Because the G-phases are periods when there is no activity related to DNA duplication, they are thought of as gaps or interruptions in DNA duplication.

The G_1 **phase** is the interval between the mitotic phase and the S phase. During G_1, the cell is metabolically active; it duplicates most of its organelles and cytosolic components but not its DNA. Virtually all the cellular activities described in this chapter happen during G_1. For a cell with a total cell cycle time of 24 hours, G_1 lasts 8 to 10 hours. The duration of this phase, however, is quite variable. It is very short in many embryonic cells or cancer cells.

The **S phase** is the interval between G_1 and G_2 and lasts about 8 hours. During the S phase, DNA replication occurs. As a result, the two daughter cells formed during cell division will have identical genetic material. The centrosome also replicates during the S phase. The G_2 **phase** is the interval between the S phase and the mitotic phase. It lasts 4 to 6 hours. During G_2, cell growth continues and enzymes and other proteins are synthesized in preparation for cell division. Cells that remain in G_1 for a very long time, perhaps destined never to divide again, are said to be in the G_0 **state.** For example, most nerve cells are in this state. Once a cell enters the S phase, however, it is committed to go through cell division.

When DNA replicates during the S phase, its helical structure partially uncoils, and the two strands separate at the points where hydrogen bonds connect base pairs (Figure 3.32). Each exposed base of the old DNA strand then pairs with the complementary base of a newly synthesized nucleotide. A new DNA strand takes shape as chemical bonds form between neighboring nucleotides. The uncoiling and complementary base pairing continues until each of the two original DNA strands is joined with a newly formed complementary DNA strand. The original DNA molecule has become two identical DNA molecules.

A microscopic view of a cell during interphase shows a clearly defined nuclear envelope, a nucleolus, and a tangled mass of chromatin (Figure 3.33a). Once a cell completes its activities during the G_1, S, and G_2 phases of interphase, the mitotic phase begins.

Mitotic Phase

The **mitotic (M) phase** of the cell cycle consists of nuclear division, or mitosis, and cytoplasmic division, or cytokinesis. The events that occur during mitosis and cytokinesis are plainly visible under a microscope because chromatin condenses into discrete chromosomes.

NUCLEAR DIVISION: MITOSIS Distribution of the two sets of chromosomes, one set into each of two separate nuclei, occurs during **mitosis.** The process results in the *exact* partitioning of genetic information. For convenience, biologists divide the process into four stages: prophase, metaphase, anaphase, and telophase.

1. Prophase. During early prophase, the chromatin fibers condense and shorten into chromosomes that are visible under the light microscope (Figure 3.33b). The condensation process may

Figure 3.32 Replication of DNA. The two strands of the double helix separate by breaking the hydrogen bonds (shown as dotted lines) between nucleotides. New, complementary nucleotides attach at the proper sites, and a new strand of DNA is synthesized alongside each of the original strands. Arrows indicate hydrogen bonds forming again between pairs of bases.

 Replication doubles the amount of DNA.

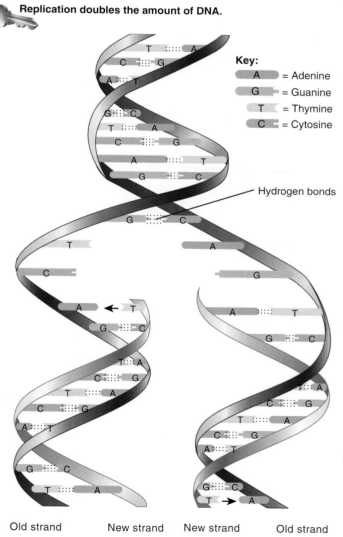

Key:
A = Adenine
G = Guanine
T = Thymine
C = Cytosine

Hydrogen bonds

Old strand   New strand   New strand   Old strand

Why is it crucial that DNA replication occurs before cytokinesis in somatic cell division?

Figure 3.33 Cell division: mitosis and cytokinesis. Begin the sequence at (a) at the top of the figure and read clockwise until you complete the process.

In somatic cell division, a single diploid cell divides to produce two identical diploid daughter cells.

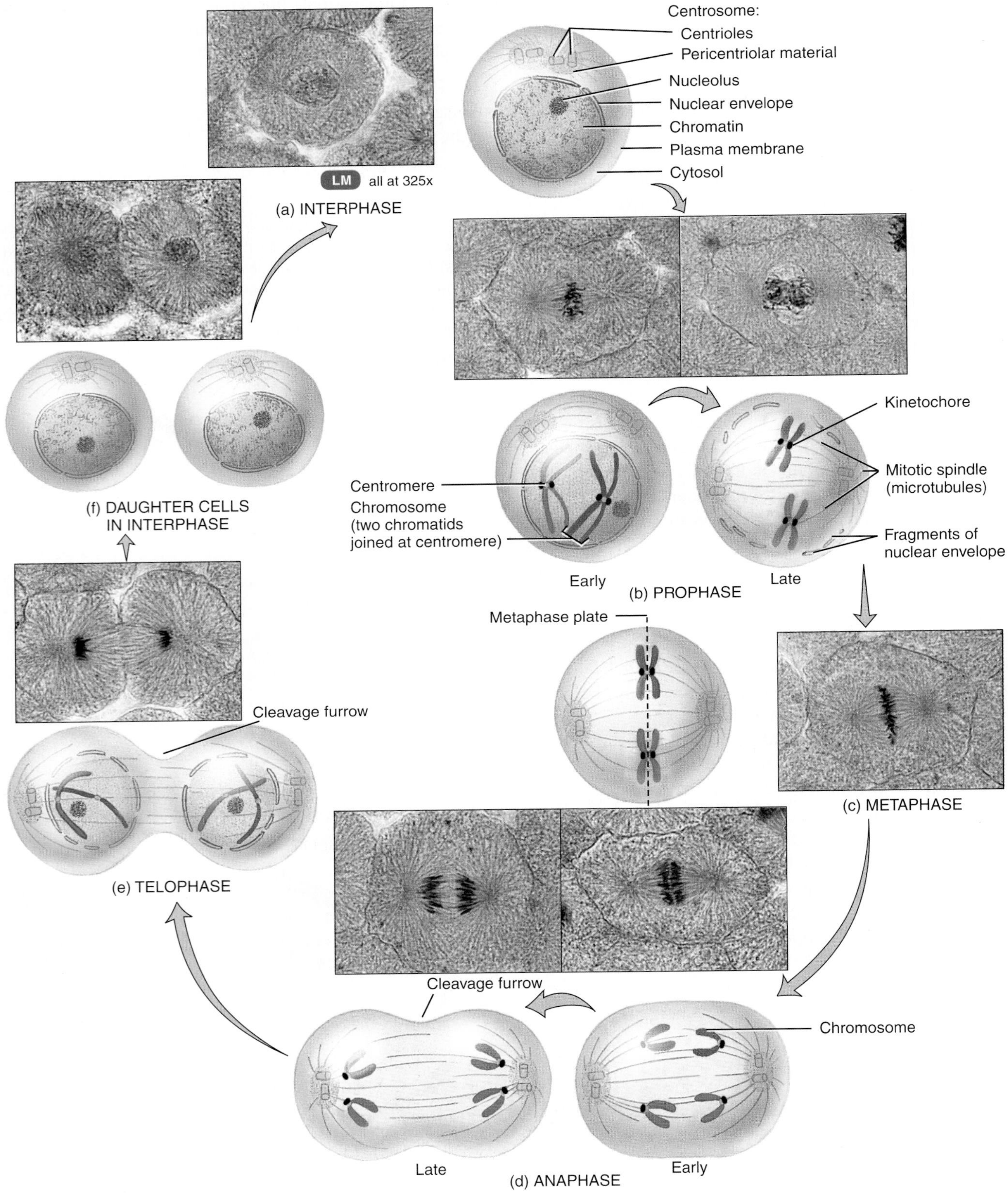

LM all at 325x

(a) INTERPHASE

Centrosome:
Centrioles
Pericentriolar material
Nucleolus
Nuclear envelope
Chromatin
Plasma membrane
Cytosol

Kinetochore

Centromere
Chromosome
(two chromatids
joined at centromere)

Mitotic spindle
(microtubules)

Fragments of
nuclear envelope

Early   Late

(b) PROPHASE

Metaphase plate

(c) METAPHASE

(f) DAUGHTER CELLS
IN INTERPHASE

Cleavage furrow

(e) TELOPHASE

Cleavage furrow

Chromosome

Late   Early

(d) ANAPHASE

When does cytokinesis begin?

prevent entangling of the long DNA strands as they move during mitosis. Because DNA replication took place during the S phase of interphase, each prophase chromosome consists of a pair of identical, double-stranded *chromatids*. A constricted region called a *centromere* holds the chromatid pair together. At the outside of each centromere is a protein complex known as the **kinetochore** (ki-NET-ō-kor). Later in prophase, tubulins in the pericentriolar material of the centrosomes start to form the **mitotic spindle,** a football-shaped assembly of microtubules that attach to the kinetochore (Figure 3.33b). As the microtubules lengthen, they push the centrosomes to the poles (ends) of the cell so that the spindle extends from pole to pole. The spindle is responsible for the separation of chromatids to opposite poles of the cell. Then, the nucleolus disappears and the nuclear envelope breaks down.

2. **Metaphase.** During metaphase, the kinetochore microtubules align the centromeres of the chromatid pairs at the exact center of the mitotic spindle (Figure 3.33c). This midpoint region is called the **metaphase plate.**

3. **Anaphase.** During anaphase, the centromeres split, separating the two members of each chromatid pair, which move toward opposite poles of the cell (Figure 3.33d). Once separated, the chromatids are termed chromosomes. As the chromosomes are pulled by the kinetochore microtubules during anaphase, they appear V-shaped because the centromeres lead the way, dragging the trailing arms of the chromosomes toward the pole.

4. **Telophase.** The final stage of mitosis, telophase, begins after chromosomal movement stops (Figure 3.33e). The identical sets of chromosomes now at opposite poles of the cell, uncoil and revert to the threadlike chromatin form. A nuclear envelope forms around each chromatin mass, nucleoli reappear in the daughter nuclei, and the mitotic spindle disappears.

CYTOPLASMIC DIVISION: CYTOKINESIS Division of a parent cell's cytoplasm and organelles into two daughter cells is called **cytokinesis** (sī′-tō-ki-NĒ-sis; *-kinesis* = motion). This process begins in late anaphase or early telophase with formation of a **cleavage furrow,** a slight indentation of the plasma membrane. The cleavage furrow usually appears midway between the centrosomes and extends around the periphery of the cell (see Figure 3.33d and e). Actin microfilaments that lie just inside the plasma membrane form a *contractile ring* that pulls the plasma membrane progressively inward. The ring constricts the center of the cell, like tightening a belt around the waist, and ultimately pinches it in two. Because the plane of the cleavage furrow is always perpendicular to the mitotic spindle, the two sets of chromosomes end up in separate daughter cells. When cytokinesis is complete, interphase begins (Figure 33.3f).

Considering the cell cycle in its entirety, the sequence of events is

$G_1 \longrightarrow$ S phase $\longrightarrow G_2$ phase $\longrightarrow$ mitosis $\longrightarrow$ cytokinesis

Table 3.3 summarizes the events of the cell cycle in somatic cells.

Table 3.3	Events of the Somatic Cell Cycle
Phase	**Activity**
INTERPHASE	Period between cell divisions; chromosomes not visible under light microscope.
G$_1$ phase	Metabolically active cell duplicates organelles and cytosolic components.
S phase	Replication of DNA and centrosomes.
G$_2$ phase	Cell growth, enzyme and protein synthesis continues.
MITOTIC PHASE	Parent cell produces daughter cells with identical chromosomes; chromosomes visible under light microscope.
Mitosis	Nuclear division; distribution of two sets of chromosomes into separate nuclei.
Prophase	Chromatin fibers condense into paired chromatids; nucleolus and nuclear envelope disappear; each centrosome moves to an opposite pole of the cell.
Metaphase	Centromeres of chromatid pairs line up at metaphase plate.
Anaphase	Centromeres split; identical sets of chromosomes move to opposite poles of cell.
Telophase	Nuclear envelopes and nucleoli reappear; chromosomes resume chromatin form; mitotic spindle disappears.
Cytokinesis	Cytoplasmic division; contractile ring forms cleavage furrow around center of cell, dividing cytoplasm into separate and equal portions.

Control of Cell Destiny

A cell has three possible destinies—to remain alive and functioning without dividing, to grow and divide, or to die. Homeostasis is maintained when there is a balance between cell proliferation and cell death. The signals that tell a cell when to exist in the G_0 phase, when to divide, and when to die have been the subjects of intense and fruitful research during the past decade. A key signal that induces cell division is **maturation promoting factor (MPF).** One component of MPF is a group of enzymes called **cdc2 proteins,** so named because they participate in the cell division cycle (cdc). Another component of MPF is a protein called **cyclin,** so named because its level rises and falls during the cell cycle. Cyclin builds up in the cell during interphase and activates cdc2 proteins and thus MPF. As a result, the cell undergoes mitosis.

Cellular death is also regulated. Throughout the lifetime of an organism, certain cells undergo **apoptosis** (ap-ō-TŌ-sis = a falling off), an orderly, genetically programmed death. In apoptosis, a triggering agent from either outside or inside the cell causes "cell-suicide" genes to produce enzymes that damage the cell in several ways, including disrupting its cytoskeleton and nucleus. As a result, the cell shrinks and pulls away from neighboring cells. The DNA within the nucleus fragments, and the cytoplasm shrinks, although the plasma membrane remains intact. Phagocytes in the vicinity then ingest the dying cell. Apoptosis is especially useful because it removes unneeded cells during development before birth, for example, the separation of webbed digits during fetal development. It continues to occur after birth to regulate the number of cells in a tissue and eliminate potentially dangerous cells such as cancer cells.

Apoptosis, a normal type of cell death, contrasts with **necrosis** (ne-KRŌ-sis = death), a pathological type of cell death that results from tissue injury. In necrosis, many adjacent cells swell, burst, and spill their cytoplasm into the interstitial fluid. The cellular debris usually stimulates an inflammatory response by the immune system, which does not occur in apoptosis.

Tumor-suppressor Genes

Abnormalities in genes that regulate the cell cycle or apoptosis are associated with many diseases. For example, some cancers are caused by damage to genes called **tumor-suppressor genes,** which produce proteins that normally inhibit cell division. Loss or alteration of a tumor-suppressor gene called *p53* on chromosome 17 is the most common genetic change leading to a wide variety of tumors, including breast and colon cancers. The normal p53 protein arrests cells in the G_1 phase, which prevents cell division. Normal p53 protein also assists in repair of damaged DNA and induces apoptosis in the cells where DNA repair was not successful. For this reason, the *p53* gene is nicknamed "the guardian angel of the genome." ■

▶ C H E C K P O I N T

21. Distinguish between the somatic and reproductive types of cell division. Why is each important?

22. Define interphase. When does DNA replicate?

23. What are the major events of each stage of the mitotic phase?

24. How are apoptosis and necrosis similar and different?

Figure 3.34 Diverse shapes and sizes of human cells. The relative difference in size between the smallest and largest cells is actually much greater than shown here.

The nearly 100 trillion cells in an average adult can be classified into about 200 different cell types.

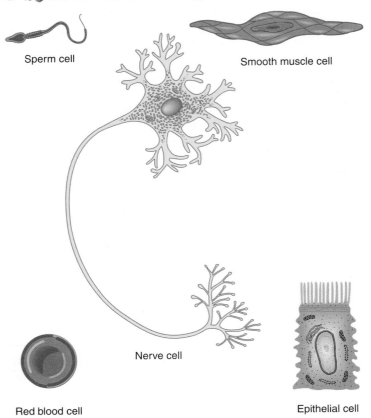

Sperm cell

Smooth muscle cell

Nerve cell

Red blood cell

Epithelial cell

Why are sperm the only body cells that need to have a flagellum?

CELLULAR DIVERSITY

The body of an average human adult is composed of nearly 100 trillion cells. All of these cells can be classified into about 200 different cell types. Cells vary considerably in size. High-powered microscopes are needed to see the smallest cells of the body. The largest cell, a single oocyte, is barely visible to the unaided eye. The sizes of cells are measured in units called *micrometers.* One micrometer (μm) is equal to 1 one-millionth of a meter, or 10^{-6} m (1/25,000 of an inch). Whereas a red blood cell has a diameter of 8 μm, an oocyte has a diameter of about 140 μm.

The shapes of cells also vary considerably (Figure 3.34). They may be round, oval, flat, cuboidal, columnar, elongated, star-shaped, cylindrical, or disc-shaped. A cell's shape is related to its function in the body. For example, a sperm cell has a long whiplike tail (flagellum) that it uses for locomotion. The disc shape of a red blood cell gives it a large surface area that enhances its ability to pass oxygen to other cells. The long, spindle shape of a relaxed smooth muscle cell shortens as it contracts. This shape change allows groups of smooth muscle cells to narrow or widen the passage for blood flowing through blood vessels. In this way, they regulate blood flow through various tissues. Some cells contain microvilli, which greatly increase their surface area. Microvilli are common in the epithelial cells that line the small intestine, where the large surface area speeds the absorption of digested food. Nerve cells have long extensions that permit them to conduct nerve impulses over great distances. As you will see, cellular diversity also permits organization of cells into more complex tissues and organs.

AGING AND CELLS

OBJECTIVE

• Describe the cellular changes that occur with aging.

Aging is a normal process accompanied by a progressive alteration of the body's homeostatic adaptive responses. It produces observable changes in structure and function and increases vulnerability to environmental stress and disease. The specialized branch of medicine that deals with the medical problems and care of elderly persons is **geriatrics** (jer′-ē-AT-riks; *ger-* = old age; *-iatrics* = medicine).

Although many millions of new cells normally are produced each minute, several kinds of cells in the body—heart muscle cells, skeletal muscle cells, and nerve cells—do not divide because they are arrested permanently in the G_0 phase. Experiments have shown that many other cell types have only a limited capability to divide. Normal cells grown outside the body divide only a certain number of times and then stop. These observations suggest that cessation of mitosis is a normal, genetically programmed event. According to this view, "aging genes" are part of the genetic blueprint at birth. These genes have an important function in normal cells but their activities slow over time. They bring about aging by slowing down or halting processes vital to life.

Another aspect of aging involves **telomeres** (TĒ-lō-merz), specific DNA sequences found only at the tips of each chromosome. These pieces of DNA protect the tips of chromosomes from erosion and from sticking to one another. However, in most normal body cells each cycle of cell division shortens the telomeres. Eventually, after many cycles of cell division, the telomeres can be completely gone and even some of the functional chromosomal material may be lost. These observations suggest that erosion of DNA from the tips of our chromosomes contributes greatly to aging and death of cells.

Glucose, the most abundant sugar in the body, plays a role in the aging process. It is haphazardly added to proteins inside and outside cells, forming irreversible cross-links between adjacent protein molecules. With advancing age, more cross-links form, which contributes to the stiffening and loss of elasticity that occur in aging tissues.

Free radicals produce oxidative damage in lipids, proteins, or nucleic acids by "stealing" an electron to accompany their unpaired electrons. Some effects are wrinkled skin, stiff joints, and hardened arteries. Normal cellular metabolism—for example, aerobic cellular respiration in mitochondria—produces some free radicals. Others are present in air pollution, radiation, and certain foods we eat. Naturally occurring enzymes in peroxisomes and in the cytosol normally dispose of free radicals. Certain dietary substances, such as vitamin E, vitamin C, beta-carotene, and selenium, are antioxidants that inhibit free radical formation.

Whereas some theories of aging explain the process at the cellular level, others concentrate on regulatory mechanisms operating within the entire organism. For example, the immune system may start to attack the body's own cells. This *autoimmune response* might be caused by changes in cell-identity markers at the surface of cells that cause antibodies to attach to and mark the cell for destruction. As changes in the proteins on the plasma membrane of cells increase, the autoimmune response intensifies, producing the well-known signs of aging. In the chapters that follow, we will discuss the effects of aging on each body system.

Progeria and Werner Syndrome

Progeria (prō-JER-ē-a) is a noninherited disease characterized by normal development in the first year of life followed by rapid aging. The condition is expressed by dry and wrinkled skin, total baldness, and birdlike facial features. Death usually occurs around age 13.

Werner syndrome is a rare, inherited disease that causes a rapid acceleration of aging, usually while the person is only in his or her twenties. It is characterized by wrinkling of the skin, graying of the hair and baldness, cataracts, muscular atrophy, and a tendency to develop diabetes mellitus, cancer, and cardiovascular disease. Most afflicted individuals die before age 50. Recently, the gene that causes Werner syndrome has been identified. Researchers hope to use the information to gain insight into the mechanisms of aging. ■

▶ **CHECKPOINT**

25. What is one reason that some tissues become stiffer as they age?

DISORDERS: HOMEOSTATIC IMBALANCES

Cancer

Cancer is a group of diseases characterized by uncontrolled cell proliferation. When cells in a part of the body divide without control, the excess tissue that develops is called a **tumor** or **neoplasm** (NĒ-ō-plazm; *neo-* = new). The study of tumors is called **oncology** (on-KOL-ō-jē; *onco-* = swelling or mass). Tumors may be cancerous and often fatal, or they may be harmless. A cancerous neoplasm is called a **malignant tumor** or **malignancy.** One property of most malignant tumors is their ability to undergo **metastasis** (me-TAS-ta-sis), the spread of cancerous cells to other parts of the body. A **benign tumor** is a neoplasm that does not metastasize. An example is a wart. A benign tumor may be surgically removed if it interferes with normal body function or becomes disfiguring.

Types of Cancer

The name of a cancer is derived from the type of tissue in which it develops. Most human cancers are **carcinomas** (kar-si-NŌ-maz; *carcin-* = cancer; *-omas* = tumors), malignant tumors that arise from epithelial cells. **Melanomas** (mel-a-NŌ-maz; *melan-* = black), for example, are cancerous growths of melanocytes, skin epithelial cells that produce the pigment melanin. **Sarcoma** (sar-KŌ-ma; *sarc-* = flesh) is a general term for any cancer arising from muscle cells or connective tissues. For example, **osteogenic sarcoma** (*osteo-* = bone; *-genic* = origin), the most frequent type of childhood cancer, destroys normal bone tissue. **Leukemia** (loo-KĒ-mē-a; *leuk-* = white; *-emia* = blood) is a cancer of blood-forming organs characterized by rapid growth of abnormal leukocytes (white blood cells). **Lymphoma** (lim-FŌ-ma) is a malignant disease of lymphatic tissue—for example, of lymph nodes.

Growth and Spread of Cancer

Cells of malignant tumors duplicate rapidly and continuously. As malignant cells invade surrounding tissues, they often trigger **angiogenesis**, the growth of new networks of blood vessels. Proteins that stimulate angiogenesis in tumors are called **tumor angiogenesis factors (TAFs).** However, inhibitors of angiogenesis may also be present. Thus, the formation of new blood vessels can occur either by overproduction of TAFs or by the lack of naturally occurring angiogenesis inhibitors. As the cancer grows, it begins to compete with normal tissues for space and nutrients. Eventually, the normal tissue decreases in size and dies. Some malignant cells may detach from the initial (primary) tumor and invade a body cavity or enter the blood or lymph, then circulate to and invade other body tissues, establishing secondary tumors. Malignant cells resist the antitumor defenses of the body. The pain associated with cancer develops when the tumor presses on nerves or blocks a passageway in an organ so that secretions build up pressure.

Causes of Cancer

Several factors may trigger a normal cell to lose control and become cancerous. One cause is environmental agents: substances in the air we breathe, the water we drink, and the food we eat. A chemical agent or radiation that produces cancer is called a **carcinogen** (car-SIN-ō-jen). Carcinogens induce **mutations,** permanent structural changes in the DNA base sequence of a gene. The World Health Organization estimates that carcinogens are associated with 60–90% of all human cancers. Examples of carcinogens are hydrocarbons found in cigarette tar, radon gas from the earth, and ultraviolet (UV) radiation in sunlight.

Intensive research efforts are now directed toward studying cancer-causing genes, or **oncogenes** (ON-kō-jēnz). These genes, when inappropriately activated, have the ability to transform a normal cell into a cancerous cell. Most oncogenes derive from normal genes called **proto-oncogenes** that regulate growth and development. The proto-oncogene undergoes some change that either causes it to produce an abnormal product or disrupts its control. It may be expressed inappropriately or make its products in excessive amounts or at the wrong time. Some oncogenes cause excessive production of growth factors, chemicals that stimulate cell growth. Others may trigger changes in a cell-surface receptor, causing it to send signals as though it were being activated by a growth factor. As a result, the growth pattern of the cell becomes abnormal.

Proto-oncogenes in every cell carry out normal cellular functions until a malignant change occurs. It appears that some proto-oncogenes are activated to oncogenes by mutations in which the DNA of the proto-oncogene is altered. Other proto-oncogenes are activated by a rearrangement of the chromosomes so that segments of DNA are exchanged. Rearrangement activates proto-oncogenes by placing them near genes that enhance their activity.

Some cancers have a viral origin. Viruses are tiny packages of nucleic acids, either RNA or DNA, that can reproduce only while inside the cells they infect. Some viruses, termed **oncogenic viruses,** cause cancer by stimulating abnormal proliferation of cells. For instance, the *human papillomavirus (HPV)* causes virtually all cervical cancers in women. The virus produces a protein that causes proteasomes to destroy the p53 protein. (Recall that p53 protein normally suppresses cell division.) In the absence of this suppressor protein, cells proliferate without control.

Carcinogenesis: A Multistep Process

Carcinogenesis (kar′-si-nō-JEN-e-sis), the process by which cancer develops, is a multistep process in which as many as 10 distinct mutations may have to accumulate in a cell before it becomes cancerous. The progression of genetic changes leading to cancer is best understood for colon (colorectal) cancer. Such cancers, as well as lung and breast cancer, take years or decades to develop. In colon cancer, the tumor begins as an area of increased cell proliferation that results from one mutation. This growth then progresses to abnormal, but noncancerous, growths called adenomas. After two or three additional mutations, a mutation of the tumor-suppressor gene *p53* occurs and a carcinoma develops. Recall that tumor-suppressor genes normally suppress unregulated cell growth. The fact that so many mutations are needed for a cancer to develop indicates that cell growth is normally controlled with many sets of checks and balances.

Treatment of Cancer

Many cancers are removed surgically. However, when cancer is widely distributed throughout the body or exists in organs such as the brain whose functioning would be greatly harmed by surgery, chemotherapy and radiation therapy may be used instead. Sometimes surgery, chemotherapy, and radiation therapy are used in combination. Chemotherapy involves administering drugs that cause death of cancerous cells. Radiation therapy breaks chromosomes, thus blocking cell division. Because cancerous cells divide rapidly, they are more vulnerable to the destructive effects of chemotherapy and radiation therapy than are normal cells. Unfortunately for the patients, hair follicle cells, red bone marrow cells, and cells lining the gastrointestinal tract also are rapidly dividing. Hence, the side effects of chemotherapy and radiation therapy include hair loss due to death of hair follicle cells, vomiting and nausea due to death of cells lining the stomach and intestines, and susceptibility to infection due to slowed production of white blood cells in red bone marrow.

Treating cancer is difficult because it is not a single disease and because the cells in a single tumor population rarely behave all in the same way. Although most cancers are thought to derive from a single abnormal cell, by the time a tumor reaches a clinically detectable size, it may contain a diverse population of abnormal cells. For example, some cancerous cells metastasize readily, and others do not. Some are sensitive to chemotherapy drugs and some are drug resistant. Because of differences in drug resistance, a single chemotherapeutic agent may destroy susceptible cells but permit resistant cells to proliferate.

MEDICAL TERMINOLOGY

Anaplasia (an′-a-PLĀ-zē-a; *an* = not; *plasia* = to shape) The loss of tissue differentiation and function that is characteristic of most malignancies.

Atrophy (AT-rō-fē; *a* = without; *-trophy* = nourishment) A decrease in the size of cells, with a subsequent decrease in the size of the affected tissue or organ; wasting away.

Dysplasia (dis-PLĀ-zē-a; *dys-* = abnormal) Alteration in the size, shape, and organization of cells due to chronic irritation or inflammation; may progress to neoplasia (tumor formation, usually malignant) or revert to normal if the irritation is removed.

Hyperplasia (hī-per-PLĀ-zē-a; *hyper-* = over) Increase in the number of cells of a tissue due to an increase in the frequency of cell division.

Hypertrophy (hī-PER-trō-fē) Increase in the size of cells without cell division.

Metaplasia (met′-a-PLĀ-zē-a; *meta-* = change) The transformation of one type of cell into another.

Progeny (PROJ-e-nē; *pro-* = forward; *-geny* = production) Offspring or descendants.

STUDY OUTLINE

INTRODUCTION (p. 60)

1. A cell is the basic, living, structural and functional unit of the body.
2. Cell biology is the scientific study of cellular structure and function.

PARTS OF A CELL (p. 60)

1. Figure 3.1 on page 60 provides an overview of the typical structures in body cells.
2. The principal parts of a cell are the plasma membrane; the cytoplasm, which consists of cytosol and organelles; and the nucleus.

THE PLASMA MEMBRANE (p. 61)

1. The plasma membrane surrounds and contains the cytoplasm of a cell.
2. The membrane is composed of a 50:50 mix by weight of proteins and lipids that are held together by noncovalent forces.
3. According to the fluid mosaic model, the membrane is a mosaic of proteins floating like icebergs in a bilayer sea of lipids.
4. The lipid bilayer consists of two back-to-back layers of phospholipids, cholesterol, and glycolipids. The bilayer arrangement occurs because the lipids are amphipathic, having both polar and nonpolar parts.
5. Integral proteins extend into or through the lipid bilayer, whereas peripheral proteins associate with membrane lipids or integral proteins at the inner or outer surface of the membrane.
6. Many integral proteins are glycoproteins, with sugar groups attached to the ends that face the extracellular fluid. Together with glycolipids, the glycoproteins form a glycocalyx on the extracellular surface of cells.
7. Membrane proteins have a variety of functions. Ion channels and transporters are membrane proteins that help specific solutes across the membrane; receptors serve as cellular recognition sites; and linkers anchor proteins in the plasma membranes to protein filaments inside and outside the cell. Some membrane proteins are enzymes and others are cell-identity markers.

8. Membrane fluidity is greater when there are more double bonds in the fatty acid tails of the lipids that make up the bilayer. Cholesterol makes the lipid bilayer stronger but less fluid at normal body temperature. Because of its fluidity, the lipid bilayer self-seals when torn or punctured.
9. The membrane's selective permeability permits some substances to pass more readily than others. The lipid bilayer is permeable to water, urea, and most nonpolar, uncharged molecules. It is impermeable to ions and charged or polar molecules other than water and urea. Channels and transporters increase the plasma membrane's permeability to small- and medium-sized polar and charged substances, including ions, that cannot cross the lipid bilayer.
10. The selective permeability of the plasma membrane supports the existence of concentration gradients, differences in the concentrations of chemicals between one side of the membrane and the other.

TRANSPORT ACROSS THE PLASMA MEMBRANE (p. 64)

1. In passive transport, a substance moves down its concentration gradient across the membrane using its own kinetic energy of motion. In active transport, cellular energy is used to drive the substance "uphill" against its concentration gradient.
2. In endocytosis and exocytosis, tiny vesicles either detach from or merge with the plasma membrane to move materials across the membrane, into or out of a cell.
3. In diffusion, molecules or ions move from an area of higher concentration to an area of lower concentration until an equilibrium is reached.
4. The rate of diffusion across a plasma membrane is affected by the steepness of the concentration gradient, temperature, mass of the diffusing substance, surface area available for diffusion, and the distance over which diffusion must occur.
5. Osmosis is the net movement of water through a selectively permeable membrane from an area of higher water concentration to an area of lower water concentration.
6. In an isotonic solution, red blood cells maintain their normal

shape; in a hypotonic solution, they undergo hemolysis; in a hypertonic solution, they undergo crenation.

7. Nonpolar, hydrophobic molecules, such as oxygen, carbon dioxide, nitrogen, steroids, fat-soluble vitamins (A, E, D, and K), small alcohols, and ammonia plus water and urea diffuse through the lipid bilayer of the plasma membrane.

8. Ion channels selective for K^+, Cl^-, Na^+, and Ca^{2+} allow these small, inorganic ions (which are too hydrophilic to penetrate the membrane's nonpolar interior) to diffuse across the plasma membrane.

9. In facilitated diffusion, a solute such as glucose binds to a specific transporter on one side of the membrane and is released on the other side after the transporter undergoes a change in shape.

10. Substances can cross the membrane against their concentration gradient by active transport. Actively transported substances include several ions, such as Na^+, K^+, H^+, Ca^{2+}, I^-, and Cl^-; amino acids; and monosaccharides.

11. Two sources of energy are used to drive active transport: Energy obtained from hydrolysis of ATP is the source in primary active transport, and energy stored in a Na^+ or H^+ concentration gradient is the source in secondary active transport.

12. The most prevalent primary active transport pump is the sodium-potassium pump, also known as Na^+/K^+ ATPase.

13. Secondary active transport mechanisms include both symporters and antiporters that are powered by either a Na^+ or H^+ concentration gradient. Symporters move two substances in the same direction, whereas antiporters move two substances in opposite directions across the membrane.

14. Receptor-mediated endocytosis is the selective uptake of large molecules and particles (ligands) that bind to specific receptors in membrane areas called clathrin-coated pits.

15. Phagocytosis is the ingestion of solid particles. Some white blood cells destroy microbes that enter the body in this way.

16. Pinocytosis is the ingestion of extracellular fluid. In this process, a pinocytic vesicle surrounds the fluid.

17. Exocytosis involves movement of secretory or waste products out of a cell by fusion of vesicles with the plasma membrane.

18. In transcytosis, vesicles undergo endocytosis on one side of a cell, move across the cell, and then undergo exocytosis on the opposite side.

CYTOPLASM (p. 75)

1. Cytoplasm is all the cellular contents between the plasma membrane and the nucleus. It consists of cytosol and organelles.

2. Cytosol is the fluid portion of cytoplasm, containing water, ions, glucose, amino acids, fatty acids, proteins, lipids, ATP, and waste products. It is the site of many chemical reactions required for a cell's existence.

3. Organelles are specialized structures with characteristic shapes that have specific functions.

4. The cytoskeleton is a network of several kinds of protein filaments that extend throughout the cytoplasm. Components of the cytoskeleton are microfilaments, intermediate filaments, and microtubules. The cytoskeleton provides a structural framework for the cell and is responsible for cell movements.

5. The centrosome consists of a pair of centrioles and pericentriolar material. The pericentriolar material organizes microtubules in nondividing cells and the mitotic spindle in dividing cells.

6. Cilia and flagella are formed by basal bodies and are motile projections of the cell surface. Cilia move fluid along the cell surface whereas flagella move an entire cell.

7. Ribosomes, composed of ribosomal RNA and ribosomal proteins, consist of two subunits made in the nucleus. They are sites of protein synthesis.

8. Endoplasmic reticulum (ER) is a network of membranes that form flattened sacs or tubules; it extends from the nuclear envelope throughout the cytoplasm.

9. Rough ER is studded with ribosomes. Proteins synthesized by ribosomes attached to rough ER enter the space within the ER for processing and sorting. It produces secretory proteins, membrane proteins, and organelle proteins. It also forms glycoproteins and attaches proteins to phospholipids, which it also synthesizes.

10. Smooth ER lacks ribosomes. It synthesizes fatty acids and steroids; inactivates or detoxifies drugs and other potentially harmful substances; removes phosphate from glucose 6-phosphate; and releases calcium ions that trigger contraction in muscle cells.

11. The Golgi complex consists of flattened sacs called cisternae. The entry, medial, and exit regions of the Golgi complex contain different enzymes that permit each to modify, sort, and package proteins for transport to different destinations.

12. Processed proteins leave the Golgi complex in secretory vesicles, membrane vesicles, or transport vesicles that will carry the proteins to another cellular destination.

13. Lysosomes are membrane-enclosed vesicles that contain digestive enzymes. Late endosomes, phagosomes, and pinocytic vesicles deliver materials to lysosomes for degradation. Lysosomes function in digestion of worn-out organelles (autophagy), digestion of a host cell (autolysis), and extracellular digestion.

14. Peroxisomes contain oxidases that oxidize amino acids, fatty acids, and toxic substances and, in the process, produce hydrogen peroxide. Catalase, also present in peroxisomes, destroys hydrogen peroxide.

15. Proteasomes continually degrade unneeded, damaged, or faulty proteins. Their proteases cut proteins into small peptides.

16. Mitochondria consist of a smooth outer membrane, an inner membrane containing cristae, and a fluid-filled cavity called the matrix. They are called "powerhouses" of the cell because they produce most of a cell's ATP.

NUCLEUS (p. 82)

1. The nucleus consists of a double nuclear envelope; nuclear pores, which control the movement of substances between the nucleus and cytoplasm; nucleoli, which produce ribosomes; and genes arranged on chromosomes.

2. Human somatic cells have 46 chromosomes, 23 inherited from each parent. The total genetic information carried in a cell or an organism is its genome.

PROTEIN SYNTHESIS (p. 85)

1. Cells make proteins by transcribing and translating the genetic information contained in DNA.

2. The genetic code is the set of rules that relate the base triplet sequences of DNA to the corresponding codons of RNA and the amino acids they specify.

3. In transcription, the genetic information in the sequence of base triplets in DNA serves as a template for copying the information into a complementary sequence of codons in messenger RNA.

Transcription begins on DNA in a region called a promoter. Regions of DNA that code for protein synthesis are called exons; those that do not are called introns.

4. Newly synthesized pre-mRNA is modified before leaving the nucleus.

5. Translation is the process by which the nucleotide sequence of mRNA specifies the amino acid sequence of a protein. In translation, mRNA binds to a ribosome, specific amino acids attach to tRNA, and anticodons of tRNA bind to codons of mRNA, thus bringing specific amino acids into position on a growing polypeptide. Translation begins at the start codon and ends at the stop codon.

CELL DIVISION (p. 90)

1. Cell division is the process by which cells reproduce themselves. It consists of nuclear division (mitosis or meiosis) and cytoplasmic division (cytokinesis).

2. Cell division that results in an increase in the number of body cells is called somatic cell division and involves a nuclear division called mitosis plus cytokinesis.

3. Cell division that results in the production of sperm and oocytes is called reproductive cell division and consists of a nuclear division called meiosis plus cytokinesis.

4. The cell cycle is an orderly sequence of events in which a somatic cell duplicates its contents and divides in two. It consists of interphase and a mitotic phase.

5. Before the mitotic phase, the chromosomes replicate so that identical chromosomes can be passed on to the next generation of cells.

6. A cell that is between divisions and is carrying on every life process except division is in interphase, which consists of three phases: G_1, S, and G_2. During the G_1 phase, the cell duplicates its organelles and cytosolic components; during the S phase, DNA and centrosome replication occur; during the G_2 phase, enzymes and other proteins are synthesized.

7. Mitosis is the replication of the chromosomes and the distribution of two identical sets of chromosomes into separate and equal nuclei; it consists of prophase, metaphase, anaphase, and telophase.

8. Cytokinesis usually begins in late anaphase and ends in telophase. A cleavage furrow forms at the cell's metaphase plate and progresses inward, pinching through the cell to form two separate portions of cytoplasm.

9. A cell can either remain alive and functioning without dividing, grow and divide, or die. Maturation promoting factor (MPF) induces cell division (both mitosis and meiosis). MPF consists of cyclin, which builds up during interphase, and cdc2 proteins, which are activated by cyclin.

10. Apoptosis is programmed cell death, a normal type of cell death. It first occurs during embryological development and continues for the lifetime of an organism.

11. Certain genes regulate both cell division and apoptosis. Abnormalities in these genes are associated with a wide variety of diseases and disorders.

CELLULAR DIVERSITY (p. 94)

1. The almost 200 different types of cells in the body vary considerably in size and shape.

2. The sizes of cells are measured in micrometers. One micrometer (μm) equals 10^{-6} m (1/25,000 of an inch).

3. A cell's shape is related to its function.

AGING AND CELLS (p. 95)

1. Aging is a normal process accompanied by progressive alteration of the body's homeostatic adaptive responses.

2. Many theories of aging have been proposed, including genetically programmed cessation of cell division, shortening of telomeres, addition of glucose to proteins, buildup of free radicals, and an intensified autoimmune response.

Q SELF-QUIZ QUESTIONS

Fill in the blanks in the following statements.

1. The three principal parts of the cell are the _____, _____, and _____.

2. _____ refers to programmed cell death, whereas _____ refers to cell death resulting from tissue injury.

3. The fluid portion of the cytoplasm is the _____.

4. List three causes of cellular aging.

Indicate whether the following statements are true or false.

5. Duplicated chromosomes held together by centromeres are called chromatids.

6. The sodium pump is an example of primary active transport.

Choose the one best answer to the following questions.

7. The basic structural unit of the plasma membrane is the (a) lipid bilayer, (b) integral protein, (c) cholesterol molecule, (d) peripheral protein, (e) glycoprotein-glycolipid complex.

8. Integral proteins can function in the cell membrane in all the following ways *except* (a) as a channel, (b) as a transporter, (c) as a receptor, (d) as an exocytosis vesicle, (e) as a cell-identity marker.

9. Which of the following factors influence the diffusion rate of substances through a plasma membrane? (1) concentration gradient, (2) diffusion distance, (3) surface area, (4) size of diffusing substance, (5) temperature. (a) 1, 2, 3, and 5, (b) 1, 2, 3, 4, and 5, (c) 2, 3, 4, and 5, (d) 1, 2, and 5, (e) 2, 3, and 5.

10. A cell would lose water volume and shrink if placed in (a) a hypertonic solution, (b) a hypotonic solution, (c) an isotonic solution, (d) a hydrophobic solution, (e) an ionic solution.

11. Which of the following statements regarding the nucleus are *true*? (1) Nucleoli within the nucleus are the sites of ribosome synthesis. (2) Most of the cell's hereditary units, called genes, are located within the nucleus. (3) The nuclear membrane is a solid, impermeable membrane. (4) Protein synthesis occurs within the nucleus. (5) In nondividing cells, DNA is found in the nucleus in the form of chromatin. (a) 1, 2, and 3, (b) 1, 2, and 4, (c) 1, 2, and 5, (d) 2, 4, and 5, (e) 2, 3, and 4.

12. Match the following:

_____ (a) mitosis
_____ (b) meiosis
_____ (c) prophase
_____ (d) metaphase
_____ (e) anaphase
_____ (f) telophase
_____ (g) cytokinesis
_____ (h) interphase

(1) cytoplasmic division
(2) somatic cell division resulting in identical daughter cells
(3) reproductive cell division that reduces the number of chromosomes by half
(4) stage of cell division when replication of DNA occurs
(5) stage when chromatin fibers condense and shorten to form chromosomes
(6) stage when centromeres split and sister chromatids move to opposite poles of the cell
(7) stage when centromeres of chromatid pairs line up at the center of the mitotic spindle
(8) stage when chromosomes uncoil and revert to chromatin

13. Match the following:

_____ (a) codon
_____ (b) RNA polymerase
_____ (c) intron
_____ (d) exon
_____ (e) transcription
_____ (f) translation
_____ (g) messenger RNA
_____ (h) transfer RNA
_____ (i) ribosomal RNA
_____ (j) snRNP
_____ (k) promotor
_____ (l) terminator

(1) DNA region that does not code for synthesis of a part of a protein
(2) DNA region that codes for synthesis of a part of a protein
(3) enzyme that removes all introns and joins remaining exons
(4) nucleotide sequence, located near the beginning of a gene, that indicates where transcription begins; RNA polymerase attachment site
(5) the copying of the DNA message onto messenger RNA
(6) joins with ribosomal proteins to make ribosomes
(7) binds to amino acids and holds them in place on a ribosome to be incorporated into a protein
(8) nucleotide sequence near the end of a gene that indicates where transcription ends; releases RNA polymerase
(9) reading of messenger RNA for protein synthesis
(10) enzyme that catalyzes transcription of DNA
(11) a transcribed sequence of three RNA nucleotides
(12) directs synthesis of a protein

14. Match the following:

_____ (a) cytoskeleton
_____ (b) centrosome
_____ (c) ribosomes
_____ (d) rough ER
_____ (e) smooth ER
_____ (f) Golgi complex
_____ (g) lysosomes
_____ (h) peroxisomes
_____ (i) mitochondria
_____ (j) cilia
_____ (k) proteasomes

(1) membrane-enclosed vesicles formed in the Golgi complex that contain strongly hydrolytic enzymes
(2) network of protein filaments that extend throughout the cytoplasm, providing cellular shape and organization
(3) sites of protein synthesis
(4) contain enzymes that break apart unneeded, damaged, or faulty proteins into small peptides
(5) site where secretory proteins and membrane molecules are synthesized
(6) membrane-enclosed vesicles that contain enzymes that oxidize various organic substances
(7) microtubular structures extending from the plasma membrane and involved in movement of materials along the cell's surface
(8) modifies, sorts, packages, and transports molecules synthesized in the rough ER
(9) an organizing center for growth of the mitotic spindle
(10) function in ATP generation
(11) functions in synthesizing fatty acids and steroids, helping liver cells release glucose into the bloodstream, and detoxification

15. Match the following:

___ (a) diffusion
___ (b) osmosis
___ (c) facilitated diffusion
___ (d) primary active transport
___ (e) secondary active transport
___ (f) vesicular transport
___ (g) phagocytosis
___ (h) pinocytosis
___ (i) exocytosis
___ (j) receptor-mediated endocytosis
___ (k) trancytosis

(1) passive transport in which a solute binds to a specific transporter on one side of the membrane and is released on the other side

(2) movement of materials out of the cell by fusing of secretory vesicles with the plasma membrane

(3) the random mixing of particles in a solution due to the kinetic energy of the particles

(4) transport of substances either into or out of the cell by means of a small, spherical membranous sac formed by budding off from existing membranes

(5) uses energy derived from hydrolysis of ATP to change the shape of a transporter protein, which "pumps" a substance across a cellular membrane against its concentration gradient

(6) vesicular movement involving endocytosis on one side of a cell and subsequent exocytosis on the opposite side of the cell

(7) type of endocytosis that involves the nonselective uptake of tiny droplets of extracellular fluid

(8) type of endocytosis in which large solid particles are taken in

(9) movement of water from an area of higher to an area of lower water concentration through a selectively permeable membrane

(10) process that allows a cell to take specific ligands from the ECF by forming vesicles

(11) indirectly uses energy obtained from the breakdown of ATP; involves symporters and antiporters

CRITICAL THINKING QUESTIONS

1. In old detective novels, a dead body is sometimes called "a stiff." But the corpse doesn't remain stiff forever. After a time, it begins to soften. What causes this change in the tissues?
HINT *Worn out organelles suffer a similar fate.*

2. Mucin is a protein present in saliva and other secretions. When mixed with water, it becomes the slippery substance known as mucus. Trace the route taken by mucin through the cell, from its synthesis to its secretion, listing all the organelles and processes involved.
HINT *All proteins are produced initially by the same organelle.*

3. If you were trying to locate the DNA sequence that coded for an mRNA, would you look at a gene's introns or exons?
HINT *The mRNA is complementary to specific parts of the DNA.*

ANSWERS TO FIGURE QUESTIONS

3.1 The three main parts of a cell are the plasma membrane, cytoplasm, and nucleus.

3.2 The glycocalyx is the sugary coat on the extracellular surface of the plasma membrane. It is composed of the carbohydrate portions of membrane glycolipids and glycoproteins.

3.3 The membrane protein that binds to insulin is a receptor.

3.4 Both the concentration gradient and the electrical gradient will favor flow of Na^+ into cells.

3.5 Transmembrane proteins allow diffusion through channels, facilitated diffusion, and active transport.

3.6 Because fever involves an increase in body temperature, the rates of all diffusion processes would increase.

3.7 No, the water concentrations can never be the same in the two arms because the left arm contains pure water and the right arm contains a solution that is less than 100% water.

3.8 A 2% solution of NaCl will cause crenation of RBCs because it is hypertonic.

3.9 A gated channel opens or closes in response to chemical or electrical changes inside or outside the cell. When the gate is open, ions can diffuse through the channel.

3.10 The rate of facilitated diffusion depends on the concentration gradient and the number of available transporters.

3.11 ATP adds a phosphate group to the pump protein, which changes the pump's three-dimensional shape.

3.12 In secondary active transport, hydrolysis of ATP is used only indirectly to drive the activity of symporter or antiporter proteins, whereas this reaction directly powers the pump protein in primary active transport.

3.13 Cholesterol, iron, vitamins, and hormones are examples of ligands that can undergo receptor-mediated endocytosis.

3.14 The binding of particles to a plasma membrane receptor triggers pseudopod formation.

3.15 Receptor-mediated endocytosis and phagocytosis involve receptor proteins, pinocytosis does not.

3.16 Centrioles, cilia, and flagella all contain microtubules.

3.17 A cell without a centrosome probably would not be able to undergo cell division.

3.18 Cilia move fluids across cell surfaces, whereas flagella move an entire cell.

3.19 Large and small ribosomal subunits are synthesized in the nucleolus in the nucleus and then join in the cytoplasm.

3.20 Rough ER has attached ribosomes, whereas smooth ER does not.

3.21 The entry face receives and modifies proteins from rough ER while the exit face modifies, sorts, and packages molecules for transport to other destinations.

3.22 Some proteins are secreted from the cell by exocytosis, some are incorporated into the plasma membrane, and some enter transfer vesicles that deliver their cargo to other destinations, such as lysosomes.

3.23 Digestion of worn-out organelles by lysosomes is called autophagy.

3.24 Mitochondrial cristae increase the surface area available for chemical reactions and contain some of the enzymes needed for ATP production.

3.25 Chromatin is a complex of DNA, proteins, and some RNA.

3.26 A nucleosome is a double-stranded DNA wrapped twice around a core of eight histones (proteins).

3.27 Proteins determine the physical and chemical characteristics of cells.

3.28 The DNA base sequence AGCT would be transcribed into the RNA base sequence UCGA. RNA polymerase catalyzes transcription of DNA.

3.29 The P site holds the tRNA attached to the growing polypeptide. The A site holds the tRNA carrying the next amino acid to be added to the growing polypeptide.

3.30 When a ribosome encounters a stop codon at the A site, the completed protein detaches from the final tRNA.

3.31 Chromosomes replicate during the S phase.

3.32 DNA replication occurs before cytokinesis so that each of the new daughter cells will have a complete genome.

3.33 Cytokinesis usually starts in late anaphase or early telophase.

3.34 Sperm, which use the flagella for locomotion, are the only body cells that need to move considerable distances.

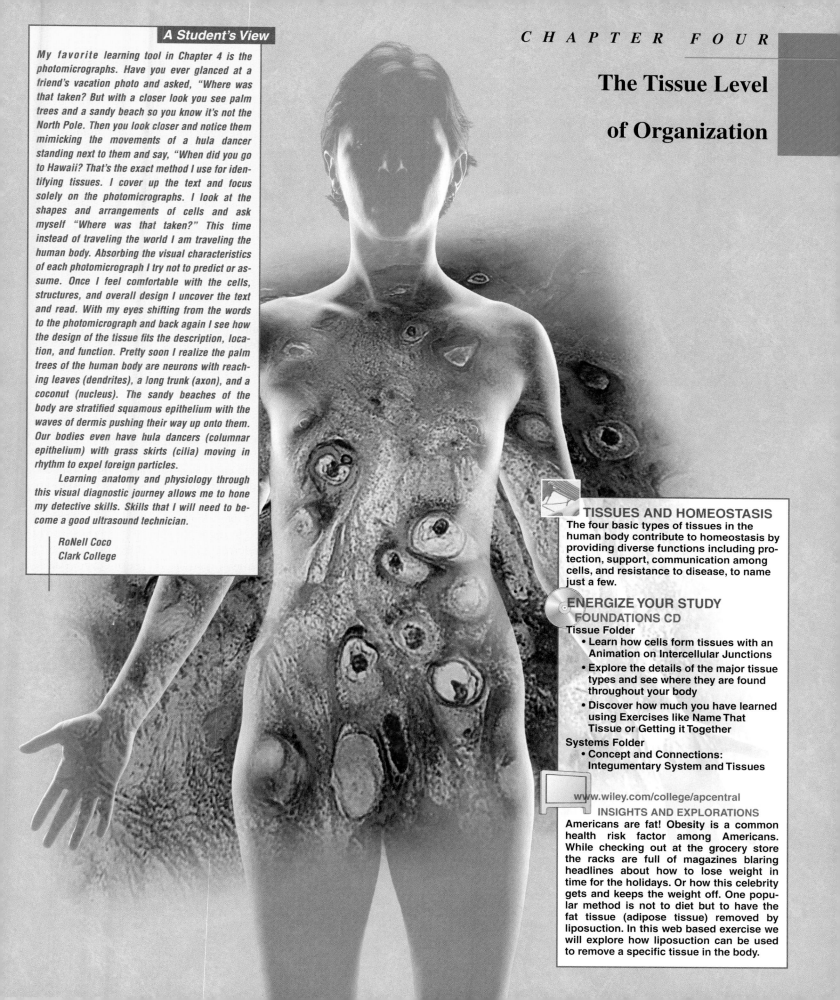

The Tissue Level
of Organization

My favorite learning tool in Chapter 4 is the photomicrographs. Have you ever glanced at a friend's vacation photo and asked, "Where was that taken? But with a closer look you see palm trees and a sandy beach so you know it's not the North Pole. Then you look closer and notice them mimicking the movements of a hula dancer standing next to them and say, "When did you go to Hawaii? That's the exact method I use for identifying tissues. I cover up the text and focus solely on the photomicrographs. I look at the shapes and arrangements of cells and ask myself "Where was that taken?" This time instead of traveling the world I am traveling the human body. Absorbing the visual characteristics of each photomicrograph I try not to predict or assume. Once I feel comfortable with the cells, structures, and overall design I uncover the text and read. With my eyes shifting from the words to the photomicrograph and back again I see how the design of the tissue fits the description, location, and function. Pretty soon I realize the palm trees of the human body are neurons with reaching leaves (dendrites), a long trunk (axon), and a coconut (nucleus). The sandy beaches of the body are stratified squamous epithelium with the waves of dermis pushing their way up onto them. Our bodies even have hula dancers (columnar epithelium) with grass skirts (cilia) moving in rhythm to expel foreign particles.

Learning anatomy and physiology through this visual diagnostic journey allows me to hone my detective skills. Skills that I will need to become a good ultrasound technician.

RoNell Coco
Clark College

TISSUES AND HOMEOSTASIS

The four basic types of tissues in the human body contribute to homeostasis by providing diverse functions including protection, support, communication among cells, and resistance to disease, to name just a few.

ENERGIZE YOUR STUDY
FOUNDATIONS CD
Tissue Folder
- **Learn how cells form tissues with an Animation on Intercellular Junctions**
- **Explore the details of the major tissue types and see where they are found throughout your body**
- **Discover how much you have learned using Exercises like Name That Tissue or Getting it Together**

Systems Folder
- **Concept and Connections: Integumentary System and Tissues**

www.wiley.com/college/apcentral
INSIGHTS AND EXPLORATIONS
Americans are fat! Obesity is a common health risk factor among Americans. While checking out at the grocery store the racks are full of magazines blaring headlines about how to lose weight in time for the holidays. Or how this celebrity gets and keeps the weight off. One popular method is not to diet but to have the fat tissue (adipose tissue) removed by liposuction. In this web based exercise we will explore how liposuction can be used to remove a specific tissue in the body.

A cell is a complex collection of compartments, each of which carries out a host of biochemical reactions that make life possible. However, a cell seldom functions as an isolated unit in the body. Instead, cells usually work together in groups called tissues. A **tissue** is a group of similar cells that usually has a common embryonic origin and functions together to carry out specialized activities. The structure and properties of a specific tissue are influenced by factors such as the nature of the extracellular material that surrounds the tissue cells and the connections between the cells that compose the tissue. Tissues may be hard, semisolid, or even liquid in their consistency, a range

exemplified by bone, fat, and blood. In addition, tissues vary greatly with respect to the kinds of cells present, how the cells are arranged, and the types of materials found between the cells. **Histology** (hiss'-TOL-ō-jē; *histo-* = tissue; *-logy* = study of) is the science that deals with the study of tissues. A **pathologist** (pa-THOL-ō-gist; *patho-* = disease) is a physician who specializes in laboratory studies of cells and tissues to help other physicians make accurate diagnoses. One of the principal functions of a pathologist is to examine tissues for any changes that might indicate disease.

TYPES OF TISSUES AND THEIR ORIGINS

▶ O B J E C T I V E

- **Name the four basic types of tissues that make up the human body and describe the characteristics of each.**

Body tissues can be classified into four basic types according to their function and structure:

1. **Epithelial tissue** covers body surfaces and lines hollow organs, body cavities, and ducts. It also forms glands.
2. **Connective tissue** protects and supports the body and its organs. Various types of connective tissue bind organs together, store energy reserves as fat, and help provide immunity to disease-causing organisms.
3. **Muscle tissue** generates the physical force needed to make body structures move.
4. **Nervous tissue** detects changes in a variety of conditions inside and outside the body and responds by generating nerve impulses that help maintain homeostasis.

Epithelial tissue and connective tissue, except for bone tissue and blood, are discussed in detail in this chapter. The general features of bone tissue and blood will be introduced here, but a detailed discussion of their characteristics and functions is presented in Chapters 6 and 19, respectively. Similarly, the structure and function of muscle tissue and nervous tissue are examined in detail in Chapters 10 and 12, respectively.

All tissues of the body develop from three **primary germ layers,** the first tissues that form in a human embryo: **ectoderm, mesoderm,** and **endoderm** (see Figure 29.7b on page 1071). All three primary germ layers contribute to epithelial tissues. Mesoderm gives rise to all connective tissues and most muscle tissues. Ectoderm develops into nervous tissue. (Table 29.1 on page 1072 provides a list of structures derived from each layer.)

Normally, most cells within a tissue remain anchored to other cells, to basement membranes (described shortly), and to

connective tissues. A few cells, such as phagocytes, move freely through the body, searching for invaders. Before birth, however, many cells migrate extensively as part of the growth and development process.

Biopsy

A **biopsy** (BĪ-op-sē; *bio-* = life, *-opsy* = to view) is the removal of a sample of living tissue for microscopic examination. This procedure is used to help diagnose many disorders, especially cancers, and to discover the cause of unexplained infections and inflammations. Both normal and potentially diseased tissues are removed for purposes of comparison. Once the tissue sample is removed, either surgically or through a needle and syringe, it may be preserved, stained to highlight special properties, or cut into thin sections for microscopic observation. Sometimes a biopsy is conducted while a patient is anesthetized during surgery to help a physician determine the most appropriate treatment. For example, if a biopsy of thyroid tissue reveals malignant cells, the surgeon can proceed with the most appropriate procedure right away. ■

▶ C H E C K P O I N T

1. Define a tissue.
2. What are the four basic types of body tissues?

CELL JUNCTIONS

▶ O B J E C T I V E

- **Describe the structure and functions of the five main types of cell junctions.**

Most epithelial cells and some muscle and nerve cells are tightly joined into functional units. **Cell junctions** are contact points between the plasma membranes of tissue cells. Depending on their structure, cell junctions may serve one of three functions: (1) forming seals between cells, like a "zip-lock" at the top of a plastic storage bag, (2) anchoring cells to one another or to extracellular material, or (3) providing channels that allow ions and molecules to pass from cell to cell within a tissue. Here we consider the five most important types of cell junctions: tight junctions, adherens junctions, desmosomes, hemidesmosomes, and gap junctions (Figure 4.1):

branes together (Figure 4.1a). Cells of epithelial tissues that line the stomach, intestines, and urinary bladder have many tight junctions. They retard the passage of substances between cells and thus prevent the contents of these organs from leaking into the blood or surrounding tissues.

Tight Junctions

Tight junctions consist of weblike strands of transmembrane proteins that fuse the outer surfaces of adjacent plasma mem-

Adherens Junctions

Adherens junctions (ad-HER-ens) contain *plaque,* a dense layer of proteins on the inside of the plasma membrane that

Figure 4.1 Cell junctions.

Most epithelial cells and some muscle and nerve cells contain cell junctions.

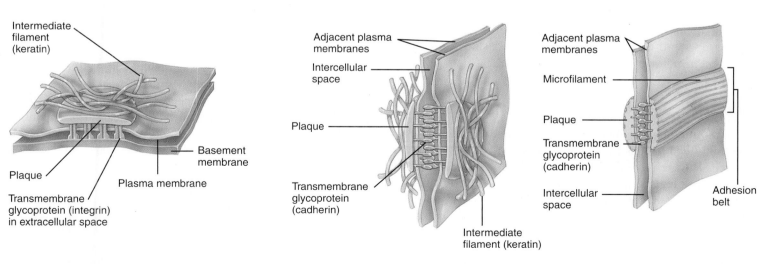

(e) Gap junction

(a) Tight junction

(d) Hemidesmosome

(c) Desmosome

(b) Adherens junction

Which type of cell junction functions in communication between adjacent cells?

attaches to both cytoskeleton proteins and membrane proteins (Figure 4.1b). Actin microfilaments extend from the plaque into the cell's cytosol. Transmembrane glycoproteins called **cadherins** insert into the plaque from the opposite side. Each cadherin partially crosses the intercellular space (the space between the cells) and connects to cadherins of the adjacent cell, thus attaching the two cells. In epithelial cells, adherens junctions often form extensive zones called *adhesion belts* that encircle the cell. Adherens junctions help epithelial surfaces resist separation.

Desmosomes

Like adherens junctions, **desmosomes** (DEZ-mō-sōms; *desmo-* = band) contain plaque and have transmembrane glycoproteins (cadherins) that extend into the intercellular space between adjacent cell membranes and attach cells to one another (Figure 4.1c). Intermediate filaments (made of keratin) extend from desmosomes on one side of the cell across the cytosol to desmosomes on the opposite side of the cell. This structural arrangement contributes to the stability of the cells and tissue. Desmosomes are common among the cells that make up the epidermis (the outermost layer of the skin) and between cardiac muscle cells in the heart. Desmosomes prevent epidermal cells from separating under tension and cardiac muscle cells from pulling apart during contraction.

Hemidesmosomes

Hemidesmosomes (*hemi-* = half) resemble desmosomes but they lack links to adjacent cells. They look like half of a desmosome, thus the name (Figure 4.1d). However, the transmembrane glycoproteins in hemidesmosomes are **integrins** rather than cadherins. On the inside of the plasma membrane, integrins attach to intermediate filaments made of the protein keratin. On the outside of the plasma membrane, the integrins attach to the protein *laminin,* which is present in the basement membrane (discussed shortly). Thus, hemidesmosomes anchor cells to the basement membrane.

Gap Junctions

At **gap junctions,** membrane proteins called **connexins** form tiny fluid-filled tunnels called **connexons** that connect neighboring cells (Figure 4.1e). The plasma membranes of gap junctions are not fused together as in tight junctions but are separated by a very narrow intercellular gap (space). Ions and small molecules can diffuse through connexons from the cytosol of one cell to another. Gap junctions allow the cells in a tissue to communicate with each other. In a developing embryo, some of the chemical and electrical signals that regulate growth and cell differentiation travel via gap junctions. Moreover, gap junctions enable nerve or muscle impulses to spread rapidly among cells, a process that is crucial for the normal operation of some parts of the nervous

system and for the contraction of muscle in the heart and gastrointestinal tract.

▶ C H E C K P O I N T
3. Which type of cell junctions allow cellular communication?
4. Which types of cell junctions are found in epithelial tissues?

EPITHELIAL TISSUE

▶ O B J E C T I V E S
• **Describe the general features of epithelial tissues.**
• **For each different type of epithelium, list its location, structure, and function.**

An **epithelial tissue** (ep-i-THĒ-lē-al), or **epithelium** (plural is *epithelia*), consists of cells arranged in continuous sheets, in either single or multiple layers. Because the cells are closely packed and are held tightly together by many cell junctions, there is little intercellular space between adjacent plasma membranes.

The various surfaces of epithelial cells often differ in structure and have specialized functions. The **apical (free) surface** of an epithelial cell faces the body surface, a body cavity, the lumen (interior space) of an internal organ, or a tubular duct that receives secretions from the cells (Figure 4.2). Apical surfaces may contain cilia or microvilli. The **lateral surfaces** of an epithelial cell face the adjacent cells on either side. As you have seen in Figure 4.1, lateral surfaces may contain tight junctions, adherens junctions, desmosomes, and gap junctions. The **basal surface** of an epithelial cell is opposite the apical surface and adheres to extracellular materials. Hemidesmosomes in the basal surfaces of epithelial cells anchor the epithelium to the basement membrane. In subsequent discussions about epithelia with multiple layers the term *apical layer* refers to the most superficial layer of cells whereas the *basal layer* is the deepest layer of cells.

The **basement membrane** is a thin extracellular layer that commonly consists of two layers, the basal lamina and reticular lamina. The *basal lamina* (*lamina* = thin layer) is closer to the epithelial cells and is secreted by them. It contains proteins such as collagen (described shortly) and laminin, as well as glycoproteins and proteoglycans (also described shortly). The laminin molecules in the basal lamina adhere to integrins in hemidesmosomes and thus attach epithelial cells to the basement membrane (see Figure 4.1d). The *reticular lamina* is closer to the underlying connective tissue and contains fibrous proteins produced by connective tissue cells called fibroblasts. Besides their function of attaching to and supporting the overlying epithelial tissue, basement membranes have other roles. They form a surface along which epithelial cells migrate during growth or wound healing, they restrict passage of larger molecules between epithelium and connective tissue, and they participate in filtration of blood in the kidneys.

Epithelial tissue is **avascular** (*a-* = without; *vascular* = vessel); that is, it lacks its own blood supply. The blood vessels that bring in nutrients and remove wastes are located in the

Figure 4.2 Surfaces of epithelial cells and the structure and location of the basement membrane.

The basement membrane is found between epithelium and connective tissue.

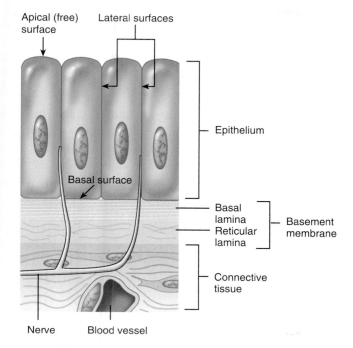

 What are the functions of the basement membrane?

adjacent connective tissue. Exchange of substances between connective tissue and epithelium occurs by diffusion. Although epithelial tissue is avascular, it has a nerve supply.

Because epithelial tissue forms boundaries between the body's organs, or between the body and the external environment, it is repeatedly subjected to physical stress and injury. A high rate of cell division, however, allows epithelial tissue to constantly renew and repair itself by sloughing off dead or injured cells and replacing them with new ones. Epithelial tissue plays many different roles in the body, the most important of which are protection, filtration, secretion, absorption, and excretion. In addition, epithelial tissue combines with nervous tissue to form special organs for smell, hearing, vision, and touch.

Epithelial tissue may be divided into two types. **Covering and lining epithelium** forms the outer covering of the skin and some internal organs. It also forms the inner lining of blood vessels, ducts, and body cavities, and the interior of the respiratory, digestive, urinary, and reproductive systems. **Glandular epithelium** constitutes the secreting portion of glands, such as the thyroid gland, adrenal glands, and sweat glands.

Covering and Lining Epithelium

The types of covering and lining epithelial tissue are classified according to two characteristics: the arrangement of cells into layers and the shapes of the cells.

1. *Arrangement of cells in layers.* The cells of covering and lining epithelia are arranged in one or more layers depending on the function the epithelium performs:

    a. *Simple epithelium* is a single layer of cells that functions in diffusion, osmosis, filtration, *secretion* (the production and release of substances such as mucus, sweat, or enzymes), and *absorption* (the intake of fluids or other substances by cells).

    b. *Stratified epithelium* consists of two or more layers of cells that protect underlying tissues in locations where there is considerable wear and tear.

    c. *Pseudostratified epithelium* contains only a single layer of cells, but it appears to have multiple layers because the cell nuclei lie at different levels and not all cells reach the apical surface. Cells that do extend to the apical surface are either ciliated or secrete mucus.

2. *Cell shapes.*

    a. *Squamous* cells (SKWĀ-mus = flat) are arranged like floor tiles and are thin, which allows for the rapid movement of substances through them.

    b. *Cuboidal* cells are as tall as they are wide and are shaped like cubes or hexagons. They may have microvilli at their apical surface and function in either secretion or absorption.

    c. *Columnar* cells are much taller than they are wide and protect underlying tissues. Their apical surface may have cilia or microvilli, and they often are specialized for secretion and absorption.

    d. *Transitional* cells change shape, from cuboidal to flat and back, as organs stretch to a larger size then collapse to a smaller size.

Considering the arrangements of layers and the cell shapes, the types of covering and lining epithelia are:

I. Simple epithelium
    A. Simple squamous epithelium
    B. Simple cuboidal epithelium
    C. Simple columnar epithelium (nonciliated and ciliated)

II. Stratified epithelium
    A. Stratified squamous epithelium (keratinized and nonkeratinized)*
    B. Stratified cuboidal epithelium*
    C. Stratified columnar epithelium*
    D. Transitional epithelium

III. Pseudostratified columnar epithelium (nonciliated and ciliated)

Each of these covering and lining epithelial tissues is described in the following sections and illustrated in (Table 4.1). The tables throughout this chapter contain figures consisting of a

---

*This classification is based on the shape of the cells at the apical surface.

photomicrograph, a corresponding diagram, and an inset that identifies a major location of the tissue in the body. Along with the illustrations are descriptions, locations, and functions of the tissues.

Simple Epithelium

SIMPLE SQUAMOUS EPITHELIUM This tissue consists of a single layer of flat cells that resembles a tiled floor when viewed from their apical surface (Table 4.1A). The nucleus of each cell is a flattened oval or sphere and is centrally located. Simple squamous epithelium is present at sites where the processes of filtration, such as blood filtration in the kidneys, or diffusion, for instance diffusion of oxygen into blood vessels of the lungs, are occurring. It is not found in body areas that are subject to wear and tear.

The simple squamous epithelium that lines the heart, blood vessels, and lymphatic vessels is known as **endothelium** (*endo-* = within; *-thelium* = covering); the type that forms the epithelial layer of serous membranes is called **mesothelium** (*meso-* = middle). Unlike other epithelial tissues, which arise from embryonic ectoderm or endoderm, endothelium and mesothelium both are derived from embryonic mesoderm.

SIMPLE CUBOIDAL EPITHELIUM The cuboidal shape of the cells in this tissue (Table 4.1B) is obvious when the tissue is sectioned and viewed from the side. Cell nuclei are usually round and centrally located. Simple cuboidal epithelium performs the functions of secretion and absorption.

SIMPLE COLUMNAR EPITHELIUM When viewed from the side, the cells appear rectangular with oval nuclei near the base of the cells. Simple columnar epithelium exists in two forms: nonciliated simple columnar epithelium and ciliated simple columnar epithelium.

Nonciliated simple columnar epithelium contains two types of cells—columnar epithelial cells with microvilli at their apical surface and goblet cells (Table 4.1C). **Microvilli** are microscopic fingerlike cytoplasmic projections that increase the surface area of the plasma membrane (see Figure 3.1 on page 60). Their presence increases the rate of absorption by the cell. **Goblet cells** are modified columnar epithelial cells that secrete mucus, a slightly sticky fluid, at their apical surfaces. Before it is released, mucus accumulates in the upper portion of the cell, causing that area to bulge out. The whole cell then resembles a goblet or wine glass. Secreted mucus serves as a lubricant for the linings of the digestive, respiratory, reproductive, and most of the urinary tracts. Mucus also helps prevent destruction of the stomach lining by acidic gastric juice.

Ciliated simple columnar epithelium contains columnar epithelial cells with cilia at their apical surface (Table 4.1D). In the airways of the upper respiratory tract, goblet cells are interspersed among ciliated columnar cells. Mucus secreted by the goblet cells forms a film over the airway surface that traps inhaled foreign particles. The cilia beat in unison, moving the mucus and any foreign particles toward the throat, where it can be coughed up and swallowed or spit out. Cilia also help move

oocytes expelled from the ovaries through the uterine (Fallopian) tubes into the uterus.

Stratified Epithelium

In contrast to simple epithelium, stratified epithelium has at least two layers of cells. Thus, it is more durable and can better protect underlying tissues. Some cells of stratified epithelia also produce secretions. The name of the specific kind of stratified epithelium depends on the shape of the apical layer of cells.

STRATIFIED SQUAMOUS EPITHELIUM Cells in the apical layer of this type of epithelium are flat, whereas in the deep layers, cells vary in shape from cuboidal to columnar (Table 4.1E). The deepest cells continually undergo cell division. As new cells grow, the cells of the deepest layer are pushed upward toward the apical layer. As they move farther from the deeper layers and from their blood supply in the underlying connective tissue, they become dehydrated, shrunken, and harder and then die. At the apical layer, the cells lose their cell junctions and are sloughed off, but they are replaced as new cells continually emerge from the deepest cells.

Stratified squamous epithelium exists in both keratinized and nonkeratinized forms. In keratinized stratified squamous epithelium, the apical layer and several layers deep to it are partially dehydrated and contain a layer of **keratin,** a tough, fibrous protein that helps protect the skin and underlying tissues from heat, microbes, and chemicals. *Nonkeratinized stratified squamous epithelium* does not contain keratin in the apical layer and several layers deep to it and remains moist.

 **Papanicolaou Test**

A **Papanicolaou test** (pa-pa-NI-kō-lō), **Pap test,** or **Pap smear** involves collecting and microscopically examining epithelial cells that have sloughed off the apical layer of a tissue. A very common type of Pap test involves examining the cells from the stratified squamous epithelium of the cervix and vagina. This type of Pap test is performed mainly to detect early changes in the cells of the female reproductive system that may indicate cancer or a precancerous condition. An annual Pap test is recommended for all women as part of a routine pelvic exam. ■

STRATIFIED CUBOIDAL EPITHELIUM This fairly rare type of epithelium sometimes consists of more than two layers of cells (Table 4.1F). Its function is mainly protective but it also functions in limited secretion and absorption.

STRATIFIED COLUMNAR EPITHELIUM Like stratified cuboidal epithelium, this type of tissue also is uncommon. Usually the basal layer consists of shortened, irregularly shaped cells; only the apical layer has cells that are columnar in form (Table 4.1G). This type of epithelium functions in protection and secretion.

(text continues on page 115)

Table 4.1 Epithelial Tissues

Covering and Lining Epithelium

A. Simple squamous epithelium

Description: Single layer of flat cells; centrally located nucleus.

Location: Lines heart, blood vessels, lymphatic vessels, air sacs of lungs, glomerular (Bowman's) capsule of kidneys, and inner surface of the tympanic membrane (eardrum); forms epithelial layer of serous membranes, such as the peritoneum.

Function: Filtration, diffusion, osmosis, and secretion in serous membranes.

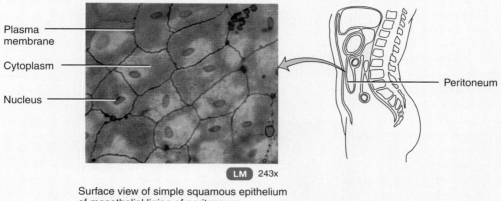

Plasma membrane
Cytoplasm
Nucleus
Peritoneum

LM 243x

Surface view of simple squamous epithelium of mesothelial lining of peritoneum

Simple squamous cell
Basement membrane
Connective tissue

Simple squamous epithelium

B. Simple cuboidal epithelium

Description: Single layer of cube-shaped cells; centrally located nucleus.

Location: Covers surface of ovary, lines anterior surface of capsule of the lens of the eye, forms the pigmented epithelium at the back of the eye, lines kidney tubules and smaller ducts of many glands, and makes up the secreting portion of some glands such as the thyroid gland and the ducts of some glands such as the pancreas.

Function: Secretion and absorption.

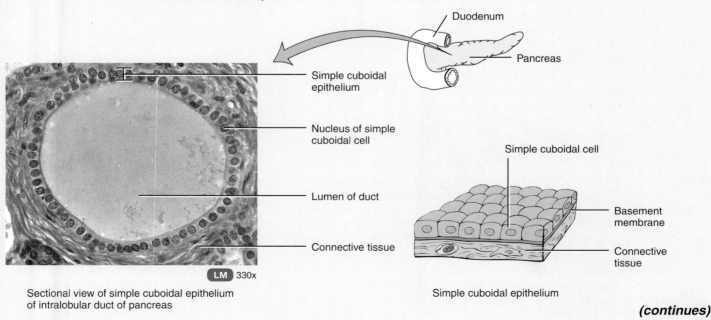

Duodenum
Pancreas
Simple cuboidal epithelium
Nucleus of simple cuboidal cell
Lumen of duct
Connective tissue

LM 330x

Sectional view of simple cuboidal epithelium of intralobular duct of pancreas

Simple cuboidal cell
Basement membrane
Connective tissue

Simple cuboidal epithelium

(continues)

Table 4.1   Epithelial Tissues *(continued)*

Covering and Lining Epithelium

C. Nonciliated simple columnar epithelium

Description: Single layer of nonciliated rectangular cells with nuclei near base of cells; contains goblet cells and cells with microvilli in some locations.

Location: Lines the gastrointestinal tract from the stomach to the anus, ducts of many glands, and gallbladder.

Function: Secretion and absorption.

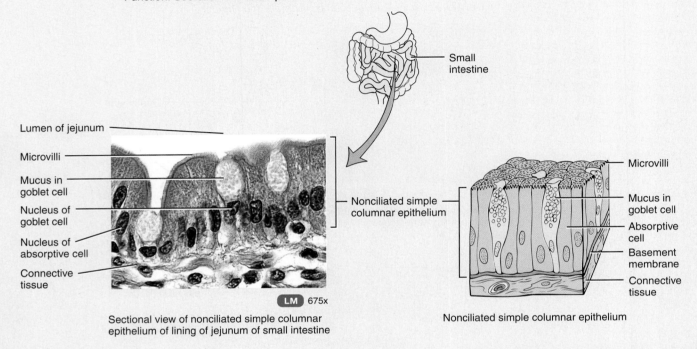

Small intestine

Lumen of jejunum
Microvilli
Mucus in goblet cell
Nucleus of goblet cell
Nucleus of absorptive cell
Connective tissue

LM 675x

Sectional view of nonciliated simple columnar epithelium of lining of jejunum of small intestine

Nonciliated simple columnar epithelium

Microvilli
Mucus in goblet cell
Absorptive cell
Basement membrane
Connective tissue

Nonciliated simple columnar epithelium

D. Ciliated simple columnar epithelium

Description: Single layer of ciliated rectangular cells with nuclei near base of cells; contains goblet cells in some locations.

Location: Lines a few portions of upper respiratory tract, uterine (Fallopian) tubes, uterus, some paranasal sinuses, and central canal of spinal cord.

Function: Moves mucus and other substances by ciliary action.

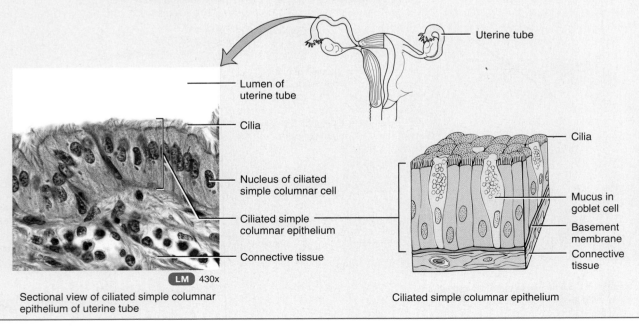

Uterine tube

Lumen of uterine tube
Cilia
Nucleus of ciliated simple columnar cell
Ciliated simple columnar epithelium
Connective tissue

LM 430x

Sectional view of ciliated simple columnar epithelium of uterine tube

Cilia
Mucus in goblet cell
Basement membrane
Connective tissue

Ciliated simple columnar epithelium

Covering and Lining Epithelium

E. Stratified squamous epithelium

Description: Several layers of cells; cuboidal to columnar shape in deep layers; squamous cells form the apical layer and several layers deep to it; cells from the basal layer replace surface cells as they are lost.

Location: Keratinized variety forms superficial layer of skin; nonkeratinized variety lines wet surfaces, such as lining of the mouth, esophagus, part of epiglottis, and vagina, and covers the tongue.

Function: Protection.

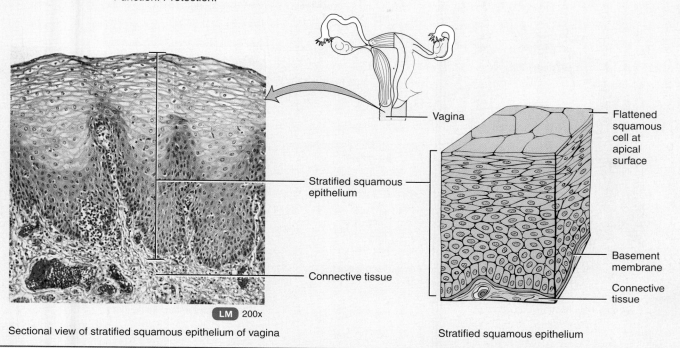

Sectional view of stratified squamous epithelium of vagina   LM 200x

Stratified squamous epithelium

F. Stratified cuboidal epithelium

Description: Two or more layers of cells in which the cells in the apical layer are cube-shaped.

Location: Ducts of adult sweat glands and esophageal glands and part of male urethra.

Function: Protection and limited secretion and absorption.

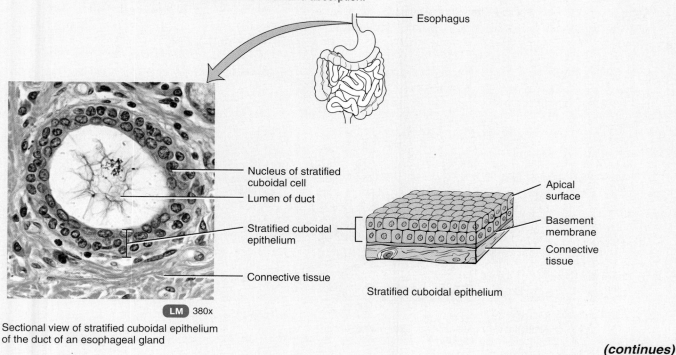

Sectional view of stratified cuboidal epithelium of the duct of an esophageal gland   LM 380x

Stratified cuboidal epithelium

(continues)

Table 4.1 Epithelial Tissues (continued)

Covering and Lining Epithelium

G. Stratified columnar epithelium

Description: Several layers of irregularly shaped cells; columnar cells are only in the apical layer.

Location: Lines part of urethra, large excretory ducts of some glands, such as esophageal glands, small areas in anal mucous membrane, and part of the conjunctiva of the eye.

Function: Protection and secretion.

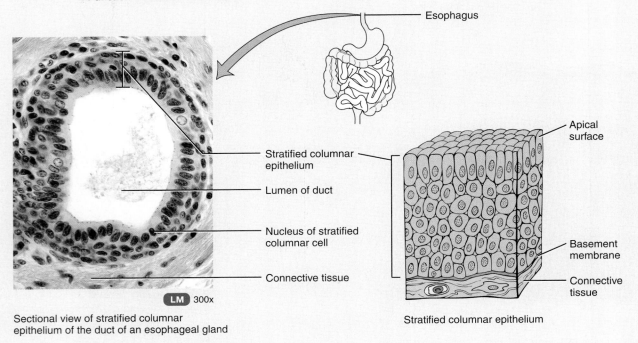

Sectional view of stratified columnar epithelium of the duct of an esophageal gland

Stratified columnar epithelium

H. Transitional epithelium

Description: Appearance is variable (transitional); shape of cells in apical layer ranges from squamous (when stretched) to cuboidal (when relaxed).

Location: Lines urinary bladder and portions of ureters and urethra.

Function: Permits distention.

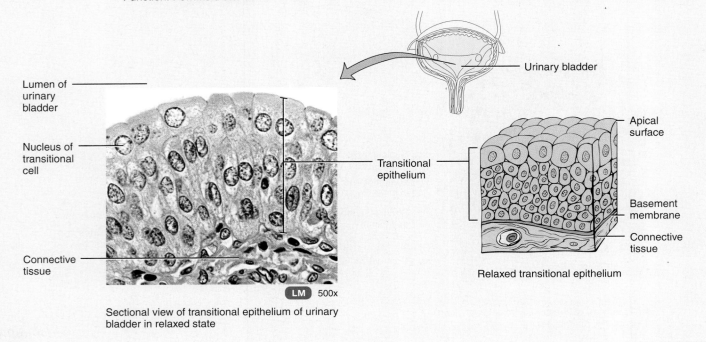

Sectional view of transitional epithelium of urinary bladder in relaxed state

Relaxed transitional epithelium

Covering and Lining Epithelium

I. Pseudostratified columnar epithelium

Description: Not a true stratified tissue; nuclei of cells are at different levels; all cells are attached to basement membrane, but not all reach the surface.

Location: Pseudostratified ciliated columnar epithelium lines the airways of most of upper respiratory tract; pseudostratified nonciliated columnar epithelium lines larger ducts of many glands, epididymis, and part of male urethra.

Function: Secretion and movement of mucus by ciliary action.

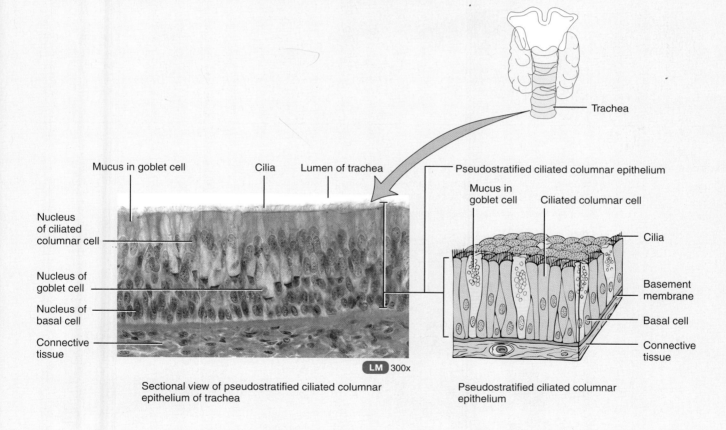

Sectional view of pseudostratified ciliated columnar epithelium of trachea

Pseudostratified ciliated columnar epithelium

(continues)

Table 4.1	Epithelial Tissues *(continued)*

Glandular Epithelium

J. Endocrine glands

Description: Secretory products (hormones) diffuse into blood after passing through interstitial fluid.

Location: Examples include pituitary gland at base of brain, pineal gland in brain, thyroid and parathyroid glands near larynx (voice box), adrenal glands superior to kidneys, pancreas near stomach, ovaries in pelvic cavity, testes in scrotum, and thymus in thoracic cavity.

Function: Produce hormones that regulate various body activities.

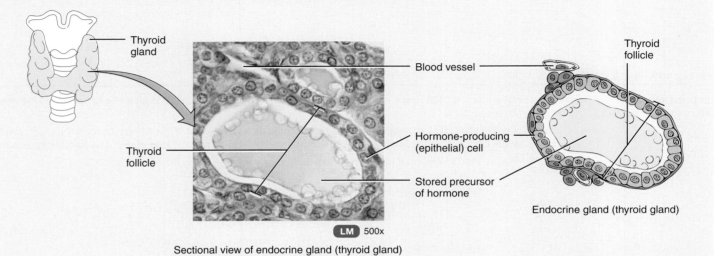

Thyroid gland

Blood vessel

Thyroid follicle

Thyroid follicle

Hormone-producing (epithelial) cell

Stored precursor of hormone

Endocrine gland (thyroid gland)

LM 500x

Sectional view of endocrine gland (thyroid gland)

K. Exocrine glands

Description: Secretory products released into ducts.

Location: Sweat, oil, and earwax glands of the skin; digestive glands such as salivary glands, which secrete into mouth cavity, and pancreas, which secretes into the small intestine.

Function: Produce mucus, sweat, oil, earwax, saliva, or digestive enzymes.

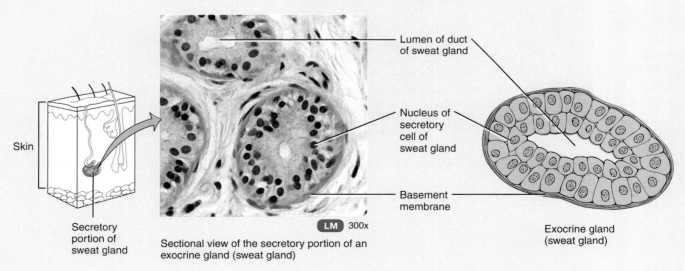

Skin

Secretory portion of sweat gland

Lumen of duct of sweat gland

Nucleus of secretory cell of sweat gland

Basement membrane

Exocrine gland (sweat gland)

LM 300x

Sectional view of the secretory portion of an exocrine gland (sweat gland)

TRANSITIONAL EPITHELIUM   This kind of stratified epithelium is present only in the urinary system and has a variable appearance. In its relaxed or unstretched state (Table 4.1H), transitional epithelium looks similar to stratified cuboidal epithelium, except that the cells in the apical layer tend to be large and rounded. As the tissue is stretched, its cells become flatter, giving the appearance of stratified squamous epithelium. Because of its elasticity, transitional epithelium is ideal for lining hollow structures that are subjected to expansion from within, such as the urinary bladder. It functions to accommodate a variable volume of fluid without rupturing.

Pseudostratified Columnar Epithelium

The third category of covering and lining epithelium is pseudostratified columnar epithelium (Table 4.1I). The nuclei of the cells are at various depths. Even though all the cells are attached to the basement membrane in a single layer, some cells do not extend to the surface. When viewed from the side, these features give the false impression of a multilayered tissue—thus the name *pseudo*stratified epithelium (*pseudo-* = false). In *pseudostratified ciliated columnar epithelium,* the cells that extend to the surface either secrete mucus (goblet cells) or bear cilia. The mucus traps foreign particles and the cilia sweep away mucus for eventual elimination from the body. *Pseudostratified nonciliated columnar epithelium* contains cells without cilia and lacks goblet cells.

Glandular Epithelium

The function of glandular epithelium is secretion, which is accomplished by glandular cells that often lie in clusters deep to the covering and lining epithelium. A **gland** may consist of a single cell or a group of cells that secrete substances into ducts (tubes), onto a surface, or into the blood. All glands of the body are classified as either endocrine or exocrine.

The secretions of **endocrine glands** (Table 4.1J) enter the interstitial fluid and then diffuse directly into the bloodstream without flowing through a duct. These secretions, called *hormones,* regulate many metabolic and physiological activities to maintain homeostasis. The pituitary, thyroid, and adrenal glands are examples of endocrine glands. Endocrine glands will be described in detail in Chapter 18.

Exocrine glands (*exo-* = outside; *-crine* = secretion; (Table 4.1K) secrete their products into ducts that empty onto the surface of a covering and lining epithelium. Thus, the product of an exocrine gland is released at the skin surface or into the lumen of a hollow organ. The secretions of exocrine glands include mucus, sweat, oil, earwax, saliva, and digestive enzymes. Examples of exocrine glands are sudoriferous (sweat) glands, which produce sweat to help lower body temperature, and salivary glands, which secrete saliva. Saliva contains mucus and digestive enzymes among other substances. As you will see later, some glands of the body, such as the pancreas, ovaries, and testes, are mixed glands that contain both endocrine and exocrine tissue.

Structural Classification of Exocrine Glands

Exocrine glands are classified into unicellular and multicellular types. **Unicellular glands** are single-celled. Goblet cells, an important unicellular exocrine gland, secrete mucus directly onto the apical surface of a lining epithelium rather than into ducts. Most glands are **multicellular glands,** composed of many cells that form a distinctive microscopic structure or macroscopic organ. Examples are sudoriferous, sebaceous (oil), and salivary glands.

Multicellular glands are categorized according to two criteria: by whether the ducts of the glands are branched or unbranched and by the shape of the secretory portions of the gland (Figure 4.3). If the duct of the gland does not branch, it is a **simple gland.** If the duct branches, it is a **compound gland.** Glands with tubular secretory parts are **tubular glands;** those with more rounded secretory portions are **acinar glands** (AS-i-nar; *acin-* = berry). **Tubuloacinar glands** have both tubular and more rounded secretory parts.

Combinations of the degree of duct branching and the shape of the secretory part are the criteria for the following structural classification scheme for multicellular exocrine glands:

I.   Simple glands
   A. **Simple tubular.** Tubular secretory part is straight and attaches to a single unbranched duct. Example: glands in the large intestine.
   B. **Simple branched tubular.** Tubular secretory part is branched and attaches to a single unbranched duct. Example: gastric glands.
   C. **Simple coiled tubular.** Tubular secretory part is coiled and attaches to a single unbranched duct. Example: sweat glands.
   D. **Simple acinar.** Secretory portion is rounded and attaches to a single unbranched duct. Example: glands of penile urethra.
   E. **Simple branched acinar.** Rounded secretory part is branched and attaches to a single unbranched duct. Example: sebaceous glands.
II. Compound glands
   A. **Compound tubular.** Secretory portion is tubular and attaches to a branched duct. Example: bulbourethral (Cowper's) glands.
   B. **Compound acinar.** Secretory portion is rounded and attaches to a branched duct. Example: mammary glands.
   C. **Compound tubuloacinar.** Secretory portion is both tubular and rounded and attaches to a branched duct. Example: acinar glands of the pancreas.

Functional Classification of Exocrine Glands

The functional classification of exocrine glands is based on how their secretion is released. In **merocrine glands** (MER-ō-krin;

Figure 4.3   Multicellular exocrine glands.   Pink represents the secretory portion; lavender represents the duct.

🔑 Structural classification of multicellular exocrine glands is based on the branching pattern of the duct and the shape of the secreting portion.

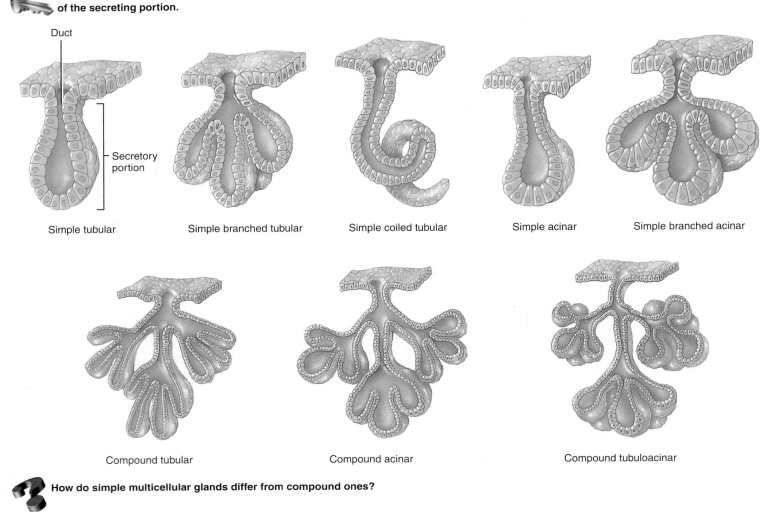

Duct

Secretory portion

Simple tubular

Simple branched tubular

Simple coiled tubular

Simple acinar

Simple branched acinar

Compound tubular

Compound acinar

Compound tubuloacinar

🔑 **How do simple multicellular glands differ from compound ones?**

mero- = a part), the secretion is synthesized on ribosomes attached to rough ER; processed, sorted, and packaged by the Golgi complex; and released from the cell in secretory vesicles via exocytosis (Figure 4.4a). Most exocrine glands of the body are merocrine glands. Examples are the salivary glands and pancreas. **Apocrine glands** (AP-ō-krin; *apo-* = from) accumulate their secretory product at the apical surface of the secreting cell. Then, that portion of the cell pinches off from the rest of the cell to release the secretion (Figure 4.4b). The remaining part of the cell repairs itself and repeats the process. Electron micrographic studies have called into question whether humans have apocrine glands. What were once thought to be apocrine glands—for example, mammary glands that secrete milk—are probably merocrine glands. The cells of **holocrine glands** (HŌ-lō-krin; *holo-* = entire) accumulate a secretory product in their cytosol. As the secretory cell matures, it ruptures and becomes the secretory prod-

uct (Figure 4.4c). The sloughed off cell is replaced by a new cell. One example of a holocrine gland is a sebaceous gland of the skin.

▶ **CHECKPOINT**

5. Describe the various layering arrangements and cell shapes of epithelium.

6. What characteristics are common to all epithelial tissues?

7. How is the structure of the following kinds of epithelium related to their functions: simple squamous, simple cuboidal, simple columnar (nonciliated and ciliated), stratified squamous (keratinized and nonkeratinized), stratified cuboidal, stratified columnar, transitional, and pseudostratified columnar (ciliated and nonciliated)?

8. Where are endothelium and mesothelium located?

9. What distinguishes endocrine glands and exocrine glands? Name and give examples of the three functional classes of exocrine glands.

Figure 4.4 Functional classification of multicellular exocrine glands.

 The functional classification of exocrine glands is based on whether a secretion is a product of a cell or consists of an entire or a partial glandular cell.

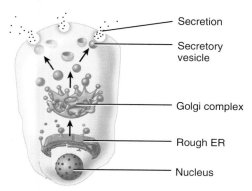

- Secretion
- Secretory vesicle
- Golgi complex
- Rough ER
- Nucleus

(a) Merocrine secretion

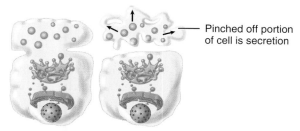

- Pinched off portion of cell is secretion

(b) Apocrine secretion

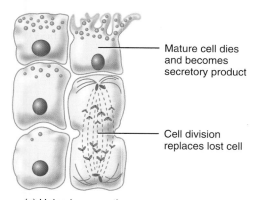

- Mature cell dies and becomes secretory product
- Cell division replaces lost cell

(c) Holocrine secretion

 What class of glands are sebaceous (oil) glands? Salivary glands?

CONNECTIVE TISSUE

OBJECTIVES

- **Describe the general features of connective tissue.**
- **Describe the structure, location, and function of the various types of connective tissues.**

Connective tissue is one of the most abundant and widely distributed tissues in the body. In its various forms, connective tissue has a variety of functions. It binds together, supports, and strengthens other body tissues; protects and insulates internal organs; compartmentalizes structures such as skeletal muscles; is the major transport system within the body (blood, a fluid connective tissue); is the major site of stored energy reserves (adipose, or fat, tissue); and is the main site of immune responses.

General Features of Connective Tissue

Connective tissue consists of two basic elements: cells and matrix. A connective tissue's **matrix** fills the wide spaces between its cells. The matrix consists of protein-based fibers and ground substance, the material between the cells and the fibers. It is usually secreted by the connective tissue cells and determines the tissue's qualities. For instance, in cartilage, the matrix is firm but pliable. In bone, by contrast, the matrix is hard and not pliable.

In contrast to epithelia, connective tissues do not usually occur on body surfaces, such as the covering or lining of internal organs, the lining of body cavities, or the external surface of the body. However, a type of connective tissue called areolar connective tissue lines joint cavities. Also, unlike epithelia, connective tissues usually are highly vascular; that is, they have a rich blood supply. Exceptions include cartilage, which is avascular, and tendons, which have a scanty blood supply. Except for cartilage, connective tissues, like epithelia, have a nerve supply.

Connective Tissue Cells

Mesodermal embryonic cells called mesenchymal cells give rise to the cells of connective tissue. Each major type of connective tissue contains an immature class of cells whose name ends in -*blast,* which means "to bud or sprout." These immature cells are called *fibroblasts* in loose and dense connective tissue, *chondroblasts* in cartilage, and *osteoblasts* in bone. Blast cells retain the capacity for cell division and secrete the matrix that is characteristic of the tissue. In cartilage and bone, once the matrix is produced, the fibroblasts differentiate into mature cells whose names end in -*cyte,* namely chondrocytes and osteocytes. Mature cells have reduced capacity for cell division and matrix formation and are mostly involved in maintaining the matrix.

The types of cells present in connective tissues vary according to the type of tissue and include the following (Figure 4.5):

1. **Fibroblasts** (FĪ-brō-blasts; *fibro-* = fibers) are large, flat cells with branching processes. They are present in several connective tissues, where they usually are the most numerous connective tissue cells. Fibroblasts migrate through the connective tissue, secreting the fibers and ground substance of the matrix.

2. **Macrophages** (MAK-rō-fā-jez; *macro-* = large; *-phages* = eaters) develop from monocytes, a type of white blood cell. Macrophages have an irregular shape with short branching projections and are capable of engulfing bacteria and cellular debris by phagocytosis. Some are *fixed macrophages,* which means they reside in a particular tissue. Examples are alveolar macrophages in the lungs or spleen macrophages in the spleen. Others are *wandering macrophages,* which roam the tissues and gather at sites of infection or inflammation.

3. **Plasma cells** are small cells that develop from a type of white blood cell called a B lymphocyte. Plasma cells secrete antibodies, proteins that attack or neutralize foreign substances in the body. Thus, plasma cells are an important part of the body's immune system. Although they are found in many places in the body, most plasma cells reside in connective tissues, especially in the gastrointestinal and respiratory tracts. They are also abundant in the salivary glands, lymph nodes, and red bone marrow.

4. **Mast cells** are abundant alongside the blood vessels that supply connective tissue. They produce histamine, a chemical that dilates small blood vessels as part of the inflammatory response, the body's reaction to injury or infection.

5. **Adipocytes,** also called fat cells or adipose cells, are connective tissue cells that store triglycerides (fats). They are found below the skin and around organs such as the heart and kidneys.

6. **White blood cells** are not found in significant numbers in normal connective tissue. However, in response to certain conditions they migrate from blood into connective tissues. For example, *neutrophils* gather at sites of infection, and *eosinophils* migrate to sites of parasitic invasions and allergic responses.

Connective Tissue Matrix

Each type of connective tissue has unique properties, due to accumulation of specific matrix materials between the cells. The matrix consists of fluid, semifluid, gelatinous, or calcified ground substance and protein fibers.

Ground Substance

As mentioned earlier, the **ground substance** is the component of a connective tissue between the cells and fibers. The ground substance supports cells, binds them together, and provides a medium through which substances are exchanged between the blood and cells. It plays an active role in how tissues develop, migrate, proliferate, and change shape, and in how they carry out their metabolic functions.

Ground substance contains water and an assortment of large molecules, many of which are complex combinations of polysaccharides and proteins. The polysaccharides include

Figure 4.5   Representative cells and fibers in connective tissues.

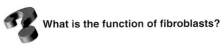 Fibroblasts are usually the most numerous connective tissue cells.

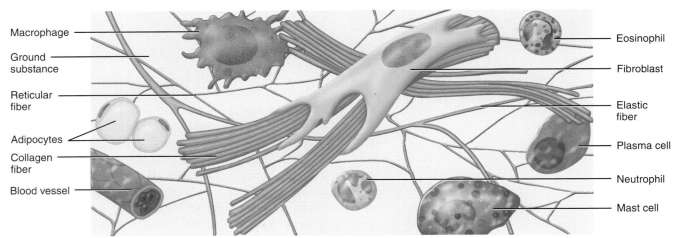

What is the function of fibroblasts?

hyaluronic acid, chondroitin sulfate, dermatan sulfate, and keratan sulfate. Collectively, they are referred to as **glycosaminoglycans** (glī-kos-a-mē′-nō-GLĪ-kans) or **GAGs.** Except for hyaluronic acid, the GAGs are associated with proteins called **proteoglycans** (prō′-tē-ō-GLĪ-kans). The proteoglycans form a core protein and the GAGs project from the protein like the bristles of a brush. One of the most important properties of GAGs is that they trap water, making the ground substance more jellylike.

Hyaluronic acid (hī′-a-loo-RON-ik) is a viscous, slippery substance that binds cells together, lubricates joints, and helps maintain the shape of the eyeballs. White blood cells, sperm cells, and some bacteria produce *hyaluronidase,* an enzyme that breaks apart hyaluronic acid, thus causing the ground substance of connective tissue to become more liquid. The ability to produce hyaluronidase helps white blood cells move more easily through connective tissues to reach sites of infection and aids penetration of an oocyte by a sperm cell during fertilization. It may also allow bacteria to spread more rapidly through connective tissues. **Chondroitin sulfate** (kon-DROY-tin) provides support and adhesiveness in cartilage, bone, skin, and blood vessels. The skin, tendons, blood vessels, and heart valves contain **dermatan sulfate,** whereas bone, cartilage, and the cornea of the eye contain **keratan sulfate.** Also present in the ground substance are **adhesion proteins,** which are responsible for linking components of the ground substance to each other and to the surfaces of cells. The main adhesion protein of connective tissue is **fibronectin,** which binds to both collagen fibers (discussed shortly) and ground substance thereby linking them together. It also attaches cells to the ground substance.

Fibers

Three types of **fibers** are embedded in the matrix between the cells: collagen fibers, elastic fibers, and reticular fibers. They function to strengthen and support connective tissues.

Collagen fibers (*colla* = glue) are very strong and resist pulling forces, but they are not stiff, which promotes tissue flexibility. Different types of collagen fibers in various tissues have slightly different properties. For example, the collagen fibers found in cartilage attract more water molecules than do the collagen fibers in bone, which gives cartilage a more cushioning consistency. Collagen fibers often occur in bundles lying parallel to one another (Figure 4.5). The bundle arrangement affords great strength. Chemically, collagen fibers consist of the protein *collagen.* This is the most abundant protein in your body, representing about 25% of the total protein. Collagen fibers are found in most types of connective tissues, especially bone, cartilage, tendons, and ligaments.

Elastic fibers, which are smaller in diameter than collagen fibers, branch and join together to form a network within a tissue. An elastic fiber consists of molecules of the protein *elastin* surrounded by a glycoprotein named *fibrillin,* which strengthens and stabilizes elastic fibers. Because of their unique molecular structure, elastic fibers are strong but can be stretched up to 150% of their relaxed length without breaking. Equally important, elastic fibers have the ability to return to their original shape after being stretched, a property called *elasticity.* Elastic fibers are plentiful in skin, blood vessel walls, and lung tissue.

Reticular fibers (*reticul-* = net), consisting of *collagen* arranged in fine bundles and a coating of glycoprotein, provide support in the walls of blood vessels and form a network around the cells in some tissues, for instance, areolar tissue, adipose tissue, and smooth muscle tissue. Produced by fibroblasts, reticular fibers are much thinner than collagen fibers and form branching networks. Like collagen fibers, reticular fibers provide support and strength. Reticular fibers are plentiful in reticular connective tissue, which forms the **stroma** (= bed or covering) or supporting framework of many soft organs, such as the spleen and lymph nodes. These fibers also help form the basement membrane.

 **Marfan Syndrome**

Marfan syndrome (MAR-fan) is an inherited disorder caused by a defective fibrillin gene. The result is abnormal development of elastic fibers. Tissues rich in elastic fibers are malformed or weakened. Structures affected most seriously are the covering layer of bones (periosteum), the ligament that suspends the lens of the eye, and the walls of the large arteries. People with Marfan syndrome tend to be tall and have disproportionately long arms, legs, fingers, and toes. A common symptom is blurred vision caused by displacement of the lens of the eye. The most life-threatening complication of Marfan syndrome is weakening of the aorta (the main artery that emerges from the heart), which can suddenly burst. ■

Classification of Connective Tissues

Classifying connective tissues can be challenging because of the diversity of cells and matrix present, and the differences in their relative proportions. Thus, the grouping of connective tissues into categories is not always clear-cut. We will classify them as follows:

I. Embryonic connective tissue
    A. Mesenchyme
    B. Mucous connective tissue
II. Mature connective tissue
    A. Loose connective tissue
        1. Areolar connective tissue
        2. Adipose tissue
        3. Reticular connective tissue

B. Dense connective tissue
   1. Dense regular connective tissue
   2. Dense irregular connective tissue
   3. Elastic connective tissue
C. Cartilage
   1. Hyaline cartilage
   2. Fibrocartilage
   3. Elastic cartilage
D. Bone tissue
E. Blood tissue
F. Lymph

Note that our classification scheme has two major sub-classes of connective tissue: embryonic and mature. **Embryonic connective tissue** is present primarily in the *embryo,* the developing human from fertilization through the first two months of pregnancy, and in the *fetus,* the developing human from the third month of pregnancy to birth.

One example of embryonic connective tissue found almost exclusively in the embryo is **mesenchyme** (MEZ-en-kīm), the tissue from which all other connective tissues eventually arise (Table 4.2A). Mesenchyme is composed of irregularly shaped cells, a semifluid ground substance, and delicate reticular fibers. Another kind of embryonic tissue is **mucous connective tissue (Wharton's jelly),** found mainly in the umbilical cord of the fetus. Mucous connective tissue is a form of mesenchyme that contains widely scattered fibroblasts, a more viscous jellylike ground substance, and collagen fibers (Table 4.2B).

The second major subclass of connective tissue, **mature connective tissue,** is present in the newborn. Its cells arise from mesenchyme. Mature connective tissue is of several types, which we explore next.

Types of Mature Connective Tissue

The six types of mature connective tissue are (1) loose connective tissue, (2) dense connective tissue, (3) cartilage, (4) bone tissue, (5) blood tissue, and (6) lymph. We now examine each in detail.

Loose Connective Tissue

In **loose connective tissue** the fibers are loosely intertwined and many cells are present. The types of loose connective tissue are areolar connective tissue, adipose tissue, and reticular connective tissue.

AREOLAR CONNECTIVE TISSUE   One of the most widely distributed connective tissues in the body is **areolar connective tissue** (a-RĒ-ō-lar; *areol-* = a small space). It contains several kinds of cells, including fibroblasts, macrophages, plasma cells, mast cells, adipocytes, and a few white blood cells (Table 4.3A). All three types of fibers—collagen, elastic, and reticular—are

arranged randomly throughout the tissue. The ground substance contains hyaluronic acid, chondroitin sulfate, dermatan sulfate, and keratan sulfate. Combined with adipose tissue, areolar connective tissue forms the *subcutaneous layer,* the layer of tissue that attaches the skin to underlying tissues and organs.

ADIPOSE TISSUE   Adipose tissue is a loose connective tissue in which the cells, called **adipocytes** (*adipo-* = fat), are specialized for storage of triglycerides (Table 4.3B). Adipocytes are derived from fibroblasts. Because the cell fills up with a single, large triglyceride droplet, the cytoplasm and nucleus are pushed to the periphery of the cell. Adipose tissue is found wherever areolar connective tissue is located. Adipose tissue is a good insulator and can therefore reduce heat loss through the skin. It is a major energy reserve and generally supports and protects various organs. As the amount of adipose tissue increases with weight gain, new blood vessels form. Thus, an obese person has many more miles of blood vessels than does a lean person, a situation that can cause high blood pressure.

Most adipose tissue in adults is **white adipose tissue,** the type just described. Another type, called **brown adipose tissue (BAT),** obtains its darker color from a very rich blood supply and numerous mitochondria, which contain colored pigments that participate in aerobic cellular respiration. Although BAT is widespread in the fetus and infant, in adults only small amounts are present. BAT generates considerable heat and probably helps to maintain body temperature in the newborn. The heat generated by the many mitochondria is carried away to other body tissues by the extensive blood supply.

Liposuction

A surgical procedure, called **liposuction** (*lip-* = fat) or **suction lipectomy** (*-ectomy* = to cut out), involves suctioning out small amounts of adipose tissue from various areas of the body. The technique can be used as a body-contouring procedure in regions such as the thighs, buttocks, arms, breasts, and abdomen. Postsurgical complications that may develop include fat emboli (clots), infection, fluid depletion, injury to internal structures, and severe postoperative pain. ■

RETICULAR CONNECTIVE TISSUE   Reticular connective tissue consists of fine interlacing reticular fibers and reticular cells (Table 4.3C). Reticular connective tissue forms the stroma (supporting framework) of the liver, spleen, and lymph nodes and helps bind together smooth muscle cells. Additionally, reticular fibers in the spleen filter blood and remove worn-out blood cells and reticular fibers in lymph nodes filter lymph and remove bacteria.

Dense Connective Tissue

Dense connective tissue contains more numerous, thicker, and denser fibers but considerably fewer cells than loose connective

Table 4.2 Embryonic Connective Tissues

A. Mesenchyme

Description: Consists of irregularly shaped mesenchymal cells embedded in a semifluid ground substance that contains reticular fibers.

Location: Under skin and along developing bones of embryo; some mesenchymal cells are found in adult connective tissue, especially along blood vessels.

Function: Forms all other kinds of connective tissue.

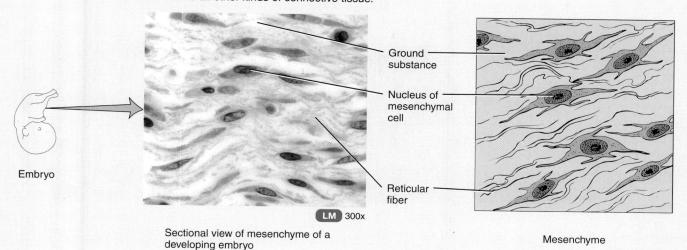

Embryo

Ground substance

Nucleus of mesenchymal cell

Reticular fiber

LM 300x

Sectional view of mesenchyme of a developing embryo

Mesenchyme

B. Mucous connective tissue

Description: Consists of widely scattered fibroblasts embedded in a viscous, jellylike ground substance that contains fine collagen fibers.

Location: Umbilical cord of fetus.

Function: Support.

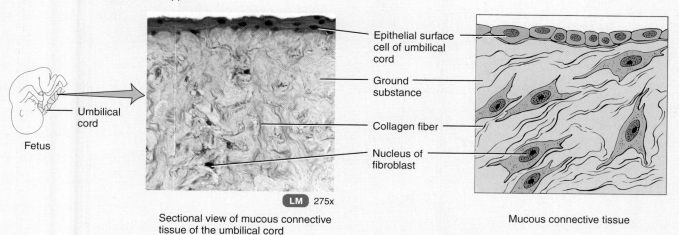

Fetus

Umbilical cord

Epithelial surface cell of umbilical cord

Ground substance

Collagen fiber

Nucleus of fibroblast

LM 275x

Sectional view of mucous connective tissue of the umbilical cord

Mucous connective tissue

Table 4.3 Mature Connective Tissues

Loose Connective Tissue

A. Areolar connective tissue

Description: Consists of fibers (collagen, elastic, and reticular) and several kinds of cells (fibroblasts, macrophages, plasma cells, adipocytes, and mast cells) embedded in a semifluid ground substance.

Location: Subcutaneous layer deep to skin; papillary (superficial) region of dermis of skin; lamina propria of mucous membranes; and around blood vessels, nerves, and body organs.

Function: Strength, elasticity, and support.

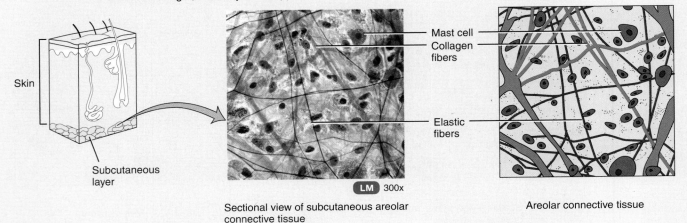

Sectional view of subcutaneous areolar connective tissue

LM 300x

Areolar connective tissue

B. Adipose tissue

Description: Consists of adipocytes, cells specialized to store triglycerides (fats) as a large centrally located droplet; nucleus and cytoplasm are peripherally located.

Location: Subcutaneous layer deep to skin, around heart and kidneys, yellow bone marrow, and padding around joints and behind eyeball in eye socket.

Function: Reduces heat loss through skin, serves as an energy reserve, supports, and protects. In newborns, brown adipose tissue generates considerable heat that helps maintain proper body temperature.

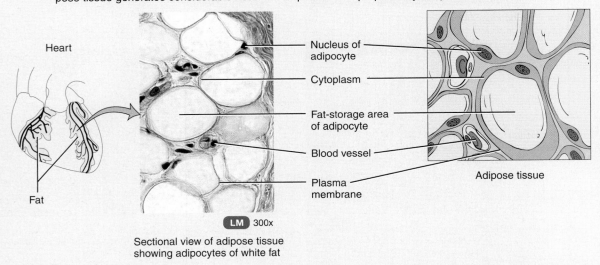

Sectional view of adipose tissue showing adipocytes of white fat

LM 300x

Adipose tissue

tissue. There are three types: dense regular connective tissue, dense irregular connective tissue, and elastic connective tissue.

DENSE REGULAR CONNECTIVE TISSUE In this tissue, bundles of collagen fibers are *regularly* arranged in parallel patterns that confer great strength (Table 4.3D). The tissue withstands pulling along the axis of the fibers. Fibroblasts, which produce the fibers and ground substance, appear in rows between the fibers. The tissue is silvery white and tough, yet somewhat pliable. Examples are tendons and most ligaments.

Loose Connective Tissue

C. Reticular connective tissue

Description: Consists of a network of interlacing reticular fibers and reticular cells.

Location: Stroma (supporting framework) of liver, spleen, lymph nodes; red bone marrow, which gives rise to blood cells; reticular lamina of the basement membrane; and around blood vessels and muscles.

Function: Forms stroma of organs; binds together smooth muscle tissue cells; filters and removes worn-out blood cells in the spleen and microbes in lymph nodes.

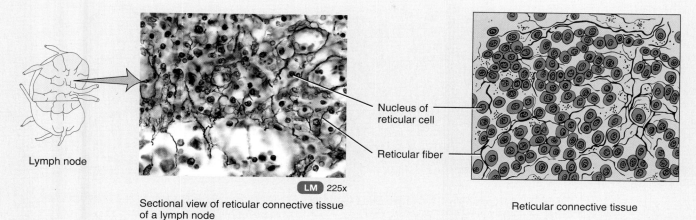

Lymph node

Nucleus of reticular cell

Reticular fiber

LM 225x

Sectional view of reticular connective tissue of a lymph node

Reticular connective tissue

Dense Connective Tissue

D. Dense regular connective tissue

Description: Matrix looks shiny white; consists mainly of collagen fibers arranged in bundles; fibroblasts present in rows between bundles.

Location: Forms tendons (attach muscle to bone), most ligaments (attach bone to bone), and aponeuroses (sheetlike tendons that attach muscle to muscle or muscle to bone).

Function: Provides strong attachment between various structures.

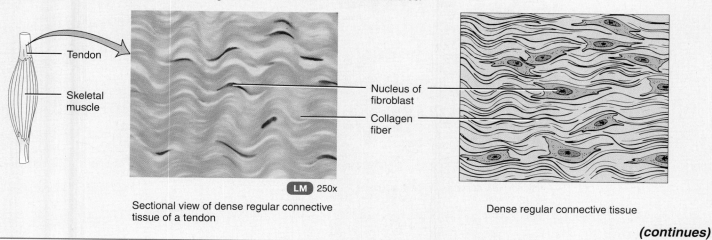

Tendon

Skeletal muscle

Nucleus of fibroblast

Collagen fiber

LM 250x

Sectional view of dense regular connective tissue of a tendon

Dense regular connective tissue

(continues)

DENSE IRREGULAR CONNECTIVE TISSUE This tissue contains collagen fibers that are packed more closely together than in loose connective tissue and are usually *irregularly* arranged (Table 4.3E). It is found in parts of the body where pulling forces are exerted in various directions. The tissue often occurs in sheets, such as in the dermis of the skin, which underlies the epidermis, or the pericardium around the heart. Heart valves, the perichondrium (the membrane surrounding cartilage), and the periosteum (the membrane surrounding bone) are dense irregular connective tissues, although they have a fairly orderly arrangement of their collagen fibers.

Table 4.3   Mature Connective Tissues *(continued)*

Dense Connective Tissue

E. Dense irregular connective tissue

Description: Consists predominantly of collagen fibers, randomly arranged, and a few fibroblasts.

Location: Fasciae (tissue beneath skin and around muscles and other organs), reticular (deeper) region of dermis of skin, periosteum of bone, perichondrium of cartilage, joint capsules, membrane capsules around various organs (kidneys, liver, testes, lymph nodes), pericardium of the heart, and heart valves.

Function: Provides strength.

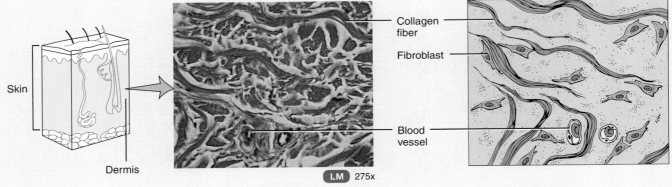

LM 275x

Sectional view of dense irregular connective tissue of reticular region of dermis

Dense irregular connective tissue

F. Elastic connective tissue

Description: Consists of predominantly freely branching elastic fibers; fibroblasts present in spaces between fibers.

Location: Lung tissue, walls of elastic arteries, trachea, bronchial tubes, true vocal cords, suspensory ligament of penis, and ligaments between vertebrae.

Function: Allows stretching of various organs.

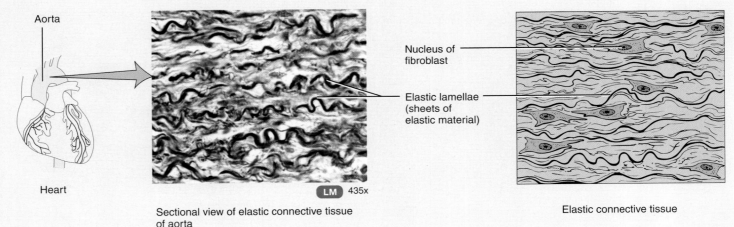

LM 435x

Sectional view of elastic connective tissue of aorta

Elastic connective tissue

ELASTIC CONNECTIVE TISSUE Branching elastic fibers predominate in elastic connective tissue (Table 4.3F), giving the unstained tissue a yellowish color. Fibroblasts are present in the spaces between the fibers. Elastic connective tissue is quite strong and can recoil to its original shape after being stretched. Elasticity is important to the normal functioning of lung tissue, which recoils as you exhale, and elastic arteries, whose recoil between heart beats helps maintain blood flow.

Cartilage

G. Hyaline cartilage

Description: Consists of a bluish-white, shiny ground substance with fine collagen fibers and many chondrocytes; most abundant type of cartilage.

Location: Ends of long bones, anterior ends of ribs, nose, parts of larynx, trachea, bronchi, bronchial tubes, and embryonic and fetal skeleton.

Function: Provides smooth surfaces for movement at joints, as well as flexibility and support.

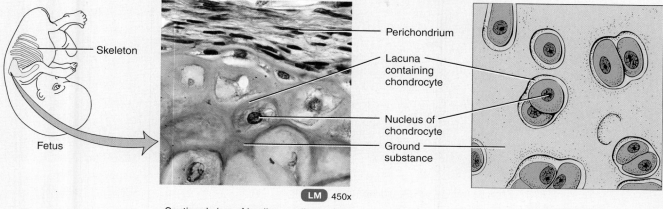

Skeleton

Fetus

Perichondrium

Lacuna containing chondrocyte

Nucleus of chondrocyte

Ground substance

LM 450x

Sectional view of hyaline cartilage of a developing fetal bone

Hyaline cartilage

H. Fibrocartilage

Description: Consists of chondrocytes scattered among bundles of collagen fibers within the matrix.

Location: Pubic symphysis (point where hip bones join anteriorly), intervertebral discs (discs between vertebrae), menisci (cartilage pads) of knee, and portions of tendons that insert into cartilage.

Function: Support and fusion.

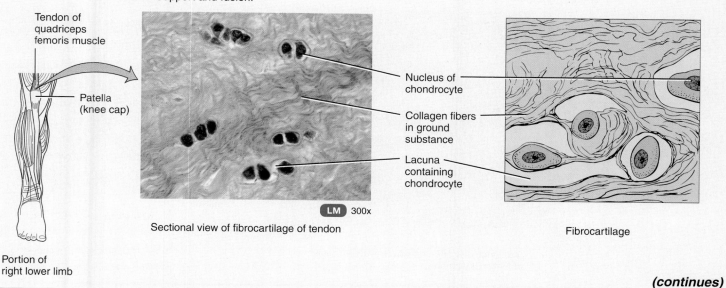

Tendon of quadriceps femoris muscle

Patella (knee cap)

Nucleus of chondrocyte

Collagen fibers in ground substance

Lacuna containing chondrocyte

LM 300x

Sectional view of fibrocartilage of tendon

Fibrocartilage

Portion of right lower limb

(continues)

Cartilage

Cartilage consists of a dense network of collagen fibers and elastic fibers firmly embedded in chondroitin sulfate, a gel-like component of the ground substance. Cartilage can endure considerably more stress than loose and dense connective tissues. Whereas the strength of cartilage is due to its collagen fibers, its resilience (ability to assume its original shape after deformation) is due to chondroitin sulfate.

Table 4.3	Mature Connective Tissues *(continued)*

Cartilage

I. Elastic cartilage

Description: Consists of chondrocytes located in a threadlike network of elastic fibers within the matrix.

Location: Lid on top of larynx (epiglottis), part of external ear (auricle), and auditory (Eustachian) tubes.

Function: Gives support and maintains shape.

Auricle of ear

Perichondrium

Nucleus of chondrocyte in lacuna

Elastic fiber in ground substance

LM 250x

Sectional view of elastic cartilage of auricle of ear

Elastic cartilage

Bone Tissue

J. Compact bone

Description: Compact bone tissue consists of osteons (Haversian systems) that contain lamellae, lacunae, osteocytes, canaliculi, and central (Haversian) canals. By contrast, spongy bone tissue (see Figure 6.3 on page 167) consists of thin columns called trabeculae; spaces between trabeculae are filled with red bone marrow.

Location: Both compact and spongy bone tissue make up the various parts of bones of the body.

Function: Support, protection, storage; houses blood-forming tissue; serves as levers that act together with muscle tissue to enable movement.

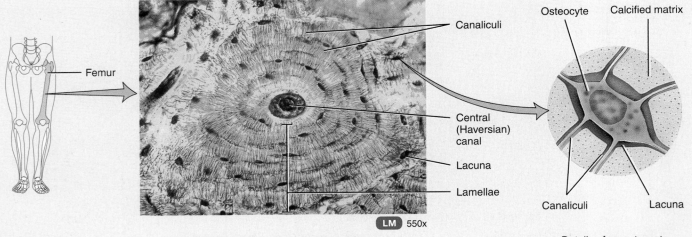

Femur

Canaliculi

Osteocyte   Calcified matrix

Central (Haversian) canal

Lacuna

Lamellae

Canaliculi   Lacuna

LM 550x

Sectional view of an osteon (Haversian system) of femur (thigh bone)

Details of an osteocyte

The cells of mature cartilage, called **chondrocytes** (KON-drō-sīts; *chondro-* = cartilage), occur singly or in groups within spaces called **lacunae** (la-KOO-nē = little lakes; singular is *lacuna*) in the matrix. A membrane of dense irregular connective tissue called the **perichondrium** (per′-i-KON-drē-um; *peri-* = around) covers the surface of most cartilage. Unlike other connective tissues, cartilage has no blood vessels or nerves, except in the perichondrium. Three kinds of cartilage are recognized: hyaline cartilage, fibrocartilage, and elastic cartilage.

HYALINE CARTILAGE   This type of cartilage contains a resilient gel as its ground substance and appears in the body as a bluish-

Blood Tissue

K. Blood

Description: Consists of blood plasma and formed elements: red blood cells (erythrocytes), white blood cells (leukocytes), and platelets (thrombocytes).

Location: Within blood vessels (arteries, arterioles, capillaries, venules, and veins) and within the chambers of the heart.

Function: Red blood cells transport oxygen and carbon dioxide; white blood cells carry on phagocytosis and are involved in allergic reactions and immune system responses; platelets are essential for the clotting of blood.

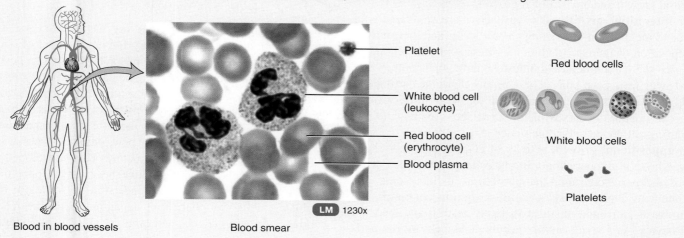

Blood in blood vessels

Blood smear

LM 1230x

Platelet

White blood cell (leukocyte)

Red blood cell (erythrocyte)

Blood plasma

Red blood cells

White blood cells

Platelets

white, shiny substance. The fine collagen fibers are not visible with ordinary staining techniques, and prominent chondrocytes are found in lacunae (Table 4.3G). Most hyaline cartilage is surrounded by a perichondrium. The exceptions are the articular cartilage in joints and at the epiphyseal plates, the regions where bones lengthen as a person grows. Hyaline cartilage is the most abundant cartilage in the body. It affords flexibility and support and, at joints, reduces friction and absorbs shock. Hyaline cartilage is the weakest of the three types of cartilage.

FIBROCARTILAGE Chondrocytes are scattered among clearly visible bundles of collagen fibers within the matrix of this type of cartilage (Table 4.3H). Fibrocartilage lacks a perichondrium. This tissue combines strength and rigidity and is the strongest of the three types of cartilage. One location of fibrocartilage is the intervertebral discs, the disk-shaped material between the backbones.

ELASTIC CARTILAGE   In this tissue, chondrocytes are located within a threadlike network of elastic fibers within the matrix (Table 4.3I). A perichondrium is present. Elastic cartilage provides strength and elasticity and maintains the shape of certain structures, such as the external ear.

REPAIR AND GROWTH OF CARTILAGE   Metabolically, cartilage is a relatively inactive tissue that grows slowly. When injured or inflamed, cartilage repair proceeds slowly, in large part because cartilage is avascular. Substances needed for repair and blood cells that participate in tissue repair must diffuse or migrate into the cartilage. The growth of cartilage follows two basic patterns: interstitial growth and appositional growth.

In **interstitial growth,** the cartilage increases rapidly in size due to the division of existing chondrocytes and the continuous deposition of increasing amounts of matrix by the chondrocytes. As the chondrocytes synthesize new matrix, they are pushed away from each other. These events cause the cartilage to expand from within, which is the reason for the term *inter*stitial. This growth pattern occurs while the cartilage is young and pliable, during childhood and adolescence.

In **appositional growth,** activity of cells in the inner chondrogenic layer of the perichondrium leads to growth. The deeper cells of the perichondrium, the fibroblasts, divide; some differentiate into chondroblasts. As differentiation continues, the chondroblasts surround themselves with matrix and become chondrocytes. As a result, matrix accumulates beneath the perichondrium on the outer surface of the cartilage, causing it to grow in width. Appositional growth starts later than interstitial growth and continues through adolescence.

Bone Tissue

Cartilage, joints, and bones make up the skeletal system. The skeletal system supports soft tissues, protects delicate structures, and works with skeletal muscles to generate movement. Bones store calcium and phosphorus; house red bone marrow, which produces blood cells; and contain yellow bone marrow, a storage site for triglycerides. Bones are organs composed of several different connective tissues, including **bone** or **osseous tissue** (OS-ē-us), the periosteum, red and yellow bone marrow, and the endosteum (a membrane that lines a space within bone that stores yellow bone marrow). Bone tissue is classified as either compact or spongy, depending on how its matrix and cells are organized.

The basic unit of **compact bone** is an **osteon** or **Haversian system** (Table 4.3J). Each osteon has four parts:

1.   The **lamellae** (la-MEL-lē = little plates) are concentric rings of matrix that consist of mineral salts (mostly calcium and phosphates), which give bone its hardness, and collagen fibers,

which give bone its strength. The lamellae are responsible for the compact nature of this type of bone tissue.

2.   **Lacunae** are small spaces between lamellae that contain mature bone cells called **osteocytes.**

3.   Projecting from the lacunae are **canaliculi** (KAN-a-lik-ū-lī = little canals), networks of minute canals containing the processes of osteocytes. Canaliculi provide routes for nutrients to reach osteocytes and for wastes to leave them.

4.   A **central (Haversian) canal** contains blood vessels and nerves.

Spongy bone lacks osteons. Rather, it consists of columns of bone called **trabeculae** (tra-BEK-ū-lē = little beams), which contain lamellae, osteocytes, lacunae, and canaliculi. Spaces between lamellae are filled with red bone marrow. Chapter 6 presents bone tissue histology in more detail.

Tissue Engineering

The technology of **tissue engineering** has allowed scientists to grow new tissues in the laboratory for replacement of damaged tissues in the body. Tissue engineers have already developed laboratory-grown versions of skin and cartilage. In the procedure, scaffolding beds of biodegradable synthetic materials or collagen are used as substrates that permit body cells such as skin cells or cartilage cells to be cultured. As the cells divide and assemble, the scaffolding degrades, and the new, permanent tissue is then implanted in the patient. Other structures being developed by tissue engineers include bones, tendons, heart valves, bone marrow, and intestines. Work is also underway to develop insulin-producing cells for diabetics, dopamine-producing cells for Parkinson disease patients, and even entire livers and kidneys. ∎

Blood Tissue

Blood tissue (or simply blood) is a connective tissue with a liquid matrix called **blood plasma,** a pale yellow fluid that consists mostly of water with a wide variety of dissolved substances — nutrients, wastes, enzymes, plasma proteins, hormones, respiratory gases, and ions (Table 4.3K). Suspended in the blood plasma are formed elements — red blood cells (erythrocytes), white blood cells (leukocytes), and platelets (thrombocytes). **Red blood cells** transport oxygen to body cells and remove carbon dioxide from them. **White blood cells** are involved in phagocytosis, immunity, and allergic reactions. **Platelets** participate in blood clotting. The details of blood are considered in Chapter 19.

Lymph

Lymph is the extracellular fluid that flows in lymphatic vessels. It is a connective tissue that consists of a clear fluid similar to blood plasma but with much less protein. In the lymph are various cells and chemicals, whose composition varies from one part of the body to another. For example, lymph leaving lymph nodes includes many lymphocytes, a type of white blood cell, whereas lymph from the small intestine has a high content of newly absorbed dietary lipids. The details of lymph are considered in Chapter 22.

10. In what ways do connective tissues differ from epithelia?

11. What are the features of the cells, ground substance, and fibers that make up connective tissue?

12. How are connective tissues classified? List the various types.

13. How are the structures of the following connective tissues related to their functions: areolar connective tissue, adipose tissue, reticular connective tissue, dense regular connective tissue, dense irregular connective tissue, elastic connective tissue, hyaline cartilage, fibrocartilage, elastic cartilage, and bone tissue?

14. How are interstitial and appositional growth of cartilage different?

MEMBRANES

► OBJECTIVES

- **Define a membrane.**
- **Describe the classification of membranes.**

Membranes are flat sheets of pliable tissue that cover or line a part of the body. The combination of an epithelial layer and an underlying connective tissue layer constitutes an **epithelial membrane.** The principal epithelial membranes of the body are mucous membranes, serous membranes, and the cutaneous membrane, or skin. (Skin is discussed in detail in Chapter 5, and will not be discussed here.) Another kind of membrane, a **synovial membrane,** lines joints and contains connective tissue but no epithelium.

Epithelial Membranes

Mucous Membranes

A **mucous membrane** or **mucosa** lines a body cavity that opens directly to the exterior. Mucous membranes line the entire digestive, respiratory, and reproductive tracts, and much of the urinary tract. They consist of both a lining layer of epithelium and an underlying layer of connective tissue.

The epithelial layer of a mucous membrane is an important feature of the body's defense mechanisms because it is a barrier that microbes and other pathogens have difficulty penetrating. Usually, tight junctions connect the cells, so materials cannot leak in between them. Goblet cells and other cells of the epithelial layer of a mucous membrane secrete mucus, and this slippery fluid prevents the cavities from drying out. It also traps particles in the respiratory passageways and lubricates food as it moves through the gastrointestinal tract. In addition, the epithelial layer secretes some of the enzymes needed for digestion and is the site of food and fluid absorption in the gastrointestinal tract. The epithelia of mucous membranes vary greatly in different parts of the body. For example, the epithelium of the mucous membrane of the small intestine is nonciliated simple columnar (see Table 4.1C), whereas the epithelium

of the large airways to the lungs is pseudostratified ciliated columnar (see Table 4.1I).

The connective tissue layer of a mucous membrane is areolar connective tissue and is called the **lamina propria** (LAM-i-na PRŌ-prē-a). The lamina propria is so named because it belongs to the mucous membrane (*propria* = one's own). The lamina propria supports the epithelium, binds it to the underlying structures, and allows some flexibility of the membrane. It also holds blood vessels in place and protects underlying muscles from abrasion or puncture. Oxygen and nutrients diffuse from the lamina propria to the epithelium covering it whereas carbon dioxide and wastes diffuse in the opposite direction.

Serous Membranes

A **serous membrane** (*serous* = watery) or **serosa** lines a body cavity that does not open directly to the exterior, and it covers the organs that lie within the cavity. Serous membranes consist of areolar connective tissue covered by mesothelium (simple squamous epithelium). Serous membranes have two layers: The layer attached to the cavity wall is called the **parietal layer** (pa-RĪ-e-tal; *pariet-* = wall); the layer that covers and attaches to the organs inside the cavity is the **visceral layer** (*viscer-* = body organ). The mesothelium of a serous membrane secretes **serous fluid,** a watery lubricating fluid that allows organs to glide easily over one another or to slide against the walls of cavities.

The serous membrane lining the thoracic cavity and covering the lungs is the **pleura.** The serous membrane lining the heart cavity and covering the heart is the **pericardium.** The serous membrane lining the abdominal cavity and covering the abdominal organs is the **peritoneum.**

Synovial Membranes

Synovial membranes (sin-Ō-vē-al; *syn-* = together, referring here to a place where bones come together) line the cavities of freely movable joints. Like serous membranes, synovial membranes line structures that do not open to the exterior. Unlike mucous, serous, and cutaneous membranes, they lack epithelium and are therefore not epithelial membranes. Synovial membranes are composed of areolar connective tissue with elastic fibers and varying numbers of adipocytes. Synovial membranes secrete **synovial fluid,** which lubricates and nourishes the cartilage covering the bones at movable joints; these membranes are called *articular synovial membranes.* Other synovial membranes line cushioning sacs, called *bursae,* and *tendon sheaths* in the hands and feet, thus easing the movement of muscle tendons.

► CHECKPOINT

15. Define the following kinds of membranes: mucous, serous, cutaneous, and synovial.

16. Where is each type of membrane located in the body? What are their functions?

MUSCLE TISSUE

OBJECTIVES

- **Describe the general features of muscle tissue.**
- **Contrast the structure, location, and mode of control of skeletal, cardiac, and smooth muscle tissue.**

Muscle tissue consists of elongated cells called *muscle fibers* that can use ATP to generate force. As a result, muscle tissue produces body movements, maintains posture, and generates heat. Based on its location and certain structural and functional features, muscle tissue is classified into three types: skeletal, cardiac, and smooth (Table 4.4).

Skeletal muscle tissue is named for its location—usually attached to the bones of the skeleton (Table 4.4A). It is also *striated;* the fibers contain alternating light and dark bands called *striations* that are visible under a light microscope. Skeletal muscle is *voluntary* because it can be made to contract or relax by conscious control. A single skeletal muscle fiber is very long, roughly cylindrical in shape, and has many nuclei located at the

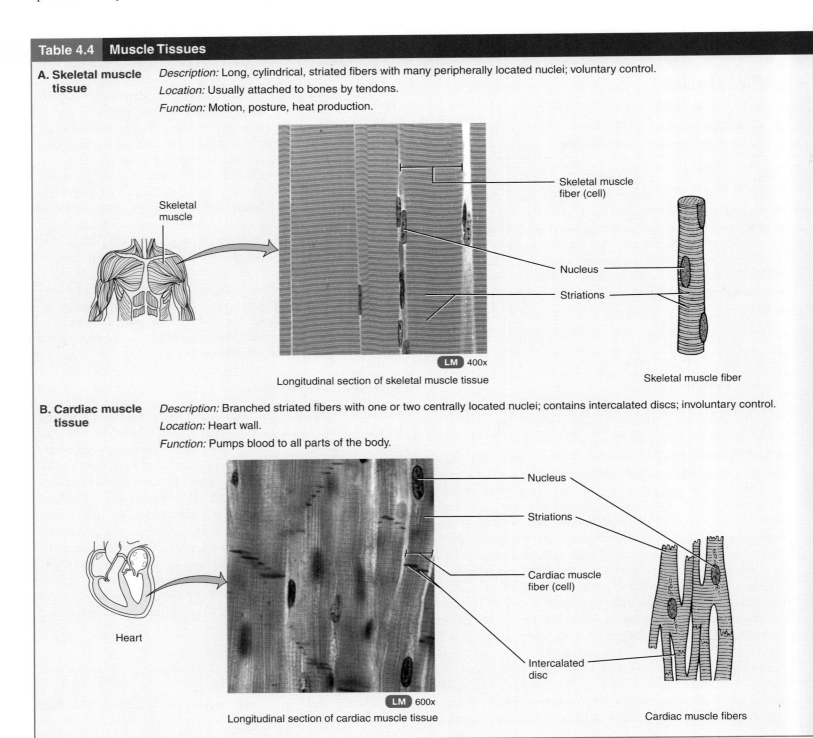

Table 4.4	Muscle Tissues

A. Skeletal muscle tissue

Description: Long, cylindrical, striated fibers with many peripherally located nuclei; voluntary control.

Location: Usually attached to bones by tendons.

Function: Motion, posture, heat production.

Skeletal muscle

Skeletal muscle fiber (cell)

Nucleus

Striations

LM 400x

Longitudinal section of skeletal muscle tissue

Skeletal muscle fiber

B. Cardiac muscle tissue

Description: Branched striated fibers with one or two centrally located nuclei; contains intercalated discs; involuntary control.

Location: Heart wall.

Function: Pumps blood to all parts of the body.

Heart

Nucleus

Striations

Cardiac muscle fiber (cell)

Intercalated disc

LM 600x

Longitudinal section of cardiac muscle tissue

Cardiac muscle fibers

periphery of the cell. Within a whole muscle, the individual muscle fibers are parallel to each other.

Cardiac muscle tissue forms most of the wall of the heart (Table 4.4B). Like skeletal muscle, it is striated. However, unlike skeletal muscle tissue, it is *involuntary;* its contraction is not consciously controlled. Cardiac muscle fibers are branched and usually have only one centrally located nucleus, although an occasional cell has two nuclei. They attach end to end to each other by transverse thickenings of the plasma membrane called **intercalated discs** (*intercalat-* = to insert between), which con-

tain both desmosomes and gap junctions. Intercalated discs are unique to cardiac muscle. The desmosomes strengthen the tissue and hold the fibers together during their vigorous contractions. The gap junctions provide a route for quick conduction of muscle action potentials throughout the heart.

Smooth muscle tissue is located in the walls of hollow internal structures such as blood vessels, airways to the lungs, the stomach, intestines, gallbladder, and urinary bladder (Table 4.4C). Its contraction helps constrict or narrow the lumen of blood vessels, physically break down and move food along

C. Smooth muscle tissue

Description: Spindle-shaped (thickest in middle and tapering at both ends), nonstriated fibers with one centrally located nucleus; involuntary control.

Location: Walls of hollow internal structures such as blood vessels, airways to the lungs, stomach, intestines, gallbladder, urinary bladder, and uterus.

Function: Motion (constriction of blood vessels and airways, propulsion of foods through gastrointestinal tract, contraction of urinary bladder and gallbladder).

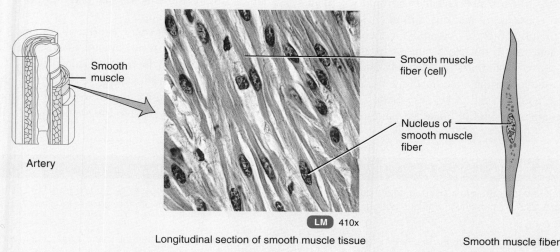

Artery

Smooth muscle

Smooth muscle fiber (cell)

Nucleus of smooth muscle fiber

LM 410x

Longitudinal section of smooth muscle tissue

Smooth muscle fiber

the gastrointestinal tract, move fluids through the body, and eliminate wastes. Smooth muscle fibers are usually *involuntary,* and they are nonstriated (lack striations), hence the term *smooth.* A smooth muscle fiber is small, thickest in the middle, and tapering at each end. It contains a single, centrally located nucleus. Gap junctions connect many individual fibers in some smooth muscle tissues, for example, in the wall of the intestines. Such muscle tissues can produce powerful contractions as many muscle fibers contract in unison. In other locations such as the iris of the eye, smooth muscle fibers contract individually, like skeletal muscle fibers, because gap junctions are absent. Chapter 10 provides a detailed discussion of muscle tissue.

▶ CHECKPOINT

17. Which muscles are striated and which are smooth?

18. Which types of muscle tissue have gap junctions?

NERVOUS TISSUE

▶ OBJECTIVE

• **Describe the structural features and functions of nervous tissue.**

Despite the awesome complexity of the nervous system, it consists of only two types of cells: neurons and neuroglia. **Neurons,** or nerve cells, are sensitive to various stimuli. They convert stimuli into nerve impulses (action potentials) and conduct these impulses to other neurons, to muscle tissue, or to glands. Most neurons consist of three basic parts: a cell body and two kinds of cell processes—dendrites and axons (Table 4.5). The **cell body** contains the nucleus and other organelles. **Dendrites** (*dendr-* = tree) are tapering, highly branched, and usually short

cell processes. They are the major receiving or input portion of a neuron. The **axon** (*axo-* = axis) of a neuron is a single, thin, cylindrical process that may be very long. It is the output portion of a neuron, conducting nerve impulses toward another neuron or to some other tissue.

Even though **neuroglia** (noo-RŌG-lē-a; *-glia* = glue) do not generate or conduct nerve impulses, these cells do have many important supportive functions (see Table 12.1 on page 392). The detailed structure and function of neurons and neuroglia are considered in Chapter 12.

▶ CHECKPOINT

19. What are the functions of the dendrites, cell body, and axon of a neuron?

TISSUE REPAIR: RESTORING HOMEOSTASIS

▶ OBJECTIVE

• **Describe the role of tissue repair in restoring homeostasis.**

Tissue repair is the process that replaces worn-out, damaged, or dead cells. New cells originate by cell division from the **stroma,** the supporting connective tissue, or from the **parenchyma,** cells that constitute the functioning part of the tissue or organ. In adults, each of the four basic tissue types (epithelial, connective, muscle, and nervous) has a different capacity for replenishing parenchymal cells lost by damage, disease, or other processes.

Epithelial cells, which endure considerable wear and tear (and even injury) in some locations, have a continuous capacity for renewal. In some cases, immature, undifferentiated cells called **stem cells** divide to replace lost or damaged cells. For example,

Table 4.5	Nervous Tissue

Description: Consists of neurons (nerve cells) and neuroglia. Neurons consist of a cell body and processes extending from the cell body (multiple dendrites and a single axon). Neuroglia do not generate or conduct nerve impulses but have other important functions.

Location: Nervous system.

Function: Exhibits sensitivity to various types of stimuli, converts stimuli into nerve impulses (action potentials), and conducts nerve impulses to other neurons, muscle fibers, or glands.

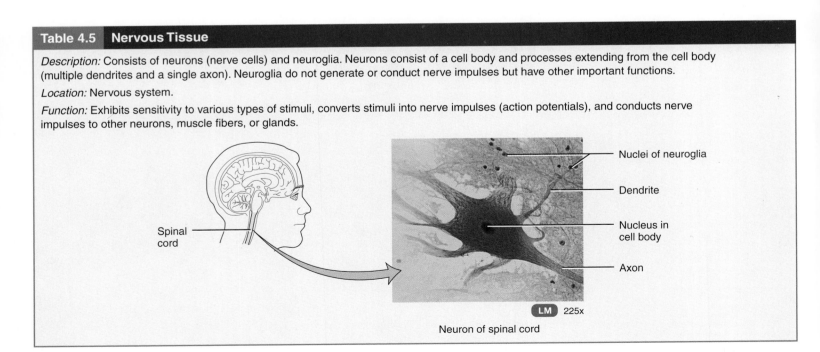

Spinal cord

Nuclei of neuroglia

Dendrite

Nucleus in cell body

Axon

LM   225x

Neuron of spinal cord

stem cells reside in protected locations in the epithelia of the skin and gastrointestinal tract to replenish cells sloughed from the apical layer, and stem cells in red bone marrow continually provide new red and white blood cells and platelets. In other cases, mature, differentiated cells can undergo cell division; examples are hepatocytes (liver cells) and endothelial cells in blood vessels.

Some connective tissues also have a continuous capacity for renewal. One example is bone, which has an ample blood supply. Other connective tissues such as cartilage can replenish cells less readily in part because of a smaller blood supply.

Muscle tissue has a relatively poor capacity for renewal of lost cells. Even though skeletal muscle tissue contains stem cells called *satellite cells,* they do not divide rapidly enough to replace extensively damaged muscle fibers. Cardiac muscle tissue lacks satellite cells, and existing cardiac muscle fibers do not undergo mitosis to form new cells. Recent evidence suggests, however, that stem cells do migrate into the heart from the blood. There, they can differentiate and replace a limited number of thyroid cardiac muscle fibers and endothelial cells in heart blood vessels. Smooth muscle fibers can proliferate to some extent, but they do so much more slowly than the cells of epithelial or connective tissues.

Nervous tissue has the poorest capacity for renewal. Although experiments have revealed the presence of some stem cells in the brain, they normally do not undergo mitosis to replace damaged neurons. Discovering why this is so is a major goal of researchers who seek ways to encourage repair of damaged or diseased nervous tissue.

The restoration of an injured tissue or organ to normal structure and function depends entirely on whether parenchymal cells are active in the repair process. If parenchymal cells accomplish the repair, **tissue regeneration** is possible, and a near-perfect reconstruction of the injured tissue may occur. However, if fibroblasts of the stroma are active in the repair, the replacement tissue will be a new connective tissue. The fibroblasts synthesize collagen and other matrix materials that aggregate to form scar tissue, a process known as **fibrosis.** Because scar tissue is not specialized to perform the functions of the parenchymal tissue, the original function of the tissue or organ is impaired.

When tissue damage is extensive, as in large, open wounds, both the connective tissue stroma and the parenchymal cells are active in repair. This repair involves the rapid cell division of many fibroblasts and the manufacture of new collagen fibers to provide structural strength. Blood capillaries also sprout new buds to supply the healing tissue the materials it needs. All these processes create an actively growing connective tissue called **granulation tissue.** This new tissue forms across a wound or surgical incision to provide a framework (stroma) that supports the epithelial cells that migrate into the open area and fill it. The newly formed granulation tissue also secretes a fluid that kills bacteria.

Three factors affect tissue repair: nutrition, blood circulation, and age. Nutrition is vital because the healing process places a great demand on the body's store of nutrients. Adequate protein in the diet is important because most of the structural components of a tissue are proteins. Several vitamins also play a direct role in wound healing and tissue repair. For example, vita-min C directly affects the normal production and maintenance of matrix materials, especially collagen. Vitamin C also strengthens and promotes the formation of new blood vessels. With vitamin C deficiency, even superficial wounds fail to heal, and the walls of the blood vessels become fragile and are easily ruptured.

In tissue repair, proper blood circulation is essential to transport oxygen, nutrients, antibodies, and many defensive cells to the injured site. The blood also plays an important role in the removal of tissue fluid, bacteria, foreign bodies, and debris, elements that would otherwise interfere with healing.

Adhesions

Scar tissue can form **adhesions,** abnormal joining of tissues. Adhesions commonly form in the abdomen around a site of previous inflammation such as an inflamed appendix, and they can develop after surgery. Although adhesions do not always cause problems, they can decrease tissue flexibility, cause obstruction (such as in the intestine), and make a subsequent operation more difficult. An *adhesiotomy,* the surgical release of adhesions, may be required. ■

AGING AND TISSUES

▶ O B J E C T I V E

- **Describe the effects of aging on tissues.**

Generally, tissues heal faster and leave less obvious scars in the young than in the aged. In fact, surgery performed on fetuses leaves no scars. The younger body is generally in a better nutritional state, its tissues have a better blood supply, and its cells have a higher metabolic rate. Thus, cells can synthesize needed materials and divide more quickly. The extracellular components of tissues also change with age. Glucose, the most abundant sugar in the body, plays a role in the aging process. Glucose is haphazardly added to proteins inside and outside cells, forming irreversible cross-links between adjacent protein molecules. With advancing age, more cross-links form, which contributes to the stiffening and loss of elasticity that occur in aging tissues. Collagen fibers, responsible for the strength of tendons, increase in number and change in quality with aging. These changes in the collagen of arterial walls are as much responsible for their loss of extensibility as are the deposits associated with atherosclerosis, the deposition of fatty materials in arterial walls. Elastin, another extracellular component, is responsible for the elasticity of blood vessels and skin. It thickens, fragments, and acquires a greater affinity for calcium with age—changes that may also be associated with the development of atherosclerosis.

▶ C H E C K P O I N T

20. How are stromal and parenchymal repair of a tissue different?

21. What is the importance of granulation tissue?

22. What common changes occur in epithelial and connective tissues with aging?

DISORDERS: HOMEOSTATIC IMBALANCES

Disorders of epithelial tissues are mainly specific to individual organs, for example, peptic ulcer disease (PUD), which erodes the epithelial lining of the stomach or small intestine. For this reason, epithelial disorders are described together with the relevant body system throughout the text. The most prevalent disorders of connective tissues are **autoimmune diseases**—diseases in which antibodies produced by the immune system fail to distinguish what is foreign from what is self and attack the body's own tissues. The most common autoimmune disorder is rheumatoid arthritis, which attacks the synovial membranes of joints. Because connective tissue is one of the most abundant and widely distributed of the four main types of tissues, its disorders often affect multiple body systems. Common disorders of muscle tissue and nervous tissue are described at the ends of Chapters 10 and 12, respectively.

Sjögren's Syndrome

Sjögren's syndrome (SHŌ-grenz) is a common autoimmune disorder that causes inflammation and destruction of exocrine glands, especially the lacrimal (tear) glands and salivary glands. Signs include dryness of the mucous membranes in the eyes and mouth and salivary gland enlargement. Systemic effects include arthritis, difficulty swallowing, pancreatitis (inflammation of the pancreas), pleuritis (inflammation of the pleurae of the lungs), and migraine headaches. The disorder affects more females than males by a ratio of 9 to 1. About 20% of older adults experience some signs of Sjögren's. Treatment is supportive, including using artificial tears to moisten the eyes, sipping fluids, chewing sugarless gum, and using a saliva substitute to moisten the mouth.

Systemic Lupus Erythematosus

Systemic lupus erythematosus (er-i-the-ma-TŌ-sus), **SLE,** or simply lupus, is a chronic inflammatory disease of connective tissue occurring mostly in nonwhite women during their childbearing years. It is an autoimmune disease that can cause tissue damage in every body system. The disease, which can range from a mild condition in most patients to a rapidly fatal disease, is marked by periods of exacerbation and remission. The prevalence of SLE is about 1 in 2000 persons, with females more likely to be afflicted than males by a ratio of 8 or 9 to 1.

Although the cause of SLE is not known, genetic, environmental, and hormonal factors are implicated. The genetic component is suggested by studies of twins and family history. Environmental factors include viruses, bacteria, chemicals, drugs, exposure to excessive sunlight, and emotional stress. With regard to hormones, sex hormones, such as estrogens, may trigger SLE.

Signs and symptoms of SLE include painful joints, low-grade fever, fatigue, mouth ulcers, weight loss, enlarged lymph nodes and spleen, sensitivity to sunlight, rapid loss of large amounts of scalp hair, and anorexia. A distinguishing feature of lupus is an eruption across the bridge of the nose and cheeks called a "butterfly rash." Other skin lesions may occur, including blistering and ulceration. The erosive nature of some SLE skin lesions was thought to resemble the damage inflicted by the bite of a wolf—thus, the term *lupus* (= wolf). The most serious complications of the disease involve inflammation of the kidneys, liver, spleen, lungs, heart, brain, and gastrointestinal tract. Because there is no cure for SLE, treatment is supportive, including anti-inflammatory drugs, such as aspirin, and immunosuppressive drugs.

MEDICAL TERMINOLOGY

Atrophy   (AT-rō-fē; *a-* = without; *-trophy* = nourishment) A decrease in the size of cells, with a subsequent decrease in the size of the affected tissue or organ.

Hypertrophy   (hī-PER-trō-fē; *hyper-* = above or excessive) Increase in the size of a tissue because its cells enlarge without undergoing cell division.

Tissue rejection   An immune response of the body directed at foreign proteins in a transplanted tissue or organ; immunosuppressive drugs, such as cyclosporine, have largely overcome tissue rejection in heart-, kidney-, and liver-transplant patients.

Tissue transplantation   The replacement of a diseased or injured tissue or organ. The most successful transplants involve use of a person's own tissues or those from an identical twin.

Xenotransplantation   (zen′-ō-trans-plan-TĀ-shun;   *xeno-* = strange, foreign) The replacement of a diseased or injured tissue or organ with cells or tissues from an animal. Porcine (from pigs) and bovine (from cows) heart valves are used for some heart-valve replacement surgeries.

STUDY OUTLINE

TYPES OF TISSUES AND THEIR ORIGINS (p. 104)

1.   A tissue is a group of similar cells that usually has a similar embryological origin and is specialized for a particular function.
2.   The various tissues of the body are classified into four basic types: epithelial, connective, muscle, and nervous.
3.   All tissues of the body develop from three primary germ layers,

the first tissues that form in a human embryo: ectoderm, mesoderm, and endoderm.

CELL JUNCTIONS (p. 104)

1.   Cell junctions are points of contact between adjacent plasma membranes.

2. Tight junctions form fluid-tight seals between cells; adherens junctions, desmosomes, and hemidesmosomes anchor cells to one another or to the basement membrane; and gap junctions permit electrical and chemical signals to pass between cells.

EPITHELIAL TISSUE (p. 106)

1. The subtypes of epithelia include covering and lining epithelia and glandular epithelia.
2. An epithelium consists mostly of cells with little extracellular material between adjacent plasma membranes. The apical, lateral, and basal surfaces of epithelial cells are modified in various ways to carry out specific functions. Epithelium is arranged in sheets and attached to a basement membrane. Although it is avascular, it has a nerve supply. Epithelia are derived from all three primary germ layers and have a high capacity for renewal.
3. Epithelial layers are simple (one layer), stratified (several layers), or pseudostratified (one layer that appears to be several). The cell shapes may be squamous (flat), cuboidal (cubelike), columnar (rectangular), or transitional (variable).
4. Simple squamous epithelium consists of a single layer of flat cells (Table 4.1A). It is found in parts of the body where filtration or diffusion are priority processes. One type, endothelium, lines the heart and blood vessels. Another type, mesothelium, forms the serous membranes that line the thoracic and abdominopelvic cavities and cover the organs within them.
5. Simple cuboidal epithelium consists of a single layer of cube-shaped cells that function in secretion and absorption (Table 4.1B). It is found covering the ovaries, in the kidneys and eyes, and lining some glandular ducts.
6. Nonciliated simple columnar epithelium consists of a single layer of nonciliated rectangular cells (Table 4.1C). It lines most of the gastrointestinal tract. Specialized cells containing microvilli perform absorption. Goblet cells secrete mucus.
7. Ciliated simple columnar epithelium consists of a single layer of ciliated rectangular cells (Table 4.1D). It is found in a few portions of the upper respiratory tract, where it moves foreign particles trapped in mucus out of the respiratory tract.
8. Stratified squamous epithelium consists of several layers of cells; cells of the apical layer and several layers deep to it are flat (Table 4.1E). A nonkeratinized variety lines the mouth; a keratinized variety forms the epidermis, the most superficial layer of the skin.
9. Stratified cuboidal epithelium consists of several layers of cells; cells at the apical layer are cube-shaped (Table 4.1F). It is found in adult sweat glands and a portion of the male urethra.
10. Stratified columnar epithelium consists of several layers of cells; cells of the apical layer have a columnar shape (Table 4.1G). It is found in a portion of the male urethra and large excretory ducts of some glands.
11. Transitional epithelium consists of several layers of cells whose appearance varies with the degree of stretching (Table 4.1H). It lines the urinary bladder.
12. Pseudostratified columnar epithelium has only one layer but gives the appearance of many (Table 4.1I). A ciliated variety contains goblet cells and lines most of the upper respiratory tract; a nonciliated variety has no goblet cells and lines ducts of many glands, the epididymis, and part of the male urethra.
13. A gland is a single cell or a group of epithelial cells adapted for secretion.
14. Endocrine glands secrete hormones into the blood (Table 4.1J).
15. Exocrine glands (mucous, sweat, oil, and digestive glands) secrete into ducts or directly onto a free surface (Table 4.1K).
16. Structural classification of exocrine glands includes unicellular and multicellular glands.
17. Functional classification of exocrine glands includes holocrine, apocrine, and merocrine glands.

CONNECTIVE TISSUE (p. 117)

1. Connective tissue is one of the most abundant body tissues.
2. Connective tissue consists of cells and a matrix of ground substance and fibers; it has abundant matrix with relatively few cells. It does not usually occur on free surfaces, has a nerve supply (except for cartilage), and is highly vascular (except for cartilage, tendons, and ligaments).
3. Cells in connective tissue are derived from mesenchymal cells.
4. Cell types include fibroblasts (secrete matrix), macrophages (perform phagocytosis), plasma cells (secrete antibodies), mast cells (produce histamine), adipocytes (store fat), and white blood cells (migrate from blood in response to infections).
5. The ground substance and fibers make up the matrix.
6. The ground substance supports and binds cells together, provides a medium for the exchange of materials, and is active in influencing cell functions.
7. Substances found in the ground substance include water and polysaccharides such as hyaluronic acid, chondroitin sulfate, dermatan sulfate, and keratan sulfate (glycosaminoglycans). Also present are proteoglycans and adhesion proteins.
8. The fibers in the matrix provide strength and support and are of three types. (a) Collagen fibers (composed of collagen) are found in large amounts in bone, tendons, and ligaments. (b) Elastic fibers (composed of elastin, fibrillin, and other glycoprotein) are found in skin, blood vessel walls, and lungs. (c) Reticular fibers (composed of collagen and glycoprotein) are found around fat cells, nerve fibers, and skeletal and smooth muscle cells.
9. The two major subclasses of connective tissue are embryonic connective tissue (found in the embryo and fetus) and mature connective tissue (present in the newborn).
10. The embryonic connective tissues are mesenchyme, which forms all other connective tissues (Table 4.2A), and mucous connective tissue, found in the umbilical cord of the fetus, where it gives support (Table 4.2B).
11. Mature connective tissue differentiates from mesenchyme. It is subdivided into several kinds: loose or dense connective tissue, cartilage, bone tissue, blood tissue, and lymph.
12. Loose connective tissue includes areolar connective tissue, adipose tissue, and reticular connective tissue.
13. Areolar connective tissue consists of the three types of fibers, several types of cells, and a semifluid ground substance (Table 4.3A). It is found in the subcutaneous layer, in mucous membranes, and around blood vessels, nerves, and body organs.
14. Adipose tissue consists of adipocytes, which store triglycerides (Table 4.3B). It is found in the subcutaneous layer, around organs, and in yellow bone marrow. Brown adipose tissue (BAT) generates heat.
15. Reticular connective tissue consists of reticular fibers and reticular cells and is found in the liver, spleen, and lymph nodes (Table 4.3C).
16. Dense connective tissue includes dense regular connective tissue, dense irregular connective tissue, and elastic connective tissue.

17. Dense regular connective tissue consists of parallel bundles of collagen fibers and fibroblasts (Table 4.3D). It forms tendons, most ligaments, and aponeuroses.
18. Dense irregular connective tissue consists of usually randomly arranged collagen fibers and a few fibroblasts (Table 4.3E). It is found in fasciae, the dermis of skin, and membrane capsules around organs.
19. Elastic connective tissue consists of branching elastic fibers and fibroblasts (Table 4.3F). It is found in the walls of large arteries, lungs, trachea, and bronchial tubes.
20. Cartilage contains chondrocytes and has a rubbery matrix (chondroitin sulfate) containing collagen and elastic fibers.
21. Hyaline cartilage is found in the embryonic skeleton, at the ends of bones, in the nose, and in respiratory structures (Table 4.3G). It is flexible, allows movement, and provides support.
22. Fibrocartilage is found in the pubic symphysis, intervertebral discs, and menisci (cartilage pads) of the knee joint (Table 4.3H).
23. Elastic cartilage maintains the shape of organs such as the epiglottis of the larynx, auditory (Eustachian) tubes, and external ear (Table 4.3I).
24. Cartilage enlarges by interstitial growth (from within) and appositional growth (from without).
25. Bone or osseous tissue consists of a matrix of mineral salts and collagen fibers that contribute to the hardness of bone, and osteocytes that are located in lacunae (Table 4.3J). It supports, protects, helps provide movement, stores minerals, and houses blood-forming tissue.
26. Blood tissue consists of blood plasma and formed elements—red blood cells, white blood cells, and platelets (Table 4.3K). Its cells function to transport oxygen and carbon dioxide, carry on phagocytosis, participate in allergic reactions, provide immunity, and bring about blood clotting.
27. Lymph is the extracellular fluid that flows in lymphatic vessels. It is a clear fluid similar to blood plasma but with less protein.

MEMBRANES (p. 129)

1. An epithelial membrane consists of an epithelial layer overlying a connective tissue layer. Examples are mucous, serous, and cutaneous membranes.
2. Mucous membranes line cavities that open to the exterior, such as the gastrointestinal tract.

3. Serous membranes line closed cavities (pleura, pericardium, peritoneum) and cover the organs in the cavities. These membranes consist of parietal and visceral layers.
4. Synovial membranes line joint cavities, bursae, and tendon sheaths and consist of areolar connective tissue instead of epithelium.

MUSCLE TISSUE (p. 130)

1. Muscle tissue consists of fibers that are specialized for contraction. It provides motion, maintenance of posture, and heat production.
2. Skeletal muscle tissue is attached to bones and is striated, and its action is voluntary (Table 4.4A).
3. Cardiac muscle tissue forms most of the heart wall and is striated, and its action is involuntary (Table 4.4B).
4. Smooth muscle tissue is found in the walls of hollow internal structures (blood vessels and viscera) and is nonstriated, and its action is involuntary (Table 4.4C).

NERVOUS TISSUE (p. 132)

1. The nervous system is composed of neurons (nerve cells) and neuroglia (protective and supporting cells) (Table 4.5).
2. Neurons are sensitive to stimuli, convert stimuli into nerve impulses, and conduct nerve impulses.
3. Most neurons consist of a cell body and two types of processes, dendrites and axons.

TISSUE REPAIR: RESTORING HOMEOSTASIS (p. 132)

1. Tissue repair is the replacement of worn-out, damaged, or dead cells by healthy ones.
2. Stem cells may divide to replace lost or damaged cells.
3. If the injury is superficial, tissue repair involves parenchymal regeneration; if damage is extensive, granulation tissue is involved.
4. Good nutrition and blood circulation are vital to tissue repair.

AGING AND TISSUES (p. 133)

1. Tissues heal faster and leave less obvious scars in the young than in the aged; surgery performed on fetuses leaves no scars.
2. The extracellular components of tissues, such as collagen and elastic fibers, also change with age.

Q SELF-QUIZ QUESTIONS

Fill in the blanks in the following statements.

1. A group of cells that usually have a common origin and a specialized function is called a _____.
2. Secretions of _____ glands enter the interstitial fluid, and then diffuse into the bloodstream.
3. The three components that can be used to classify connective tissue are _____, _____, and _____.

Indicate whether the following statements are true or false.

4. Connective tissue fibers that are arranged in bundles and lend strength and flexibility to a tissue are collagen fibers.

5. Due to its ample blood supply, damaged cartilage is quickly repaired.

Choose the one best answer to the following questions.

6. The tissue type that can detect changes in the internal and external environments and respond to the changes is (a) nervous tissue, (b) muscle tissue, (c) connective tissue, (d) epithelial tissue.
7. Which of the following statements are *true* of epithelium? (1) The cells are arranged in continuous single- or multiple-layer sheets. (2) The attachment between the basal layer and the connective tissue is called the basement membrane. (3) This tissue has an

extensive blood supply. (4) This tissue has a high rate of cell division. (5) This tissue functions in protection, secretion, absorption, and excretion. (a) 1, 2, 3, and 4, (b) 2, 3, 4, and 5, (c) 1, 2, 4, and 5, (d) 1, 3, and 5, (e) 2, 4, and 5.

8. The type of exocrine gland that forms its secretory product and simply releases it from the cell by exocytosis is the (a) apocrine gland, (b) merocrine gland, (c) holocrine gland, (d) endocrine gland, (e) apical gland.

9. The connective tissue cells responsible for secreting fibers and ground substance are (a) macrophages, (b) mast cells, (c) fibroblasts, (d) adipocytes, (e) plasma cells.

10. The membrane lining a body cavity that opens directly to the exterior is a (a) serous membrane, (b) mucous membrane, (c) synovial membrane, (d) plasma membrane, (e) basement membrane.

11. The muscle tissue that forms the bulk of the wall of the heart is (a) skeletal muscle, (b) smooth muscle, (c) involuntary, non-striated muscle, (d) cardiac muscle, (e) striated, voluntary muscle.

12. Match the following:

_____ (a) contains a single layer of flat cells; found in the body where filtration (kidney) or diffusion (lungs) are priority processes

_____ (b) found in the superficial part of skin; provides protection from heat, microbes, and chemicals

_____ (c) contains cube-shaped cells; found in the kidney that function in secretion and absorption

_____ (d) lines the upper respiratory tract and uterine tubes; wavelike motion of cilia propel material through the lumen

_____ (e) found in the urinary bladder; contains cells that can change shape (stretch or relax)

_____ (f) contains cells with microvilli and goblet cells; found in linings of the digestive, reproductive and urinary tracts

_____ (g) contains cells that are all attached to the basement membrane, although some do not reach the surface; those cells that do extend to the surface secrete mucus or bear cilia at their apical surface

_____ (h) a fairly rare type of epithelium that has a mainly protective function

(1) pseudostratified ciliated columnar epithelium
(2) ciliated simple columnar epithelium
(3) transitional epithelium
(4) simple squamous epithelium
(5) simple cuboidal epithelium
(6) nonciliated simple columnar epithelium
(7) stratified cuboidal epithelium
(8) keratinized stratified squamous epithelium

13. Match the following:

_____ (a) the tissue from which all other connective tissues eventually arise

_____ (b) connective tissue with a clear, liquid matrix that flows in lymphatic vessels

_____ (c) connective tissue consisting of several kinds of cells, containing all three fiber types randomly arranged, and found in the subcutaneous layer of the skin

_____ (d) a loose connective tissue specialized for triglyceride storage

_____ (e) tissue that contains reticular fibers and reticular cells and forms the stroma of certain organs such as the spleen

_____ (f) tissue with irregularly arranged collagen fibers found in the dermis of the skin

_____ (g) tissue found in the lungs that is strong and can recoil back to its original shape after being stretched

_____ (h) tissue that affords flexibility at joints and reduces joint friction

_____ (i) tissue that provides strength and rigidity and is the strongest of the three types of cartilage

_____ (j) bundles of collagen arranged in parallel patterns; compose tendons and ligaments

_____ (k) tissue that forms the internal framework of the body and works with skeletal muscle to generate movement

_____ (l) connective tissue with formed elements suspended in a liquid matrix called plasma

(1) blood
(2) fibrocartilage
(3) mesenchyme
(4) dense regular connective tissue
(5) lymph
(6) hyaline cartilage
(7) dense irregular connective tissue
(8) areolar connective tissue
(9) reticular connective tissue
(10) bone (osseous tissue)
(11) elastic connective tissue
(12) adipose tissue

14. Match the following:

_____ (a) prevents organ contents from leaking into the blood or surrounding
tissues

_____ (b) forms adhesion belts that help epithelial surfaces resist separation

_____ (c) makes tissues stable by linking the
cytoskeletons of cells together

_____ (d) anchors cells to basement membrane

_____ (e) allows epithelial cells in a tissue to
communicate; enables nerve or
muscle impulses to spread rapidly
between cells

(1) gap
junction
(2) tight
junction
(3) desmosome
(4) hemidesmo-
some
(5) adherens
junction

15. Match the following:

_____ (a) result of repair of extensive
tissue damage by both fibroblasts and parenchymal cells;
provides a supporting framework for new tissue

_____ (b) near perfect repair of damaged tissue by parenchymal
cells

_____ (c) replacement of damaged tissue by collagen fibers and
other matrix materials, forming scar tissue

(1) fibrosis
(2) tissue regeneration
(3) granulation tissue

CRITICAL THINKING QUESTIONS

1. Imagine that you live 50 years in the future, and you can custom-
design a human to suit the environment. Your assignment is to customize the tissue makeup for life on a large planet with strong
gravity; a cold, dry climate; and a thin atmosphere. What adaptations would you incorporate into the structure and/or amount of
your tissues, and why?
HINT *Think of the adaptations seen in people and animals that
live in the Arctic versus those that live near the equator.*

2. The neighborhood kids are walking around with straight pins and
sewing needles stuck into their fingertips. There is no visible
bleeding. What type of tissue have they pierced? What if they
stuck a needle straight into the fingertip?
HINT *If you pierced an earlobe, it would bleed.*

3. Kendra was reading her *Principles of Anatomy and Physiology* text
at the college cafeteria's all-you-can-eat salad bar. She was having
trouble visualizing tissues, until her roommate plopped down a
bowl of gelatin salad. It was pale pink with grapes, shredded carrots, and coconut. As the salad quivered in the bowl, Kendra suddenly understood connective tissue. Relate the structure of the
salad to the structure of connective tissue.
HINT *The fruit is suspended in the jellylike gelatin.*

ANSWERS TO FIGURE QUESTIONS

4.1 Gap junctions allow the spread of electrical and chemical signals
between adjacent cells.

4.2 The basement membrane provides physical support of epithelium,
serves as a filter in the kidneys, and guides cell migration during
development and tissue repair.

4.3 Simple multicellular exocrine glands have a nonbranched duct;
compound multicellular exocrine glands have a branched duct.

4.4 Sebaceous (oil) glands are holocrine glands, and salivary glands
are merocrine glands.

4.5 Fibroblasts secrete the fibers and ground substance of the matrix.

The Integumentary System

THE INTEGUMENTARY SYSTEM AND HOMEOSTASIS

The integumentary system contributes to homeostasis by protecting the body and helping regulate body temperature. It also allows you to sense pleasurable, painful, and other stimuli in your external environment.

ENERGIZE YOUR STUDY

FOUNDATIONS CD

Anatomy Overview
- The Integumentary System

Animations
- Negative Feedback Control of Body Temperature
- Systems Contributions to Homeostasis

Concept and Connections
- The Integumentary System and Tissues

www.wiley.com/college/apcentral

INSIGHTS AND EXPLORATIONS

We have seen the pictures of people being rescued from burning buildings, trucks, cars and other accidents. According to the American Burn Association, 1.25 million people seek medical attention for burns every year in the United States. The standard treatment of third-degree burns (burns that extend into the deepest layers of the skin) is to remove and replace the damaged skin with new skin, ideally from other areas of the burned patient. For patients who are burned over 80 to 90 percent of their body this is not possible. Come explore how scientists are growing skin from burn patients to use in skin grafts.

Skin and its accessory structures—hair and nails, along with various glands, muscles, and nerves—make up the **integumentary system** (in-teg-ū-MEN-tar-ē; *inte-* = whole; *-gument* = body covering). The integumentary system helps protect the body, helps maintain a constant body temperature, and provides sensory information about the surrounding environment. Of all the body's organs, none is more easily inspected or more exposed to infection, disease, and injury than the skin. Although the skin's location makes it vulnerable to damage from trauma, sunlight, microbes, and pollutants in the environment, it has protective features that ward off such damage. Because of its visibility, skin reflects our emotions and some aspects of normal physiology, as evidenced by frowning, blushing, and sweating. Changes in skin color may indicate homeostatic imbalances in the body; for example, a bluish skin color is one sign of heart failure. Abnormal skin eruptions or rashes such as chickenpox, cold sores, or measles may reveal systemic infections or diseases of internal organs. Other conditions may involve just the skin itself, such as warts, age spots, or pimples. So important is the skin to one's image that many people spend much time and money to restore it to a more normal or youthful appearance. **Dermatology** (der′-ma-TOL-ō-jē; *dermato-* = skin; *-logy* = study of) is the medical specialty for diagnosis and treatment of integumentary system disorders.

STRUCTURE OF THE SKIN

OBJECTIVES

- Describe the layers of the epidermis and the cells that compose them.
- Compare the composition of the papillary and reticular regions of the dermis.
- Explain the basis for different skin colors.

The **skin** or **cutaneous membrane** covers the external surface of the body. It is the largest organ of the body in surface area and weight. In adults, the skin covers an area of about 2 square meters (22 square feet) and weighs 4.5–5 kg (10–11 lb), about 16% of total body weight. It ranges in thickness from 0.5 mm (0.02 in.) on the eyelids to 4.0 mm (0.16 in.) on the heels. However, over most of the body it is 1–2 mm (0.04–0.08 in.) thick. Structurally, the skin consists of two main parts (Figure 5.1). The superficial, thinner portion, which is composed of *epithelial tissue,* is the **epidermis** (ep′-i-DERM-is; *epi-* = above). The deeper, thicker *connective tissue* part is the **dermis.**

Deep to the dermis, but not part of the skin, is the **subcutaneous (subQ) layer.** Also called the **hypodermis** (*hypo-* = below), this layer consists of areolar and adipose tissues. Fibers that extend from the dermis anchor the skin to the subcutaneous layer, which, in turn, attaches to underlying tissues and organs. The subcutaneous layer serves as a storage depot for fat and contains large blood vessels that supply the skin. This region (and sometimes the dermis) also contains nerve endings called **lamellated (Pacinian) corpuscles** (pa-SIN-ē-an) that are sensitive to pressure (Figure 5.1).

Epidermis

The **epidermis** is keratinized stratified squamous epithelium. It contains four principal types of cells: keratinocytes, melanocytes, Langerhans cells, and Merkel cells (Figure 5.2 on page 142). About 90% of epidermal cells are **keratinocytes** (ker-a-TIN-ō-sīts; *keratino-* = hornlike; *-cytes* = cells), which are arranged in four or five layers and produce the protein **keratin** (Figure 5.2a). Recall from Chapter 4 that keratin is a tough, fibrous protein that helps protect the skin and underlying tissues from heat, microbes, and chemicals. Keratinocytes also produce lamellar granules, which release a water-repellent sealant.

About 8% of the epidermal cells are **melanocytes** (MEL-a-nō-sīts; *melano-* = black), which produce the pigment melanin (Figure 5.2b). Their long, slender projections extend between the keratinocytes and transfer melanin granules to them. **Melanin** is a brown-black pigment that contributes to skin color and absorbs damaging ultraviolet (UV) light. Once inside keratinocytes, the melanin granules cluster to form a protective veil over the nucleus, on the side toward the skin surface. In this way, they shield the nuclear DNA from being damaged by UV light. Although keratinocytes gain some protection from melanin granules, melanocytes themselves are particularly susceptible to damage by UV light.

Langerhans cells (LANG-er-hans) arise from red bone marrow and migrate to the epidermis (Figure 5.2c), where they constitute a small fraction of the epidermal cells. They participate in immune responses mounted against microbes that invade the skin, and are easily damaged by UV light.

Merkel cells are the least numerous of the epidermal cells. They are located in the deepest layer of the epidermis, where they contact the flattened process of a sensory neuron (nerve cell), a structure called a **tactile (Merkel) disc** (Figure 5.2d). Merkel cells and tactile discs detect different aspects of touch sensations (see Table 15.2 on page 507).

Several distinct layers of keratinocytes in various stages of development form the epidermis (Figure 5.3 on page 143). In most regions of the body the epidermis has four strata or layers—stratum basale, stratum spinosum, stratum granulosum, and a thin stratum corneum. Where exposure to friction is greatest, such as in the fingertips, palms, and soles, the epidermis has five layers—stratum basale, stratum spinosum, stratum granulosum, stratum lucidum, and a thick stratum corneum.

Stratum Basale

The deepest layer of the epidermis is the **stratum basale** (*basal-* = base), composed of a single row of cuboidal or columnar keratinocytes. Some cells in this layer are *stem cells* that un-

Figure 5.1 Components of the integumentary system. The skin consists of a thin, superficial epidermis and a deep, thicker dermis. Deep to the skin is the subcutaneous layer, which attaches the dermis to underlying organs and tissues.

The integumentary system includes the skin and its accessory structures—hair, nails, and skin glands—along with associated smooth muscles and nerves.

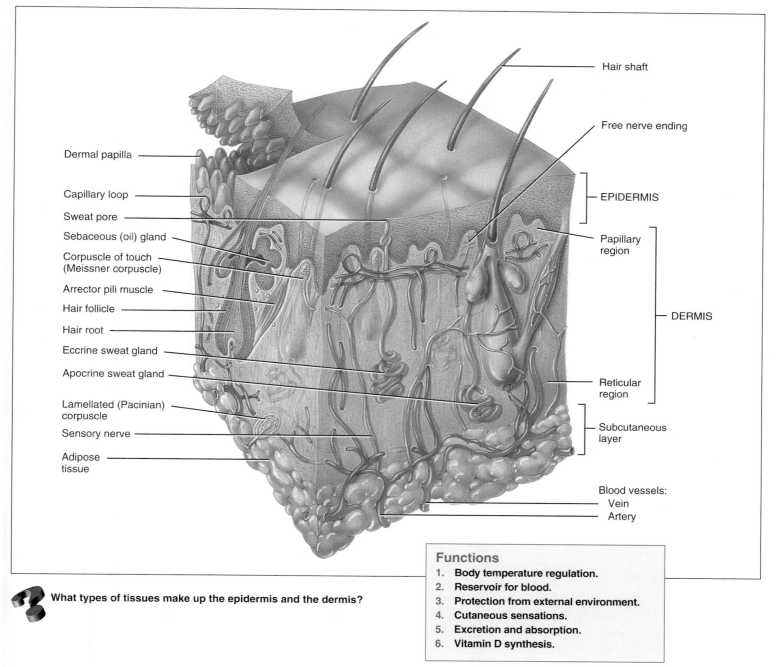

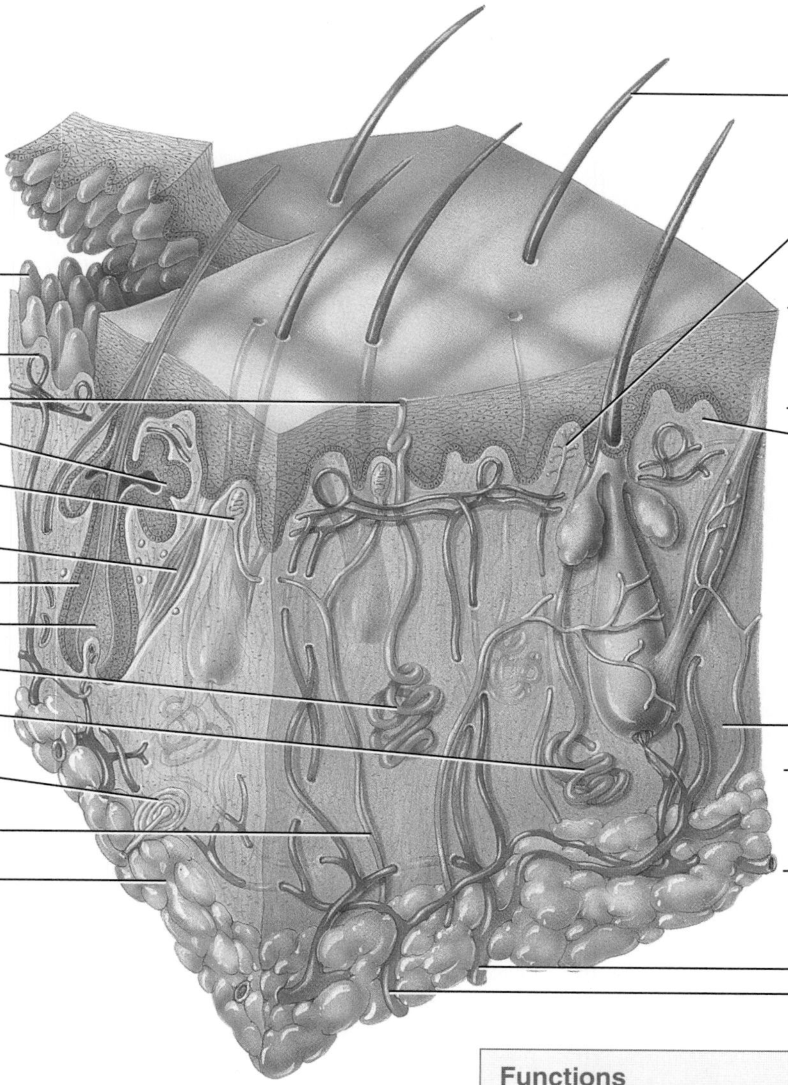
Hair shaft
Free nerve ending
Dermal papilla
Capillary loop
Sweat pore
Sebaceous (oil) gland
Corpuscle of touch (Meissner corpuscle)
Arrector pili muscle
Hair follicle
Hair root
Eccrine sweat gland
Apocrine sweat gland
Lamellated (Pacinian) corpuscle
Sensory nerve
Adipose tissue
EPIDERMIS
Papillary region
DERMIS
Reticular region
Subcutaneous layer
Blood vessels:
Vein
Artery

Functions
1. **Body temperature regulation.**
2. **Reservoir for blood.**
3. **Protection from external environment.**
4. **Cutaneous sensations.**
5. **Excretion and absorption.**
6. **Vitamin D synthesis.**

What types of tissues make up the epidermis and the dermis?

dergo cell division to continually produce new keratinocytes. The nuclei of keratinocytes in the stratum basale are large, and their cytoplasm contains many ribosomes, a small Golgi complex, a few mitochondria, and some rough endoplasmic reticulum. The cytoskeleton within keratinocytes of the stratum basale includes scattered intermediate filaments, called tonofilaments. The tonofilaments are composed of a protein that will later form the keratin in more superficial epidermal layers. Tonofilaments attach to desmosomes, which bind cells of the stratum basale to each other and to the cells of the adjacent stratum spinosum, and to hemidesmosomes, which bind the keratinocytes to the basement membrane between the epidermis and the dermis. Keratin

Figure 5.2 Types of cells in the epidermis. Besides keratinocytes, the epidermis contains melanocytes, which produce the pigment melanin; Langerhans cells, which participate in immune responses; and Merkel cells, which function in the sensation of touch.

Most of the epidermis consists of keratinocytes, which produce the protein keratin (protects underlying tissues) and lamellar granules (contain a waterproof sealant).

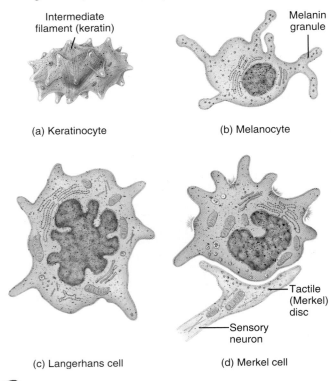

(a) Keratinocyte

(b) Melanocyte

(c) Langerhans cell

(d) Merkel cell

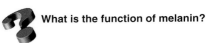

What is the function of melanin?

also protects deeper layers from injury. Melanocytes, Langerhans cells, and Merkel cells with their associated tactile discs are scattered among the keratinocytes of the basal layer. The stratum basale is also known as the **stratum germinativum** (jer′-mi-na-TĒ-vum; *germ-* = sprout), a name that emphasizes its role in forming new cells.

Skin Grafts

New skin cannot regenerate if an injury destroys the stratum basale and its stem cells. Skin wounds of this magnitude require skin grafts in order to heal. A **skin graft** involves covering the wound with a patch of healthy skin taken from a donor site. To avoid tissue rejection, the transplanted skin is usually taken from the same individual (*autograft)* or an identical twin (*isograft).* If skin damage is so extensive that an autograft would cause harm, a self-donation procedure called *autologous skin transplantation* (aw-TOL-ō-gus) may be used. In this procedure, performed most often for severely burned patients, small amounts of an individual's epidermis are removed, and the keratinocytes are cultured in the laboratory to produce thin sheets of skin. The new skin is transplanted back to the patient so that it covers the

burn wound and generates a permanent skin. Also available as skin grafts for wound coverage are products developed from the foreskins of circumsized infants (Apligraft and Transite) that are grown in the laboratory. ■

Stratum Spinosum

Superficial to the stratum basale is the **stratum spinosum** (*spinos-* = thornlike), where 8 to 10 layers of many-sided keratinocytes fit closely together. Cells in the more superficial portions of this layer become somewhat flattened. These keratinocytes have the same organelles as cells of the stratum basale, and some cells in this layer may retain their ability to undergo cell division. When cells of the stratum spinosum are prepared for microscopic examination, they shrink and pull apart such that they appear to be covered with thornlike spines (see Figure 5.2a), although they are rounded and larger in living tissue. Each spiny projection in a prepared tissue section is a point where bundles of tonofilaments are inserting into a desmosome, tightly joining the cells to one another. This arrangement provides both strength and flexibility to the skin. Projections of both Langerhans cells and melanocytes also appear in this stratum.

Stratum Granulosum

At the middle of the epidermis, the **stratum granulosum** (*granulos-* = little grains) consists of three to five layers of flattened keratinocytes that are undergoing apoptosis. (Recall from Chapter 3 that apoptosis is an orderly, genetically programmed cell death in which the nucleus fragments before the cells die.) The nuclei and other organelles of these cells begin to degenerate, and tonofilaments become more apparent. A distinctive feature of cells in this layer is the presence of darkly staining granules of a protein called **keratohyalin** (ker′-a-tō-HĪ-a-lin), which converts the tonofilaments into keratin. Also present in the keratinocytes are membrane-enclosed **lamellar granules,** which release a lipid-rich secretion. This secretion fills the spaces between cells of the stratum granulosum, stratum lucidum, and stratum corneum. The lipid-rich secretion acts as a water-repellent sealant, retarding loss of body fluids and entry of foreign materials. As their nuclei break down during apoptosis, the keratinocytes of the stratum granulosum can no longer carry on vital metabolic reactions, and they die. Thus, the stratum granulosum marks the transition between the deeper, metabolically active strata and the dead cells of the more superficial strata.

Stratum Lucidum

The **stratum lucidum** (*lucid-* = clear) is present only in the thick skin of the fingertips, palms, and soles. It consists of three to five layers of flattened clear, dead keratinocytes that contain large amounts of keratin and thickened plasma membranes.

Stratum Corneum

The **stratum corneum** (*corne-* = horn or horny) consists of 25 to 30 layers of flattened dead keratinocytes. These cells are con-

Figure 5.3   Layers of the epidermis.

 The epidermis consists of keratinized stratified squamous epithelium.

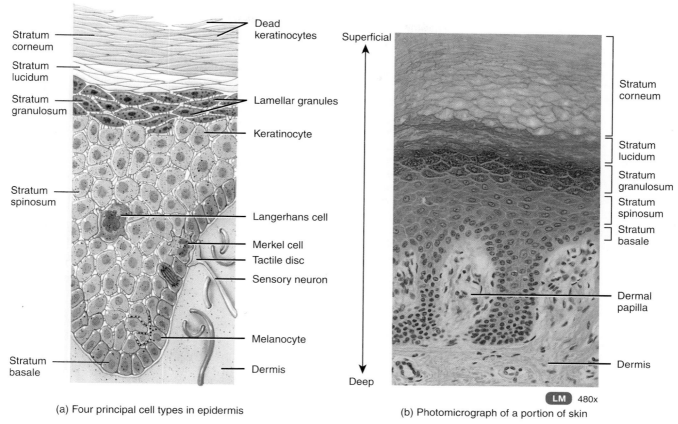

(a) Four principal cell types in epidermis

(b) Photomicrograph of a portion of skin

Which epidermal layer includes stem cells that continually undergo cell division?

tinuously shed and replaced by cells from the deeper strata. The interior of the cells contains mostly keratin. Between the cells are lipids from lamellar granules that help make this layer an effective water-repellent barrier. Its multiple layers of dead cells also help to protect deeper layers from injury and microbial invasion. Constant exposure of skin to friction stimulates the formation of a *callus,* an abnormal thickening of the stratum corneum.

Keratinization and Growth of the Epidermis

Newly formed cells in the stratum basale are slowly pushed to the surface. As the cells move from one epidermal layer to the next, they accumulate more and more keratin, a process called **keratinization** (ker′-a-tin-i-ZĀ-shun). Then they undergo apoptosis. Eventually the keratinized cells slough off and are replaced by underlying cells that, in turn, become keratinized. The whole process by which cells form in the stratum basale, rise to the surface, become keratinized, and slough off takes about four

weeks in an average epidermis of 0.1 mm (0.004 in.) thickness. The rate of cell division in the stratum basale increases when the outer layers of the epidermis are stripped away, as occurs in abrasions and burns. The mechanisms that regulate this remarkable growth are not well understood, but hormonelike proteins such as **epidermal growth factor (EGF)** play a role.

Table 5.1 summarizes the distinctive features of the epidermal strata.

 Psoriasis

Psoriasis is a common and chronic skin disorder in which keratinocytes divide and move more quickly than normal from the stratum basale to the stratum corneum. They are shed prematurely in as little as 7 to 10 days. The immature keratinocytes make an abnormal keratin, which forms flaky, silvery scales at the skin surface, most often on the knees, elbows, and scalp (dandruff). Effective treatments—various topical ointments and UV phototherapy—suppress cell division, decrease the rate of cell growth, or inhibit keratinization. ■

Table 5.1	Summary of Epidermal Strata
Stratum	**Description**
Basale	Deepest layer, composed of a single row of cuboidal or columnar keratinocytes that contain scattered tonofilaments (intermediate filaments); stem cells undergo cell division to produce new keratinocytes; melanocytes, Langerhans cells, and Merkel cells associated with tactile discs are scattered among the keratinocytes.
Spinosum	Eight to ten rows of many-sided keratinocytes with bundles of tonofilaments; includes projections of melanocytes and Langerhans cells.
Granulosum	Three to five rows of flattened keratinocytes, in which organelles are beginning to degenerate; cells contain the protein keratohyalin, which converts tonofilaments into keratin, and lamellar granules, which release a lipid-rich, water-repellent secretion.
Lucidum	Present only in skin of fingertips, palms, and soles; consists of three to five rows of clear, flat, dead keratinocytes with large amounts of keratin.
Corneum	Twenty-five to thirty rows of dead, flat keratinocytes that contain mostly keratin.

Dermis

The second, deeper part of the skin, the **dermis,** is composed mainly of connective tissue containing collagen and elastic fibers. The few cells present in the dermis include fibroblasts, macrophages, and some adipocytes. Blood vessels, nerves, glands, and hair follicles are embedded in dermal tissue. Based on its tissue structure, the dermis can be divided into a papillary region and a reticular region.

The **papillary region** is the superficial part of the dermis, about one-fifth of the thickness of the total layer (see Figure 5.1). It consists of areolar connective tissue containing fine elastic fibers. Its surface area is greatly increased by small, fingerlike projections called **dermal papillae** (pa-PIL-ē = nipples). These nipple-shaped structures indent the epidermis and some contain capillary loops. Other dermal papillae also contain tactile receptors called **corpuscles of touch** or **Meissner corpuscles,** nerve endings that are sensitive to touch. Also present in the dermal papillae are **free nerve endings,** dendrites that lack any apparent structural specialization. Different free nerve endings initiate signals that give rise to sensations of warmth, coolness, pain, tickling, and itching.

The deeper part of the dermis is called the **reticular region** (*reticul-* = netlike). It consists of dense irregular connective tissue containing bundles of collagen and some coarse elastic fibers. The bundles of collagen fibers in the reticular region interlace in a netlike manner. A few adipose cells, hair follicles, nerves, sebaceous (oil) glands, and sudoriferous (sweat) glands occupy the spaces between fibers.

The combination of collagen and elastic fibers in the reticular region provides the skin with strength, *extensibility* (ability to stretch), and *elasticity* (ability to return to original shape after stretching). The extensibility of skin can readily be seen in pregnancy and obesity. Extreme stretching, however, may produce small tears in the dermis, causing *striae* (STRĪ-ē = streaks), or stretch marks, that are visible as red or silvery white streaks on the skin surface.

The surfaces of the palms, fingers, soles, and toes have a series of ridges and grooves. They appear either as straight lines or as a pattern of loops and whorls, as on the tips of the digits. These **epidermal ridges** develop during the third and fourth fetal months as the epidermis conforms to the contours of the underlying dermal papillae of the papillary region. The ridges increase the surface area of the epidermis and thus increase the grip of the hand or foot by increasing friction. Because the ducts of sweat glands open on the tops of the epidermal ridges as sweat pores, the sweat and ridges form fingerprints (or footprints) upon touching a smooth object. The epidermal ridge pattern is genetically determined and is unique for each individual. Normally, the ridge pattern does not change during life, except to enlarge, and thus can serve as the basis for identification through fingerprints or footprints. The study of the pattern of epidermal ridges is called **dermatoglyphics** (der′-ma-tō-GLIF-iks; *glyphe* = carved work).

Table 5.2 summarizes the structural features of the papillary and reticular regions of the dermis.

 ## Photodamage and Photosensitivity Reactions

Although basking in the warmth of the sun may feel good, it is not a healthy practice. Both the longer-wavelength ultraviolet A (UVA) rays and the shorter-wavelength UVB rays cause **photodamage** of the skin. Light-skinned and dark-skinned individuals alike experience the effect of acute overexposure to UVB rays—sunburn. Even if sunburn does not occur, the UVB rays can damage DNA, producing genetic mutations in epidermal cells. In turn, the mutations can cause skin cancer. As UVA rays penetrate

Table 5.2	Summary of Papillary and Reticular Regions of the Dermis
Region	**Description**
Papillary	The superficial portion of the dermis (about one-fifth); consists of areolar connective tissue with elastic fibers; contains dermal papillae that house capillaries, corpuscles of touch, and free nerve endings.
Reticular	The deeper portion of the dermis (about four-fifths); consists of dense irregular connective tissue with bundles of collagen and some coarse elastic fibers. Spaces between fibers contain some adipose cells, hair follicles, nerves, sebaceous glands, and sudoriferous glands.

to the dermis, they produce oxygen free radicals that disrupt collagen and elastic fibers in the extracellular matrix. This is the main reason for the severe wrinkling that occurs in those who spend a great deal of time in the sun without protection.

Besides causing photodamage to the skin, exposure to direct sunlight may also produce **photosensitivity reactions** in some individuals taking certain oral medications, such as erythromycin, or herbal supplements, such as St. John's wort. The reactions may include redness, peeling, hives, blisters, and even shock. ∎

The Structural Basis of Skin Color

Melanin, carotene, and hemoglobin are three pigments that impart a wide variety of colors to skin. The amount of **melanin** causes the skin's color to vary from pale yellow to tan to black. Melanocytes are most plentiful in the epidermis of the penis, nipples of the breasts, area just around the nipples (areolae), face, and limbs. They are also present in mucous membranes. Because the *number* of melanocytes is about the same in all people, differences in skin color are due mainly to the *amount of pigment* the melanocytes produce and disperse to keratinocytes. In some people, melanin accumulates in patches called *freckles*. As one grows older, *age (liver) spots* may develop. These flat blemishes look like freckles and range in color from light brown to black. Like freckles, age spots are accumulations of melanin.

Melanocytes synthesize melanin from the amino acid *tyrosine* in the presence of an enzyme called *tyrosinase.* Synthesis occurs in an organelle called a **melanosome.** Exposure to UV light increases the enzymatic activity within melanosomes and thus stimulates melanin production. Both the amount and darkness of melanin increase, which gives the skin a tanned appearance and further protects the body against UV radiation. Thus, within limits, melanin serves a protective function. Nevertheless, repeatedly exposing the skin to UV light causes skin cancer. A tan is lost when the melanin-containing keratinocytes are shed from the stratum corneum.

Carotene (KAR-ō-tēn; *carot* = carrot), a yellow-orange pigment, is the precursor of vitamin A, which is used to synthesize pigments needed for vision. Carotene is found in the stratum corneum and fatty areas of the dermis and subcutaneous layer.

When little melanin or carotene is present, the epidermis appears translucent. Thus, the skin of white people appears pink to red, depending on the amount and oxygen content of the blood moving through capillaries in the dermis. The red color is due to **hemoglobin,** the oxygen-carrying pigment inside red blood cells.

Albinism (AL-bin-izm; *albin-* = white) is the inherited inability of an individual to produce melanin. Most **albinos** (al-BĪ-nōs), people affected by albinism, have melanocytes that are unable to synthesize tyrosinase. Melanin is missing from their hair, eyes, and skin. In another condition, called **vitiligo** (vit-i-LĪ-gō), the partial or complete loss of melanocytes from patches of skin produces irregular white spots. The loss of melanocytes may be related to an immune system malfunction in which antibodies attack the melanocytes.

Skin Color Clues

The color of skin and mucous membranes can provide clues for diagnosing certain conditions. When blood is not picking up an adequate amount of oxygen in the lungs, such as in someone who has stopped breathing, the mucous membranes, nail beds, and skin appear bluish or **cyanotic** (sī-a-NOT-ik; *cyan-* = blue). This occurs because hemoglobin that is depleted of oxygen looks deep, purplish blue. **Jaundice** (JON-dis; *jaund-* = yellow) is due to a buildup of the yellow pigment bilirubin in the blood. This condition gives a yellowish appearance to the whites of the eyes and the skin. Jaundice usually indicates liver disease. **Erythema** (er-i-THĒ-ma; *eryth-* = red), redness of the skin, is caused by engorgement of capillaries in the dermis with blood due to skin injury, exposure to heat, infection, inflammation, or allergic reactions. All skin color changes are observed most readily in people with lighter-colored skin and may be more difficult to discern in people with darker skin. ∎

Types of Skin

Although the skin over the entire body is similar in structure, there are quite a few local variations related to thickness of the epidermis, strength, flexibility, degree of keratinization, distribution and type of hair, density and types of glands, pigmentation, vascularity (blood supply), and innervation (nerve supply). Based on structural and functional properties, we recognize two major types of skin: thin (hairy) and thick (hairless).

Thin skin covers all parts of the body except for the palms, palmar surfaces of the digits, and the soles. Its epidermis is thin, just 0.10–0.15 mm (0.004–0.006 in.). A distinct stratum lucidum is lacking, and the strata spinosum and corneum are relatively thin. Thin skin has lower, broader, and fewer dermal papillae, and thus lacks epidermal ridges. Additionally, even though thin skin has hair follicles, arrector pili muscles, and sebaceous (oil) glands, it has fewer sweat glands than thick skin. Finally, thin skin has a sparser distribution of sensory receptors than thick skin.

Thick skin covers the palms, palmar surfaces of the fingers, and soles. Its epidermis is relatively thick, 0.6–4.5 mm (0.024–0.18 in.), and features a distinct stratum lucidum as well as thicker strata spinosum and corneum. The dermal papillae of thick skin are higher, narrower, and more numerous than those in thin skin, and thus thick skin has epidermal ridges. Thick skin lacks hair follicles, arrector pili muscles, and sebaceous glands, and has more sweat glands than thin skin. Sensory receptors are more densely clustered in thick skin.

Table 5.3 summarizes the features of thin and thick skin.

▶ CHECKPOINT

1. What structures are included in the integumentary system?

2. How does the subcutaneous layer relate to the skin?

3. How does the process of keratinization occur?

4. What are the structural and functional differences between the epidermis and dermis?

5. How are epidermal ridges formed?

6. In what ways are thin and thick skin similar and different?

Table 5.3	Comparison of Thin and Thick Skin	
Feature	**Thin Skin**	**Thick Skin**
Distribution	All parts of the body except palms, palmar surface of digits, and soles.	Palms, palmar surface of digits, and soles.
Epidermal thickness	0.10–0.15 mm (0.004–0.006 in.).	0.6–4.5 mm (0.024–0.18 in.).
Epidermal strata	Stratum lucidum essentially lacking; thinner strata spinosum and corneum.	Thick strata lucidum, spinosum, and corneum.
Epidermal ridges	Lacking due to poorly developed and fewer dermal papillae.	Present due to well-developed and more numerous dermal papillae.
Hair follicles and arrector pili muscles	Present.	Absent.
Sebaceous glands	Present.	Absent.
Sudoriferous glands	Fewer.	More numerous.
Sensory receptors	Sparser.	Denser.

ACCESSORY STRUCTURES OF THE SKIN

▶ O B J E C T I V E

- **Contrast the structure, distribution, and functions of hair, skin glands, and nails.**

Accessory structures of the skin—hair, skin glands, and nails—develop from the embryonic epidermis. They have a host of important functions. For example, hair and nails protect the body, and sweat glands help regulate body temperature.

Hair

Hairs, or *pili* (PI-lē), are present on most skin surfaces except the palms, palmar surfaces of the fingers, soles, and plantar surfaces of the feet. In adults, hair usually is most heavily distributed across the scalp, in the eyebrows, in the axillae (armpits), and around the external genitalia. Genetic and hormonal influences largely determine hair thickness and the pattern of distribution.

Anatomy of a Hair

Each hair is composed of columns of dead, keratinized cells bonded together by extracellular proteins. The **shaft** is the superficial portion of the hair, most of which projects from the surface of the skin (Figure 5.4a). The shaft of straight hair is round in transverse section, that of wavy hair is oval, and that of curly hair is kidney-shaped. The **root** is the portion of the hair deep to the shaft that penetrates into the dermis, and sometimes into the subcutaneous layer. The shaft and root both consist of three concen-

tric layers (Figure 5.4c, d). The inner *medulla,* which may be lacking in thinner hair, is composed of two or three rows of irregularly shaped cells containing pigment granules and air spaces. The middle *cortex* consists of elongated cells that contain pigment granules in dark hair but mostly air in gray or white hair. The *cuticle of the hair,* the outermost layer, consists of a single layer of cells. In the deeper part of the hair root, the cells of the cuticle of the hair are nucleated. In the upper part of the root and in the shaft they are scale-like and lack nuclei. Cuticle cells on the shaft are arranged like shingles on the side of a house, with their free edges pointing toward the end of the hair, and they are the most heavily keratinized of hair cells (Figure 5.4b).

Surrounding the root of the hair is the **hair follicle,** which is made up of an external root sheath and an internal root sheath (Figure 5.4c d). The *external root sheath* is a downward continuation of the epidermis. Near the surface, it contains all the epidermal layers. At the base of the hair follicle, the external root sheath contains only the stratum basale. The *internal root sheath* is produced by the matrix (described shortly) and forms a cellular tubular sheath of epithelium between the external root sheath and the hair.

The base of each hair follicle is an onion-shaped structure, the **bulb** (Figure 5.4c). This structure houses a nipple-shaped indentation, the **papilla of the hair,** which contains areolar connective tissue and many blood vessels that nourish the growing hair follicle. The bulb also contains a germinal layer of cells called the **matrix.** The matrix cells arise from the stratum basale, the site of cell division. Hence, matrix cells are responsible for the growth of existing hairs, and they produce new hairs when old hairs are shed. This replacement process occurs within the same follicle. Matrix cells also give rise to the cells of the internal root sheath.

Sebaceous (oil) glands (discussed shortly) and a bundle of smooth muscle cells are also associated with hairs (Figure 5.4a). The smooth muscle is the **arrector pili** (a-REK-tor PI-lē; *arrect* = to raise). It extends from the superficial dermis of the skin to the side of the hair follicle. In its normal position, hair emerges at an angle to the surface of the skin. Under physiologic or emotional stress, such as cold or fright, autonomic nerve endings stimulate the arrector pili muscles to contract, which pulls the hair shafts perpendicular to the skin surface. This action causes "goose bumps" or "gooseflesh" because the skin around the shaft forms slight elevations.

Surrounding each hair follicle are dendrites of neurons, called **hair root plexuses,** that are sensitive to touch (Figure 5.4a). The hair root plexuses generate nerve impulses if their hair shaft is moved.

Hair Removal

A substance that removes hair is called a **depilatory.** It dissolves the protein in the hair shaft, turning it into a gelatinous mass than can be wiped away. Because the hair root is not affected, regrowth of the hair occurs. In **electrolysis,** an electric current is used to destroy the hair matrix so the hair cannot regrow. ■

Figure 5.4   Hair.

Hairs are growths of epidermis composed of dead, keratinized cells.

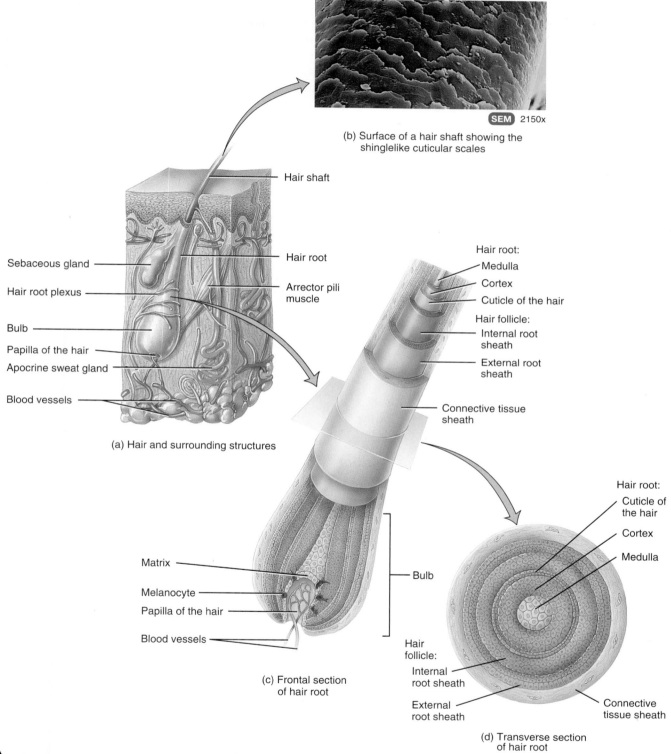

(b) Surface of a hair shaft showing the shinglelike cuticular scales

(a) Hair and surrounding structures

(c) Frontal section of hair root

(d) Transverse section of hair root

 Why does it hurt when you pluck a hair out but not when you have a haircut?

Hair Growth

Each hair follicle goes through a growth cycle, which consists of a growth stage and a resting stage. During the **growth stage,** cells of the matrix differentiate, keratinize, and die. This process forms the root sheath and hair shaft. As new cells are added at the base of the hair root, the hair grows longer. In time, the growth of the hair stops and the **resting stage** begins. After the resting stage, a new growth cycle begins. The old hair root falls out or is pushed out of the hair follicle, and a new hair begins to grow in its place. Scalp hair grows for 2 to 6 years and rests for about 3 months. At any time, about 85% of scalp hairs are in the growth stage. Visible hair is dead, but until the hair is pushed out of its follicle by a new hair, portions of its root within the scalp are alive.

Normal hair loss in the adult scalp is about 100 hairs per day. Both the rate of growth and the replacement cycle may be altered by illness, radiation therapy, chemotherapy, age, genetics, gender, and severe emotional stress. Rapid weight-loss diets that severely restrict calories or protein increase hair loss. The rate of shedding also increases for three to four months after childbirth. **Alopecia** (al′-o-PĒ-shē-a), the partial or complete lack of hair, may result from genetic factors, aging, endocrine disorders, chemotherapy for cancer, or skin disease.

Types of Hairs

Hair follicles develop between the ninth and twelfth weeks after fertilization as downgrowths of the stratum basale of the epidermis into the dermis. Usually by the fifth month of development, the follicles produce very fine, nonpigmented hairs called **lanugo** (la-NOO-gō = wool or down) that cover the body of the fetus. This hair is shed before birth, except in the scalp, eyebrows, and eyelashes. A few months after birth, slightly thicker hairs replace these downy hairs. Over the remainder of the body of an infant, a new growth of short, fine hair occurs. These hairs are known as **vellus hairs** (VEL-us = fleece), commonly called "peach fuzz." In response to hormones (androgens) secreted at puberty, coarse pigmented and frequently curly hair develops in the axillae (armpits) and pubic region. In males, these hairs also appear on the face and other parts of the body. The hairs that develop at puberty, together with those of the head, eyebrows, and eyelashes are called terminal hairs. About 95% of body hair on males is terminal hair (5% vellus hair), whereas about 35% of body hair on females is terminal (65% vellus hair).

Hair Color

The color of hair is due primarily to the amount and type of melanin in its keratinized cells. Melanin is synthesized by melanocytes scattered in the matrix of the bulb and passes into cells of the cortex and medulla (Figure 5.4c). Dark-colored hair contains mostly true melanin, whereas blond and red hair contain variants of melanin in which there is iron and more sulfur. Graying hair occurs because of a progressive decline in tyrosinase, whereas white hair results from accumulation of air bubbles in the medullary shaft.

Functions of Hair

Although the protection it offers is limited, hair on the head guards the scalp from injury and the sun's rays. It also decreases heat loss from the scalp. Eyebrows and eyelashes protect the eyes from foreign particles, as does hair in the nostrils and in the external ear canal. Touch receptors associated with hair follicles (hair root plexuses) are activated whenever a hair is even slightly moved. Thus, hairs function in sensing light touch.

 Hair and Hormones

At puberty, when the testes begin secreting significant quantities of androgens (masculinizing sex hormones), males develop the typical male pattern of hair growth, including a beard and a hairy chest. In females at puberty, the ovaries and the adrenal glands produce small quantities of androgens, which promote hair growth in the axillae and pubic region. Occasionally, a tumor of the adrenal glands, testes, or ovaries produces an excessive amount of androgens. The result in females or prepubertal males is **hirsutism** (HER-soo-tizm; *hirsut-* = shaggy), a condition of excessive body hair.

Surprisingly, androgens also must be present for occurrence of the most common form of baldness, **androgenic alopecia** or **male-pattern baldness.** In genetically predisposed adults, androgens inhibit hair growth. On men, hair loss is most obvious at the temples and crown. Women are more likely to have thinning of hair on top of the head. The first drug approved for enhancing scalp hair growth was minoxidil (Rogaine). It causes vasodilation (widening of blood vessels), thus increasing circulation. In about a third of the people who try it, minoxidil improves hair growth, causing scalp follicles to enlarge and lengthening the growth cycle. For many, however, the hair growth is meager. Minoxidil does not help people who already are bald. ■

Skin Glands

Several kinds of exocrine glands are associated with the skin: sebaceous (oil) glands, sudoriferous (sweat) glands, and ceruminous glands. Mammary glands, which are specialized sudoriferous glands that secrete milk, are discussed in Chapter 28 along with the female reproductive system.

Sebaceous Glands

Sebaceous glands (se-BĀ-shus; *sebace-* = greasy) or **oil glands** are simple, branched acinar glands. With few exceptions, they are connected to hair follicles (see Figures 5.1 and 5.4a). The secreting portion of a sebaceous gland lies in the dermis and usually opens into the neck of a hair follicle. In some locations, such as the lips, glans penis, labia minora, and tarsal glands of the eyelids, sebaceous glands open directly onto the surface of the skin. Absent in the palms and soles, sebaceous glands are small in most areas of the trunk and limbs, but large in the skin of the breasts, face, neck, and upper chest.

Sebaceous glands secrete an oily substance called **sebum** (SĒ-bum), which is a mixture of triglycerides, cholesterol, pro-

teins, and inorganic salts. Sebum coats the surface of hairs and helps keep them from drying and becoming brittle. Sebum also prevents excessive evaporation of water from the skin, keeps the skin soft and pliable, and inhibits the growth of certain bacteria.

Acne

Acne is an inflammation of sebaceous glands that usually begins at puberty, when the sebaceous glands grow in size and increase their production of sebum. Androgens from the testes, ovaries, and adrenal glands play the greatest role in stimulating sebaceous glands. Acne occurs predominantly in sebaceous follicles that have been colonized by bacteria, some of which thrive in the lipid-rich sebum. The infection may cause a cyst or sac of connective tissue cells to form, which can destroy and displace epidermal cells. This condition, called **cystic acne,** can permanently scar the epidermis. ∎

Sudoriferous Glands

There are three to four million **sweat glands,** or **sudoriferous glands** (soo´-dor-IF-er-us; *sudori-* = sweat; *-ferous* = bearing). The cells of sweat glands release their secretions by exocytosis and empty them into hair follicles or onto the skin surface through pores. They are divided into two main types, eccrine and apocrine, based on their structure, location, and type of secretion.

Eccrine sweat glands (*eccrine* = secreting outwardly) are simple, coiled tubular glands and are much more common than apocrine sweat glands. They are distributed throughout the skin, except for the margins of the lips, nail beds of the fingers and toes, glans penis, glans clitoris, labia minora, and eardrums. Eccrine sweat glands are most numerous in the skin of the forehead, palms, and soles; their density can be as high as 450 per square centimeter (3000 per square inch) in the palms. The secretory portion of eccrine sweat glands is located mostly in the deep dermis (sometimes in the upper subcutaneous layer). The excretory duct projects through the dermis and epidermis and ends as a pore at the surface of the epidermis (see Figure 5.1).

The sweat produced by eccrine sweat glands (about 600 mL per day) consists of water, ions (mostly Na^+ and Cl^-), urea, uric acid, ammonia, amino acids, glucose, and lactic acid. The main function of eccrine sweat glands is to help regulate body temperature through evaporation. As sweat evaporates, large quantities of heat energy leave the body surface. Sweat, or perspiration, usually occurs first on the forehead and scalp, extends to the face and the rest of the body, and occurs last on the palms and soles. Under conditions of emotional stress, however, the palms, soles, and axillae are the first surfaces to sweat. Eccrine sweat also plays a small role in eliminating wastes such as urea, uric acid, and ammonia. Sweat that evaporates from the skin before it is perceived as moisture is termed **insensible perspiration.** Sweat that is excreted in larger amounts and is seen as moisture on the skin is called **sensible perspiration.**

Apocrine sweat glands are also simple, coiled tubular glands. They are found mainly in the skin of the axilla (armpit),

groin, areolae (pigmented areas around the nipples) of the breasts, and bearded regions of the face in adult males. These glands were once thought to release their secretions in an apocrine manner—by pinching off a portion of the cell. We now know, however, that their secretion is released by exocytosis, which is characteristic of the release of secretions by merocrine glands (see Figure 4.4a on page 117). Although in this sense they are merocrine glands, the term *apocrine* is still used. The secretory portion of these sweat glands is located mostly in the subcutaneous layer, and the excretory duct opens into hair follicles (see Figure 5.1). Their secretory product is slightly viscous compared to eccrine secretions and contains the same components as eccrine sweat plus lipids and proteins. In women, cells of apocrine sweat glands enlarge about the time of ovulation and shrink during menstruation. Whereas eccrine sweat glands start to function soon after birth, apocrine sweat glands do not begin to function until puberty. Apocrine sweat glands are stimulated during emotional stress and sexual excitement; these secretions are commonly known as a "cold sweat."

Table 5.4 presents a comparison of eccrine and apocrine sweat glands.

Ceruminous Glands

Modified sweat glands in the external ear, called **ceruminous glands** (se-ROO-mi-nus; *cer-* = wax), produce a waxy secretion. The secretory portions of ceruminous glands lie in the subcutaneous layer, deep to sebaceous glands. Their excretory ducts

Table 5.4	Comparison of Eccrine and Apocrine Sweat Glands	
Feature	**Eccrine Sweat Glands**	**Apocrine Sweat Glands**
Distribution	Throughout skin, except for margins of lips, nail beds, glans penis and clitoris, labia minora, and eardrums.	Skin of the axilla, groin, areolae, and bearded regions of the face.
Location of secretory portion	Mostly in deep dermis.	Mostly in subcutaneous layer.
Termination of excretory duct	Surface of epidermis.	Hair follicle.
Secretion	Less viscous; consists of water, ions (Na^+, Cl^-), urea, uric acid, ammonia, amino acids, glucose, and lactic acid.	More viscous; consists of the same components as eccrine sweat glands plus lipids and proteins.
Functions	Regulation of body temperature and waste removal.	Stimulated during emotional stress and sexual excitement.
Onset of function	Soon after birth.	Puberty.

open either directly onto the surface of the external auditory canal (ear canal) or into ducts of sebaceous glands. The combined secretion of the ceruminous and sebaceous glands is called **cerumen,** or earwax. Cerumen, together with hairs in the external auditory canal, provides a sticky barrier that impedes the entrance of foreign bodies.

Impacted Cerumen

Some people produce an abnormally large amount of cerumen in the external auditory canal. The cerumen may accumulate until it becomes impacted (firmly wedged), which prevents sound waves from reaching the eardrum. The treatment for **impacted cerumen** is usually periodic ear irrigation with enzymes to dissolve the wax or removal of wax with a blunt instrument by trained medical personnel. The use of cotton-tipped swabs or sharp objects is not recommended for this purpose because they may push the cerumen farther into the external auditory canal and damage the eardrum. ■

Nails

Nails are plates of tightly packed, hard, keratinized epidermal cells. The cells form a clear, solid covering over the dorsal surfaces of the distal portions of the digits. Each nail consists of a nail body, a free edge, and a nail root (Figure 5.5). The **nail body** is the portion of the nail that is visible, the **free edge** is the part that may extend past the distal end of the digit, and the **nail root** is the portion that is buried in a fold of skin. Most of the nail body appears pink because of blood flowing through underlying capillaries. The free edge is white because there are no

underlying capillaries. The whitish, crescent-shaped area of the proximal end of the nail body is called the **lunula** (LOO-noo-la = little moon). It appears whitish because the vascular tissue underneath does not show through due to the thickened stratum basale in the area. Beneath the free edge is a thickened region of stratum corneum called the **hyponychium** (hī′-pō-NIK-ē-um; *hypo-* = below), which secures the nail to the fingertip. The **eponychium** (ep′-ō-NIK-ē-um; *ep-* = above; *-onych* = nail) or **cuticle** is a narrow band of epidermis that extends from and adheres to the margin (lateral border) of the nail wall. It occupies the proximal border of the nail and consists of stratum corneum.

The epithelium deep to the nail root is the **nail matrix,** where cells divide by mitosis to produce growth. Nail growth occurs by the transformation of superficial cells of the matrix into nail cells. In the process, the harder outer layer is pushed forward over the stratum basale. The growth rate of nails is determined by the rate of mitosis in matrix cells, which is influenced by factors such as a person's age, health, and nutritional status. Nail growth also varies according to the season, the time of day, and environmental temperature. The average growth in the length of fingernails is about 1 mm (0.04 in.) per week. The growth rate is somewhat slower in toenails. The longer the digit, the faster the nail grows.

Functionally, nails help us grasp and manipulate small objects in various ways, provide protection against trauma to the ends of the digits, and allow us to scratch various parts of the body.

▶   C H E C K P O I N T

7.   Describe the structure of a hair. What causes "goose bumps"?

8.   Contrast the locations and functions of sebaceous (oil) glands, sudoriferous (sweat) glands, and ceruminous glands.

9.   Describe the parts of a nail.

Figure 5.5   Nails.   Shown is a fingernail.

🔑 **Nail cells arise by transformation of superficial cells of the nail matrix.**

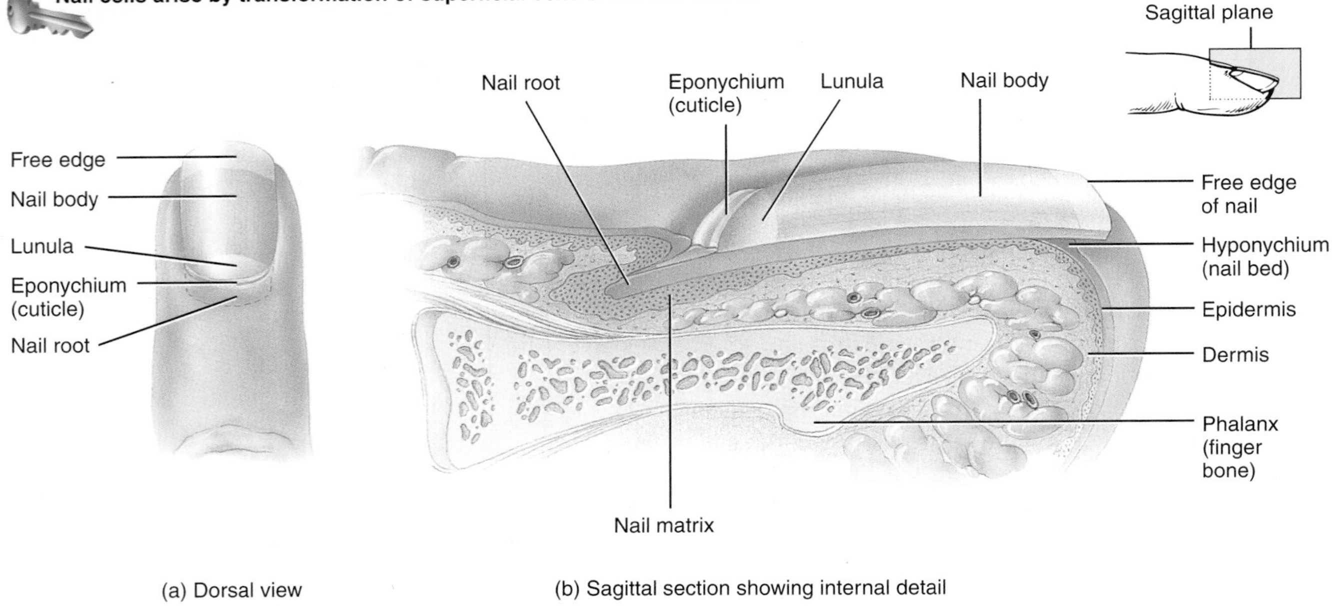

Sagittal plane

Nail root   Eponychium (cuticle)   Lunula   Nail body

Free edge
Nail body
Lunula
Eponychium (cuticle)
Nail root

Free edge of nail
Hyponychium (nail bed)
Epidermis
Dermis
Phalanx (finger bone)

Nail matrix

(a) Dorsal view   (b) Sagittal section showing internal detail

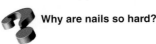 **Why are nails so hard?**

FUNCTIONS OF THE SKIN

> OBJECTIVE

- **Describe how the skin contributes to regulation of body temperature, protection, sensation, excretion and absorption, and synthesis of vitamin D.**

The skin helps regulate body temperature, serves as a water-repellent and protective barrier between the external environment and internal tissues, contains sensory nerve endings, excretes a small amount of salts and several organic compounds and has some capacity to absorb substances, and helps to synthesize the active form of vitamin D.

Thermoregulation

The skin contributes to **thermoregulation,** the homeostatic regulation of body temperature, in two ways: by liberating sweat at its surface and by adjusting the flow of blood in the dermis. In response to high environmental temperature or heat produced by exercise, the evaporation of sweat from the skin surface helps lower body temperature. In response to low environmental temperature, production of sweat is decreased, which helps conserve heat. During moderate exercise, the flow of blood through the skin increases, which increases the amount of heat radiated from the body.

The dermis houses an extensive network of blood vessels that carry 8–10% of the total blood flow in a resting adult. For this reason, the skin acts as a *blood reservoir.* During very strenuous exercise, skin blood vessels constrict (narrow) somewhat, diverting more blood to contracting skeletal muscles and the heart. Because of this shunting of blood away from the skin, however, less heat is lost from the skin, and body temperature may rise.

Protection

The skin covers the body and provides protection in various ways. Keratin in the skin protects underlying tissues from microbes, abrasion, heat, and chemicals and the tightly interlocked keratinocytes resist invasion by microbes. Lipids released by lamellar granules retard evaporation of water from the skin surface, thus protecting the body from dehydration, and they also retard entry of water across the skin surface when we take a shower or go swimming. The oily sebum from the sebaceous glands also protects skin and hairs from drying out and contains bactericidal chemicals that kill surface bacteria. The acidic pH of perspiration retards the growth of some microbes. The pigment melanin provides some protection against the damaging effects of UV light. Two types of cells carry out protective functions that are immunological in nature. Epidermal Langerhans cells alert the immune system to the presence of potentially harmful microbial invaders by recognizing and processing them, and macrophages in the dermis phagocytize bacteria and viruses that manage to penetrate the skin surface.

Cutaneous Sensations

Cutaneous sensations are those that arise in the skin. These include tactile sensations—touch, pressure, vibration, and tickling—as well as thermal sensations such as warmth and coolness. Another cutaneous sensation, pain, usually is an indication of impending or actual tissue damage. Some of the wide variety of abundantly distributed nerve endings and receptors in the skin include the tactile discs of the epidermis, the corpuscles of touch in the dermis, and hair root plexuses around each hair follicle. Chapter 15 provides more details on the topic of cutaneous sensations.

Excretion and Absorption

The skin normally has a small role in *excretion,* the elimination of substances from the body, and *absorption,* the passage of materials from the external environment into body cells. Despite the almost waterproof nature of the stratum corneum, about 400 mL of water evaporates through it daily. A sedentary person loses an additional 200 mL per day of water as sweat, whereas a physically active person loses much more. Besides removing water and heat (by evaporation), sweat also is the vehicle for excretion of small amounts of salts, carbon dioxide, and two organic molecules that result from the breakdown of proteins—ammonia and urea.

The absorption of water-soluble substances through the skin is negligible, but certain lipid-soluble materials do penetrate the skin. These include fat-soluble vitamins (A, D, E, and K), certain drugs, and the gases oxygen and carbon dioxide. Toxic materials that can be absorbed through the skin include organic solvents such as acetone (in some nail polish removers) and carbon tetrachloride (dry-cleaning fluid); salts of heavy metals such as lead, mercury, and arsenic; and the toxins in poison ivy and poison oak. Since topical (applied to the skin) steroids, such as cortisone, are lipid-soluble, they move easily into the papillary region of the dermis. Here, they exert their anti-inflammatory properties by inhibiting histamine production by mast cells. Recall that histamine contributes to inflammation.

Synthesis of Vitamin D

Synthesis of vitamin D requires activation of a precursor molecule in the skin by UV rays in sunlight. Enzymes in the liver and kidneys then modify the activated molecule, finally producing *calcitriol,* the most active form of vitamin D. Calcitriol is a hormone that aids in the absorption of calcium in foods from the gastrointestinal tract into the blood.

 **Transdermal Drug Administration**

Most drugs are either absorbed into the body through the digestive system or injected into subcutaneous tissue or muscle. An alternative route, **transdermal drug administration,** enables a drug contained within an adhesive skin patch to pass across the epidermis and into the blood vessels of the dermis. The drug is released at a controlled rate over one to several days. Because the major barrier to penetration of most drugs is the stratum corneum, transdermal absorption is most rapid in regions of the skin where this layer is thin, such as the scrotum, face, and scalp. A growing number of drugs are available for transdermal administration. These include nitroglycerin, for prevention of angina pectoris (chest pain associated with heart disease); scopolamine, for motion sickness; estradiol, used for estrogen-replacement therapy during menopause; and nicotine, used to help people stop smoking. ■

▶ CHECKPOINT

10. In what two ways does the skin help regulate body temperature?
11. In what ways does the skin serve as a protective barrier?
12. What sensations arise from stimulation of neurons in the skin?
13. What types of molecules can penetrate the stratum corneum?

MAINTAINING HOMEOSTASIS: SKIN WOUND HEALING

▶ OBJECTIVE
• **Explain how epidermal wounds and deep wounds heal.**

Skin damage sets in motion a sequence of events that repairs the skin to its normal (or near-normal) structure and function. Two kinds of wound-healing processes can occur, depending on the depth of the injury. Epidermal wound healing occurs following wounds that affect only the epidermis; deep wound healing occurs following wounds that penetrate the dermis.

Epidermal Wound Healing

Even though the central portion of an epidermal wound may extend to the dermis, the edges of the wound usually involve only slight damage to superficial epidermal cells. Common types of epidermal wounds include abrasions, in which a portion of skin has been scraped away, and minor burns.

In response to an epidermal injury, basal cells of the epidermis surrounding the wound break contact with the basement membrane. The cells then enlarge and migrate across the wound (Figure 5.6a). The cells appear to migrate as a sheet until advancing cells from opposite sides of the wound meet. When epidermal cells encounter each other, they stop migrating due to a cellular response called **contact inhibition.** Migration of the epidermal cells stops completely when each is finally in contact with other epidermal cells on all sides.

While some basal epidermal cells migrate, epidermal growth factor, a hormone, stimulates basal stem cells to divide and replace the ones that have moved into the wound. Replacement of migrating basal cells continues until the wound is resurfaced (Figure 5.6b). Following this, the relocated cells divide to build new strata, thus thickening the new epidermis.

Deep Wound Healing

Deep wound healing occurs when an injury extends to the dermis and subcutaneous layer. Because multiple tissue layers must be repaired, the healing process is more complex than in epidermal wound healing. In addition, scar tissue is formed, and the healed tissue loses some of its normal function. Deep wound healing occurs in four phases: an inflammatory phase, a migratory phase, a proliferative phase, and a maturation phase.

During the **inflammatory phase,** a blood clot forms in the wound and loosely unites the wound edges (Figure 5.7a). This phase of deep wound healing involves **inflammation,** a vascular and cellular response that helps eliminate microbes, foreign material, and dying tissue in preparation for repair. The vasodilation and increased permeability of blood vessels associated with inflammation enhance delivery of helpful cells. These include phagocytic white blood cells called neutrophils, monocytes, which develop into macrophages that phagocytize microbes, and mesenchymal cells, which develop into fibroblasts (Figure 5.7a).

Figure 5.6 Epidermal wound healing.

 In an epidermal wound, the injury does not extend into the dermis.

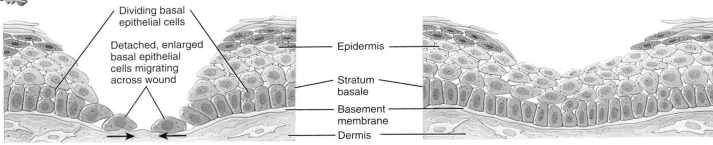

Dividing basal epithelial cells

Detached, enlarged basal epithelial cells migrating across wound

Epidermis
Stratum basale
Basement membrane
Dermis

(a) Division of basal epithelial cells and migration across wound

(b) Thickening of epidermis

Would you expect an epidermal wound to bleed? Why or why not?

Figure 5.7 Deep wound healing. The initial inflammatory phase (a) is followed by a migratory phase, a proliferative phase, and finally (b) a maturation phase.

In a deep wound, the injury extends deep to the epidermis.

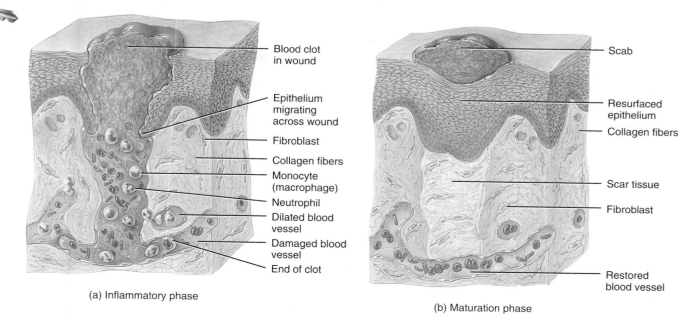

(a) Inflammatory phase

- Blood clot in wound
- Epithelium migrating across wound
- Fibroblast
- Collagen fibers
- Monocyte (macrophage)
- Neutrophil
- Dilated blood vessel
- Damaged blood vessel
- End of clot

(b) Maturation phase

- Scab
- Resurfaced epithelium
- Collagen fibers
- Scar tissue
- Fibroblast
- Restored blood vessel

Deep wound healing

Why is the arrival of phagocytic white blood cells (neutrophils and macrophages) at the scene of a wound helpful?

The three phases that follow do the work of repairing the wound. In the **migratory phase,** the clot becomes a scab, and epithelial cells migrate beneath the scab to bridge the wound. Fibroblasts migrate along fibrin threads and begin synthesizing scar tissue (collagen fibers and glycoproteins), and damaged blood vessels begin to regrow. During this phase, the tissue filling the wound is called **granulation tissue.** The **proliferative phase** is characterized by extensive growth of epithelial cells beneath the scab, deposition by fibroblasts of collagen fibers in random patterns, and continued growth of blood vessels. Finally, during the **maturation phase,** the scab sloughs off once the epidermis has been restored to normal thickness. Collagen fibers become more organized, fibroblasts decrease in number, and blood vessels are restored to normal (Figure 5.7b).

The process of scar tissue formation is called **fibrosis.** Sometimes, so much scar tissue is formed during deep wound healing that a raised scar—one that is elevated above the normal epidermal surface—results. If such a scar remains within the boundaries of the original wound, it is a **hypertrophic scar.** If it extends beyond the boundaries into normal surrounding tissues, it is a **keloid scar.** Scar tissue differs from normal skin in that its collagen fibers are more densely arranged. Scar tissue also has fewer blood vessels and might not contain the same number of hairs, skin glands, or sensory structures as undamaged skin. Because of the arrangement of collagen fibers and the scarcity of blood vessels, scars usually are lighter in color than normal skin.

▶ C H E C K P O I N T

14. Why doesn't epidermal wound healing result in scar formation?

DEVELOPMENT OF THE INTEGUMENTARY SYSTEM

▶ O B J E C T I V E

- **Describe the development of the epidermis, its accessory structures, and the dermis.**

At the end of many chapters throughout this book, we will discuss the development of body systems. About 15 days after the fertilization of an oocyte, a portion of the developing embryo differentiates into three **primary germ layers: ectoderm** (*ecto-* = outside), **mesoderm** (*meso-* = middle), and **endoderm** (*endo-* = within). They are the embryonic tissues from which all tissues and organs of the body eventually develop (see Table 29.1 on page 1072).

The *epidermis* is derived from the **ectoderm.** At the beginning of the eighth week after fertilization the ectoderm consists of simple cuboidal epithelium. These cells become flattened and are known as the **periderm.** By the fourth month all layers of the epidermis are formed and each layer assumes its characteristic structure.

The *dermis* is derived from **mesodermal cells** in a zone beneath the ectoderm. There they undergo a process that changes them into the connective tissues that begin to form the dermis at about 11 weeks.

Nails are developed at about 10 weeks. Initially they consist of a thick layer of epithelium called the **primary nail field.** The nail itself is keratinized epithelium and grows distally from its base. It is not until the ninth month that the nails actually reach the tips of the digits.

As noted earlier, *hair follicles* develop between the ninth and twelfth weeks as downgrowths of the stratum basale of the epidermis into the deeper dermis. The downgrowths soon differentiate into the bulb (dividing epithelial cells) and the papilla (connective tissue) of the hair, beginnings of the epithelial portions of sebaceous glands, and other structures associated with hair follicles. By the fifth month, the follicles produce **lanugo** (delicate fetal hair). It is produced first on the head and then on other parts of the body. The lanugo is usually shed prior to birth.

The epithelial (secretory) portions of *sebaceous glands* develop from the sides of the hair follicles at about four months and remain connected to the follicles. The epithelial portions of *sudoriferous glands* are also derived from downgrowths of the stratum basale of the epidermis into the dermis. They appear at about five months on the palms and soles and a little later in other regions. The connective tissue and blood vessels associated with the glands develop from **mesoderm.**

By the fifth month of development, secretions from sebaceous glands mix with sloughed off epidermal cells and hairs to form a fatty substance called **vernix caseosa** (VER-niks KĀ-sē-ō-sa; *vernix* = varnish, *caseosa* = cheese). This substance covers and protects the skin of the fetus from the constant exposure to the amniotic fluid in which it is bathed. In addition, the vernix caseosa facilitates the birth of the fetus because of its slippery nature and protects the skin from being damaged by the nails.

▶ C H E C K P O I N T
15. Which structures develop as downgrowths of the stratum basale?

AGING AND THE INTEGUMENTARY SYSTEM

▶ O B J E C T I V E
• **Describe the effects of aging on the integumentary system.**

The pronounced effects of skin aging do not become noticeable until people reach their late forties. Most of the age-related changes occur in the dermis. Collagen fibers in the dermis begin to decrease in number, stiffen, break apart, and disorganize into a shapeless, matted tangle. Elastic fibers lose some of their elasticity, thicken into clumps, and fray, an effect that is greatly accelerated in the skin of smokers. Fibroblasts, which produce both collagen and elastic fibers, decrease in number. As a result, the skin forms the characteristic crevices and furrows known as wrinkles.

With further aging, Langerhans cells dwindle in number and macrophages become less-efficient phagocytes, thus decreasing the skin's immune responsiveness. Moreover, decreased size of sebaceous glands leads to dry and broken skin that is more susceptible to infection. Production of sweat diminishes, which probably contributes to the increased incidence of heat stroke in the elderly. There is a decrease in the number of functioning melanocytes, resulting in gray hair and atypical skin pigmentation. An increase in the size of some melanocytes produces pigmented blotching (age spots). Walls of blood vessels in the dermis become thicker and less permeable, and subcutaneous adipose tissue is lost. Aged skin (especially the dermis) is thinner than young skin, and the migration of cells from the basal layer to the epidermal surface slows considerably. With the onset of old age, skin heals poorly and becomes more susceptible to pathological conditions such as skin cancer and pressure sores.

Growth of nails and hair slows during the second and third decades of life. The nails also may become more brittle with age, often due to dehydration or repeated use of cuticle remover or nail polish.

• • •

To appreciate the many ways that skin contributes to homeostasis of other body systems, examine *Focus on Homeostasis: The Integumentary System.* This focus box is the first of ten, found at the end of selected chapters, that explain the how the body system under consideration contributes to the homeostasis of all the other body systems. The *Focus on Homeostasis* feature will help you understand how the individual body systems interact to contribute to the homeostasis of the entire body. Next, in Chapter 6, we will explore the how bone tissue is formed and how bones are assembled into the skeletal system, which protects many of our internal organs.

DISORDERS: HOMEOSTATIC IMBALANCES

Skin Cancer

Excessive exposure to the sun has caused virtually all of the one million cases of **skin cancer** diagnosed annually in the United States. There are three common forms of skin cancer. **Basal cell carcinomas** account for about 78% of all skin cancers. The tumors arise from cells in the stratum basale of the epidermis and rarely metastasize. **Squamous cell carcinomas,** which account for about 20% of all skin cancers, arise from squamous cells of the epidermis, and they have a variable tendency to metastasize. Most arise from preexisting lesions of damaged tissue on sun-exposed skin. Basal and squamous cell carcinomas are together known as *nonmelanoma skin cancer.* They are 50% more common in males than in females.

Malignant melanomas arise from melanocytes and account for about 2% of all skin cancers. They are the most prevalent life-threatening cancer in young women. The estimated lifetime risk of developing melanoma is now 1 in 75, double the risk only 15 years ago. In part, this increase is due to depletion of the ozone layer, which absorbs some UV light high in the atmosphere. But the main reason for the increase is that more people are spending more time in the sun and in tanning beds.

(text continues on page 156)

Body System	Contribution of the Integumentary System
For all body systems	Skin and hair provide barriers that protect all internal organs from damaging agents in the external environment; sweat glands and skin blood vessels help regulate body temperature, needed for proper functioning of other body systems.
Skeletal system	Skin helps activate vitamin D, needed for proper absorption of dietary calcium and phosphorus to build and maintain bones.
Muscular system 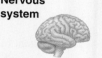	Skin helps provide calcium ions, needed for muscle contraction.
Nervous system	Nerve endings in skin and subcutaneous tissue provide input to the brain for touch pressure, thermal, and pain sensations.
Endocrine system	Keratinocytes in skin help activate vitamin D to calcitriol, a hormone that aids absorption of dietary calcium and phosphorus.
Cardiovascular system	Local chemical changes in dermis cause widening and narrowing of skin blood vessels, which help adjust blood flow to the skin.
Lymphatic and immune system	Skin is "first line of defense" in immunity, providing mechanical barriers and chemical secretions that discourage penetration and growth of microbes; Langerhans cells in epidermis participate in immune responses by recognizing and processing foreign antigens; macrophages in the dermis phagocytize microbes that penetrate the skin surface.
Respiratory system	Hairs in nose filter dust particles from inhaled air; stimulation of pain nerve endings in skin may alter breathing rate.
Digestive system	Skin helps activate vitamin D to the hormone calcitriol, which promotes absorption of dietary calcium and phosphorus in the small intestine.
Urinary system	Kidney cells receive partially activated vitamin D hormone from skin and convert it to calcitriol; some waste products are excreted from body in sweat, contributing to excretion by urinary system.
Reproductive systems	Nerve endings in skin and subcutaneous tissue respond to erotic stimuli, thereby contributing to sexual pleasure; suckling of a baby stimulates nerve endings in skin, leading to milk ejection; mammary glands (modified sweat glands) produce milk; skin stretches during pregnancy as fetus enlarges.

155

Malignant melanomas metastasize rapidly and can kill a person within months of diagnosis.

The key to successful treatment of malignant melanoma is early detection. The early warning signs of malignant melanoma are identified by the acronym ABCD (Figure 5.8). *A* is for *asymmetry;* malignant melanomas tend to lack symmetry. *B* is for *border;* malignant melanomas have irregular—notched, indented, scalloped, or indistinct—borders. *C* is for *color;* malignant melanomas have uneven coloration and may contain several colors. *D* is for *diameter;* ordinary moles typically are smaller than 6 mm (0.25 in.), about the size of a pencil eraser. Once a malignant melanoma has the characteristics of A, B, and C, it is usually larger than 6 mm.

Among the risk factors for skin cancer are the following:

1. *Skin type.* Individuals with light-colored skin who never tan but always burn are at high risk.

2. *Sun exposure.* People who live in areas with many days of sunlight per year and at high altitudes (where ultraviolet light is more intense) have a higher risk of developing skin cancer. Likewise, people who engage in outdoor occupations and those who have suffered three or more severe sunburns have a higher risk.

3. *Family history.* Skin cancer rates are higher in some families than in others.

4. *Age.* Older people are more prone to skin cancer owing to longer total exposure to sunlight.

5. *Immunological status.* Individuals who are immunosuppressed have a higher incidence of skin cancer.

Burns

A **burn** is tissue damage caused by excessive heat, electricity, radioactivity, or corrosive chemicals that denature the proteins in the skin cells. Burns destroy some of the skin's important contributions to homeostasis—protection against microbial invasion and desiccation, and thermoregulation.

Burns are graded according to their severity. A *first-degree burn* involves only the epidermis. It is characterized by mild pain and erythema (redness) but no blisters. Skin functions remain intact. Immediate flushing with cold water may lessen the pain and damage caused by a first-degree burn. Generally, healing of a first-degree burn will occur in 3 to 6 days and may be accompanied by flaking or peeling. One example of a first-degree burn is mild sunburn.

A *second-degree burn* destroys a portion of the epidermis and possibly parts of the dermis. Some skin functions are lost. In a second-degree burn, redness, blister formation, edema, and pain result. In a blister the epidermis has separated from the dermis due to the accumulation of tissue fluid between them. Associated structures, such as hair follicles, sebaceous glands, and sweat glands, usually are not injured. If there is no infection, second-degree burns heal without skin grafting in about 3 to 4 weeks, but scarring may result. First- and second-degree burns are collectively referred to as *partial-thickness burns.*

A *third-degree burn* or *full-thickness burn* destroys a portion of the epidermis, the underlying dermis, and associated structures. Most skin functions are lost. Such burns vary in appearance from marble-white to mahogany colored to charred, dry wounds. There is marked edema, and the burned region is numb because sensory nerve endings have been destroyed. Regeneration occurs slowly, and much granulation tissue forms before being covered by epithelium. Skin grafting may be required to promote healing and to minimize scarring.

The injury to the skin tissues directly in contact with the damaging agent is the *local effect* of a burn. Generally, however, the *systemic effects* of a major burn are a greater threat to life. The systemic effects of a burn may include (1) a large loss of water, plasma, and plasma proteins, which causes shock; (2) bacterial infection; (3) reduced circulation of blood; (4) decreased production of urine; and (5) diminished immune responses.

The seriousness of a burn is determined by its depth and extent of area involved, as well as the person's age and general health. According to the American Burn Association's classification of burn injury, a major burn includes third-degree burns over 10% of body surface area; or second-degree burns over 25% of body surface area; or any third-degree burns on the face, hands, feet, or perineum (per-i-NĒ-um, which includes the anal and urogenital regions). When the burn area exceeds 70%, more than half the victims die. A quick means for estimating the surface area affected by a burn in an adult is the **rule of nines** (Figure 5.9):

Figure 5.9   Rule of nines method for determining the extent of a burn. The percentages are the approximate proportions of the body surface area.

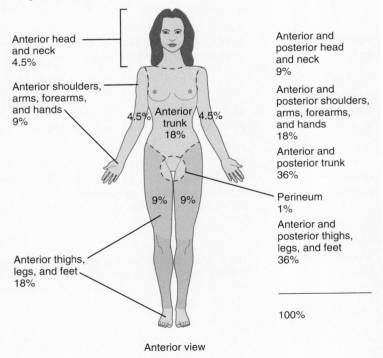

Anterior head and neck 4.5%

Anterior shoulders, arms, forearms, and hands 9%

4.5% Anterior trunk 18% 4.5%

9%   9%

Anterior thighs, legs, and feet 18%

Anterior and posterior head and neck 9%

Anterior and posterior shoulders, arms, forearms, and hands 18%

Anterior and posterior trunk 36%

Perineum 1%

Anterior and posterior thighs, legs, and feet 36%

100%

Anterior view

Figure 5.8   Comparison of a normal nevus (mole) and a malignant melanoma.

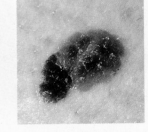

(a) Normal nevus (mole)   (b) Malignant melanoma

1. Count 9% if the anterior and posterior surfaces of the head and neck are affected.

2. Count 9% for the anterior and posterior surfaces of each upper limb (total of 18% for both upper limbs).

3. Count four times nine or 36% for the anterior and posterior surfaces of the trunk, including the buttocks.

4. Count 9% for the anterior and 9% for the posterior surfaces of each lower limb as far up as the buttocks (total of 36% for both lower limbs).

5. Count 1% for the perineum.

A more accurate way to estimate the amount of surface area affected by a burn is the **Lund–Browder method,** which estimates the extent by comparing the areas affected to the percentage of total surface area for body parts (Table 5.5). Because most of the proportions of the body change with growth, some percentages vary for different ages.

Pressure Ulcers

Pressure ulcers, also known as *decubitus ulcers* (dē-KŪ-bi-tus) or *bedsores,* are caused by a constant deficiency of blood flow to tissues. Typically the affected tissue overlies a bony projection that has been subjected to prolonged pressure against an object such as a bed, cast, or splint. If the pressure is relieved in a few hours, redness occurs but no lasting tissue damage results. Blistering of the affected area may indicate superficial damage, whereas a reddish-blue discoloration may indicate deep tissue damage. Prolonged pressure causes tissue ulceration. Small breaks in the epidermis become infected, and the sensitive subcutaneous layer and deeper tissues are damaged. Eventually, the tissue dies. Pressure ulcers occur most often in bedridden patients. With proper care, pressure ulcers are preventable.

Table 5.5   Lund–Browder Burn Assessment Chart

Area	Age (years)				
	0–1	1–4	5–9	10–15	Adult
Head	19	17	13	10	7
Neck	2	2	2	2	2
Anterior trunk	13	13	13	13	13
Posterior trunk	13	13	13	13	13
Right buttock	$2\frac{1}{2}$	$2\frac{1}{2}$	$2\frac{1}{2}$	$2\frac{1}{2}$	$2\frac{1}{2}$
Left buttock	$2\frac{1}{2}$	$2\frac{1}{2}$	$2\frac{1}{2}$	$2\frac{1}{2}$	$2\frac{1}{2}$
Genitalia	1	1	1	1	1
Right arm	4	4	4	4	4
Left arm	4	4	4	4	4
Right forearm	3	3	3	3	3
Left forearm	3	3	3	3	3
Right hand	$2\frac{1}{2}$	$2\frac{1}{2}$	$2\frac{1}{2}$	$2\frac{1}{2}$	$2\frac{1}{2}$
Left hand	$2\frac{1}{2}$	$2\frac{1}{2}$	$2\frac{1}{2}$	$2\frac{1}{2}$	$2\frac{1}{2}$
Right thigh	$5\frac{1}{2}$	$6\frac{1}{2}$	$8\frac{1}{2}$	$8\frac{1}{2}$	$9\frac{1}{2}$
Left thigh	$5\frac{1}{2}$	$6\frac{1}{2}$	$8\frac{1}{2}$	$8\frac{1}{2}$	$9\frac{1}{2}$
Right leg	5	5	$5\frac{1}{2}$	6	7
Left leg	5	5	$5\frac{1}{2}$	6	7
Right foot	$3\frac{1}{2}$	$3\frac{1}{2}$	$3\frac{1}{2}$	$3\frac{1}{2}$	$3\frac{1}{2}$
Left foot	$3\frac{1}{2}$	$3\frac{1}{2}$	$3\frac{1}{2}$	$3\frac{1}{2}$	$3\frac{1}{2}$

MEDICAL TERMINOLOGY

Cold sore   A lesion, usually in oral mucous membrane, caused by Type 1 herpes simplex virus (HSV) transmitted by oral or respiratory routes. The virus remains dormant until triggered by factors such as ultraviolet light, hormonal changes, and emotional stress. Also called a *fever blister.*

Contact dermatitis   (der-ma-TĪ-tis; *dermat-* = skin; *-itis* = inflammation of) Inflammation of the skin characterized by redness, itching, and swelling and caused by exposure of the skin to chemicals that bring about an allergic reaction, such as poison ivy toxin.

Corn   A painful conical thickening of the stratum corneum of the epidermis found principally over toe joints and between the toes, often caused by friction or pressure. Corns may be hard or soft, depending on their location. Hard corns are usually found over toe joints, and soft corns are usually found between the fourth and fifth toes.

Cutaneous anthrax   One of the three forms of anthrax that accounts for 95% of all naturally occurring anthrax. Endospores (spores) introduced at the site of a cut or abrasion produce a lesion that turns into a black scab and falls off in a week or two. The disease is characterized by low-grade fever, malaise (general feeling of discomfort), and enlarged lymph nodes. Without antibiotic treatment the mortality rate is as high as 20%.

Hemangioma   (he-man'-jē-Ō-ma; *hem-* = blood; *-angi-* = blood vessel; *-oma* = tumor) Localized tumor of the skin and subcutaneous layer that results from an abnormal increase in blood vessels. One type is a **portwine stain,** a flat, and pink, red, or purple lesion present at birth, usually at the nape of the neck.

Hives   Condition of the skin marked by reddened elevated patches that are often itchy. Most commonly caused by infections, physical trauma, medications, emotional stress, food additives, and certain food allergies. Also called *urticaria* (yur-ti- KAR-ē-a).

Impetigo   (im'-pe-TĪ-gō) Superficial skin infection caused by *Staphylococcus* bacteria; most common in children.

Intradermal   (in-tra-DER-mal; *intra-* = within) Within the skin. Also called *intracutaneous.*

Laceration   (las-er-Ā-shun; *lacer-* = torn) An irregular tear of the skin.

Nevus   (NĒ-vus) A round, flat, or raised area of pigmented skin that may be present at birth or may develop later. Varies in color from yellow-brown to black. Also called a *mole* or *birthmark.*

Pruritus   (proo-RĪ-tus; *pruri-* = to itch) Itching, one of the most common dermatological disorders. It may be caused by skin disorders (infections), systemic disorders (cancer, kidney failure), psychogenic factors (emotional stress), or allergic reactions.

Topical   In reference to a medication, applied to the skin surface rather than ingested or injected.

Wart   Mass produced by uncontrolled growth of epithelial skin cells; caused by a papillomavirus. Most warts are noncancerous.

STUDY OUTLINE

STRUCTURE OF THE SKIN (p. 140)

1. The integumentary system consists of the skin and its accessory structures—hair, nails, glands, muscles, and nerves.
2. The skin is the largest organ of the body in surface area and weight. The principal parts of the skin are the epidermis (superficial) and dermis (deep).
3. The subcutaneous layer (hypodermis) is deep to the dermis and not part of the skin. It anchors the dermis to underlying tissues and organs, and it contains lamellated (Pacinian) corpuscles.
4. The types of cells in the epidermis are keratinocytes, melanocytes, Langerhans cells, and Merkel cells.
5. The epidermal layers, from deep to superficial, are the stratum basale, stratum spinosum, stratum granulosum, stratum lucidum (in thick skin only), and stratum corneum. Stem cells in the stratum basale undergo continuous cell division, producing keratinocytes for the other layers.
6. The dermis consists of papillary and reticular regions. The papillary region is composed of areolar connective tissue containing fine elastic fibers, dermal papillae, and Meissner corpuscles. The reticular region is composed of dense irregular connective tissue containing interlaced collagen and coarse elastic fibers, adipose tissue, hair follicles, nerves, sebaceous (oil) glands, and ducts of sudoriferous (sweat) glands.
7. Epidermal ridges provide the basis for fingerprints and footprints.
8. Thin skin covers all parts of the body except for the palms, palmar surfaces of the digits, and the soles. Thick skin covers the palms, palmar surfaces of the digits, and soles.
9. The color of skin is due to melanin, carotene, and hemoglobin.

ACCESSORY STRUCTURES OF THE SKIN (p. 146)

1. Accessory structures of the skin—hair, skin glands, and nails—develop from the embryonic epidermis.
2. A hair consists of a shaft, most of which is superficial to the surface, a root that penetrates the dermis and sometimes the subcutaneous layer, and a hair follicle.
3. Associated with each hair follicle is a sebaceous (oil) gland, an arrector pili muscle, and a hair root plexus.
4. New hairs develop from division of matrix cells in the bulb; hair replacement and growth occur in a cyclic pattern consisting of alternating growth and resting stages.
5. Hairs offer a limited amount of protection—from the sun, heat loss, and entry of foreign particles into the eyes, nose, and ears. They also function in sensing light touch.
6. Sebaceous (oil) glands are usually connected to hair follicles; they are absent in the palms and soles. Sebaceous glands produce sebum, which moistens hairs and waterproofs the skin. Clogged sebaceous glands may produce acne.
7. There are two types of sudoriferous (sweat) glands: eccrine and apocrine. Eccrine sweat glands have an extensive distribution; their ducts terminate at pores at the surface of the epidermis. Apocrine sweat glands are limited to the skin of the axillae, groin, and areolae; their ducts open into hair follicles. They begin functioning at puberty and are stimulated during emotional stress and sexual excitement.

8. Ceruminous glands are modified sudoriferous glands that secrete cerumen. They are found in the external auditory canal (ear canal).
9. Lanugo of the fetus is shed before birth, except in the scalp, eyebrows, and eyelashes. Most body hair on males is terminal (coarse, pigmented); most body hair on females is vellus (fine).
10. Nails are hard, keratinized epidermal cells over the dorsal surfaces of the distal portions of the digits.
11. The principal parts of a nail are the nail body, free edge, nail root, lunula, eponychium, and matrix. Cell division of the matrix cells produces new nails.

FUNCTIONS OF THE SKIN (p. 151)

1. Skin functions include body temperature regulation, protection, sensation, excretion and absorption, and synthesis of vitamin D.
2. The skin participates in thermoregulation by liberating sweat at its surface and by adjusting the flow of blood in the dermis.
3. The skin provides physical, chemical, and biological barriers that help protect the body.
4. Cutaneous sensations include tactile sensations, thermal sensations, and pain.

MAINTAINING HOMEOSTASIS: SKIN WOUND HEALING (p. 152)

1. In an epidermal wound, the central portion of the wound usually extends down to the dermis, whereas the wound edges involve only superficial damage to the epidermal cells.
2. Epidermal wounds are repaired by enlargement and migration of basal cells, contact inhibition, and division of migrating and stationary basal cells.
3. During the inflammatory phase of deep wound healing, a blood clot unites the wound edges, epithelial cells migrate across the wound, vasodilation and increased permeability of blood vessels enhance delivery of phagocytes, and mesenchymal cells develop into fibroblasts.
4. During the migratory phase, fibroblasts migrate along fibrin threads and begin synthesizing collagen fibers and glycoproteins.
5. During the proliferative phase, epithelial cells grow extensively.
6. During the maturation phase, the scab sloughs off, the epidermis is restored to normal thickness, collagen fibers become more organized, fibroblasts begin to disappear, and blood vessels are restored to normal.

DEVELOPMENT OF THE INTEGUMENTARY SYSTEM (p. 154)

1. The epidermis develops from the embryonic ectoderm, and the accessory structures of the skin (hair, nails, and skin glands) are epidermal derivatives.
2. The dermis is derived from mesodermal cells.

AGING AND THE INTEGUMENTARY SYSTEM (p. 154)

1. Most effects of aging begin to occur when people reach their late forties.
2. Among the effects of aging are wrinkling, loss of subcutaneous adipose tissue, atrophy of sebaceous glands, and decrease in the number of melanocytes and Langerhans cells.

SELF-QUIZ QUESTIONS

Fill in the blanks in the following statements.

1. Thick skin can be found on the _____, _____ and _____.

2. The combination of collagen and elastic fibers in the reticular region of the dermis provides the skin with _____, _____, and _____.

3. Modified sweat glands in the ear are called _____ glands.

4. The pigments that give skin a wide variety of colors are _____, _____, and _____.

Choose the one best answer to the following questions.

5. The substance that prevents excessive evaporation of water from the skin, keeps the skin soft and pliable, and inhibits the growth of bacteria is (a) sebum, (b) sweat, (c) cerumen, (d) carotene, (e) melatonin.

6. Epidermal ridges (a) indicate the predominant direction of underlying collagen fiber bundles, (b) increase the grip of the hand or foot, (c) provide the pigments of skin color, (d) synthesize vitamin D in the epidermis of the skin.

7. To permanently remove a hair, you must destroy the (a) matrix, (b) shaft, (c) cuticle, (d) arrector pili, (e) external root sheath.

8. Which of the following statements are *true*? (1) Nails are composed of tightly packed, hard, keratinized cells of the epidermis that form a clear, solid covering over the dorsal surface of the terminal end of digits. (2) The free edge of the nail is white due to absence of capillaries. (3) Nails help us grasp and manipulate small objects. (4) Nails protect the ends of digits from trauma. (5) Nail color is due to a combination of melanin and carotene. (a) 1, 2, and 3, (b) 1, 3, and 4, (c) 1, 2, 3, and 4, (d) 2, 3, and 4, (e) 1, 3, and 5.

9. When performing surgery, the physician's scalpel would first cut through which layer of the epidermis? (a) stratum granulosum, (b) stratum basale, (c) stratum corneum, (d) stratum spinosum, (e) stratum lucidum.

10. Which of the following statements are *true*? (1) A first-degree burn involves only the surface epidermis. (2) In a second-degree burn, no skin function is lost. (3) First- and second-degree burns are collectively referred to as partial thickness burns. (4) A third-degree burn destroys the epidermis, dermis, and epidermal derivatives. (5) When burns exceed 20% of body surface area, more than half the victims die. (a) 1, 2, and 3, (b) 2, 3, and 4, (c) 3, 4, and 5, (d) 1, 2, and 4, (e) 1, 3, and 4.

11. Match the following:

    ____ (a) deep region of the dermis composed primarily of dense irregular connective tissue

    ____ (b) composed of keratinized stratified squamous epithelial tissue

    ____ (c) not considered part of the skin, it contains areolar and adipose tissues and blood vessels; attaches skin to underlying tissues and organs

    ____ (d) superficial region of the dermis; composed of areolar connective tissue

    (1) subcutaneous layer (hypodermis)
    (2) papillary region
    (3) reticular region
    (4) epidermis

12. Match the following:

    ____ (a) a skin disorder in which keratinocytes divide and move more quickly to the skin surface, are shed prematurely, and make an abnormal keratin

    ____ (b) an inherited inability to produce melanin

    ____ (c) a condition in which a partial or complete loss of melanocytes from patches of skin produce irregular white spots

    ____ (d) a yellowed appearance of the whites of the eyes and of light colored skin, usually indicating liver disease

    ____ (e) a bluish appearance of mucous membranes, nail beds, and light-colored skin due to oxygen depletion

    ____ (f) redness of the skin caused by engorgement of capillaries in the dermis with blood

    ____ (g) partial or complete lack of hair

    ____ (h) an inflammation of sebaceous glands

    ____ (i) a precancerous skin lesion induced by sunlight

    ____ (j) the most common form of skin cancer

    ____ (k) a skin sore caused by long-standing, pressure-induced deficiency of blood flow to tissues overlying a bony projection

    ____ (l) the process of scar-tissue formation

    ____ (m) a rapidly metastasizing, potentially fatal form of skin cancer.

    (1) fibrosis
    (2) erythema
    (3) vitiligo
    (4) cyanosis
    (5) psoriasis
    (6) alopecia
    (7) jaundice
    (8) acne
    (9) malignant melanoma
    (10) solar keratosis
    (11) decubitus ulcer
    (12) basal cell carcinoma
    (13) albinism

13. Match the following:

_____ (a) produce the protein that helps protect the skin and underlying tissues from light, heat, microbes, and many chemicals

_____ (b) produce the pigment that contributes to skin color and absorbs ultraviolet light

_____ (c) cells that arise from red bone marrow, migrate to the epidermis, and participate in immune responses

_____ (d) cells thought to function in the sensation of touch

_____ (e) an abnormal thickening of the epidermis

_____ (f) release a lipid-rich secretion that functions as a water-repellent sealant in the stratum granulosum

_____ (g) pressure-sensitive cells found mostly in the subcutaneous layer

_____ (h) a fatty substance that covers and protects the skin of the fetus from the constant exposure to amniotic fluid

(1) Merkel cells
(2) callus
(3) keratinocytes
(4) Langerhans cells
(5) melanocytes
(6) lamellar granules
(7) lamellated (Pacinian) corpuscles
(8) vernix caseosa

14. Match the following:

_____ (a) liberation of sweat at the surface and adjustment of blood flow in the dermis

_____ (b) provision of a chemical barrier and resistance to microbial invasion

_____ (c) input of touch, pressure, pain, and heat and cold information

_____ (d) elimination of usually unneeded substances and passage of materials from the external environment into body cells

_____ (e) production of calcitriol

(1) protection
(2) sensation
(3) thermoregulation
(4) vitamin D synthesis
(5) excretion and absorption

15. Match the following and place the phases in the correct order:

_____ (a) epithelial cells migrate under scab to bridge the wound; formation of granulation tissue

_____ (b) sloughing of scab; reorganization of collagen fibers; blood vessels return to normal

_____ (c) vasodilation and increased permeability of blood vessels to deliver cells involved in phagocytosis; clot formation

_____ (d) extensive growth of epithelial cells beneath scab; random deposition of collagen fibers; continued growth of blood vessels

(1) proliferative phase
(2) inflammatory phase
(3) maturation phase
(4) migratory phase

Correct order of phases:

1) _____, 2) _____, 3) _____, 4) _____,

CRITICAL THINKING QUESTIONS

1. The amount of dust that collects in a house with an assortment of dogs, cats, and people is truly amazing. A lot of these dust particles had a previous "life" as part of the home's living occupants. Where did the dust originate on the human body?
   HINT *Dogs, cats, and humans all shed in their own way.*

2. Kiko is fastidious about personal hygiene. She's convinced that body secretions on the skin are unsanitary and wants to have all her exocrine glands removed. Is this wise?
   HINT *There are about 3 to 4 million sudoriferous glands alone.*

3. Your nephew has been learning about cells in science class, and now he refuses to take a bath. He asks, "If all cells have a semipermeable membrane, and the skin is made out of cells, then won't I swell up and pop if I take a bath?" Explain the relevant anatomy to your nephew before he starts attracting flies.
   HINT *Semipermeable membranes function in living cells.*

ANSWERS TO FIGURE QUESTIONS

5.1 The epidermis is epithelial tissue, whereas the dermis is connective tissue.

5.2 Melanin protects the DNA of the nucleus of keratinocytes from damage from UV light.

5.3 The stratum basale is the layer of the epidermis that contains stem cells.

5.4 Plucking a hair stimulates hair root plexuses in the dermis, some of which are sensitive to pain. Because the cells of a hair shaft are already dead and the hair shaft lacks nerves, cutting hair is not painful.

5.5 Nails are hard because they are composed of tightly packed, hard, keratinized epidermal cells.

5.6 Epidermal wounds do not bleed because there are no blood vessels in the epidermis.

5.7 In addition to phagocytizing microbes, neutrophils and macrophages can help clean up cellular debris that results from the wound.

Appendix A
Measurements

U.S. Customary System

Parameter	Unit	Relation to Other U.S. Units	SI (Metric) Equivalent
Length	inch	1/12 foot	2.54 centimeters
foot	12 inches	0.305 meter	
yard	36 inches	9.144 meters	
mile	5,280 feet	1.609 kilometers	
Mass	grain	1/1000 pound	64.799 milligrams
dram	1/16 ounce	1.772 grams	
ounce	16 drams	28.350 grams	
pound	16 ounces	453.6 grams	
ton	2,000 pounds	907.18 kilograms	
Volume (Liquid)	ounce	1/16 pint	29.574 milliliters
pint	16 ounces	0.473 liter	
quart	2 pints	0.946 liter	
gallon	4 quarts	3.785 liters	
Volume (Dry)	pint	1/2 quart	0.551 liter
quart	2 pints	1.101 liters	
peck	8 quarts	8.810 liters	
bushel	4 pecks	35.239 liters	

International System (SI)

BASE UNITS

Unit	Quantity	Symbol
meter	length	M
kilogram	mass	Kg
second	time	S
liter	volume	L
mole	amount of matter	Mol

PREFIXES

Prefix	Multiplier	Symbol
tera-	$10^{12} = 1,000,000,000,000$	T
giga-	$10^9 = 1,000,000,000$	G
mega-	$10^6 = 1,000,000$	M
kilo-	$10^3 = 1,000$	k
hecto-	$10^2 = 100$	h
deca-	$10^1 = 10$	da
deci-	$10^{-1} = 0.1$	d
centi-	$10^{-2} = 0.01$	c
milli-	$10^{-3} = 0.001$	m
micro-	$10^{-6} = 0.000,001$	μ
nano-	$10^{-9} = 0.000,000,001$	n
pico-	$10^{-12} = 0.000,000,000,001$	p

Temperature Conversion

Fahrenheit (F) To Celsius (C)

$$°C = (°F - 32) ÷ 1.8$$

Celsius (C) To Fahrenheit (F)

$$°F = (°C × 1.8) + 32$$

U.S TO SI (Metric) Conversion

When you know	Multiply by	To find
inches	2.54	centimeters
feet	30.48	centimeters
yards	0.91	meters
miles	1.61	kilometers
ounces	28.35	grams
pounds	0.45	kilograms
tons	0.91	metric tons
fluid ounces	29.57	milliliters
pints	0.47	liters
quarts	0.95	liters
gallons	3.79	liters

SI (Metric) TO U.S. Conversion

When you know	Multiply by	To find
millimeters	0.04	inches
centimeters	0.39	inches
meters	3.28	feet
kilometers	0.62	miles
liters	1.06	quarts
cubic meters	35.32	cubic feet
grams	0.035	ounces
kilograms	2.21	pounds

Appendix B
Periodic Table

The periodic table lists the known **chemical elements,** the basic units of matter. The elements in the table are arranged left-to-right in rows in order of their **atomic number,** the number of protons in the nucleus. Each horizontal row, numbered from 1 to 7, is a **period.** All elements in a given period have the same number of electron shells as their period number. For example, an atom of hydrogen or helium each has one electron shell, while an atom of potassium or calcium each has four electron shells. The elements in each column, or **group,** share chemical properties. For example, the elements in column IA are very chemically reactive, whereas the elements in column VIIIA have full electron shells and thus are chemically inert.

Scientists now recognize 113 different elements; 92 occur naturally on Earth, and the rest are produced from the natural elements using particle accelerators or nuclear reactors. Elements are designated by **chemical symbols,** which are the first one or two letters of the element's name in English, Latin, or another language.

Twenty-six of the 92 naturally occurring elements normally are present in your body. Of these, just four elements—oxygen (O), carbon (C), hydrogen (H), and nitrogen (N) (coded blue)—constitute about 96% of the body's mass. Eight others—calcium (Ca), phosphorus (P), potassium (K), sulfur (S), sodium (Na), chlorine (Cl), magnesium (Mg), and iron (Fe) (coded pink)—contribute 3.8% of the body's mass. An additional 14 elements, called **trace elements** because they are present in tiny amounts, account for the remaining 0.2% of the body's mass. The trace elements are aluminum, boron, chromium, cobalt, copper, fluorine, iodine, manganese, molybdenum, selenium, silicon, tin, vanadium, and zinc (coded yellow). Table 2.1 on page 28 provides information about the main chemical elements in the body.

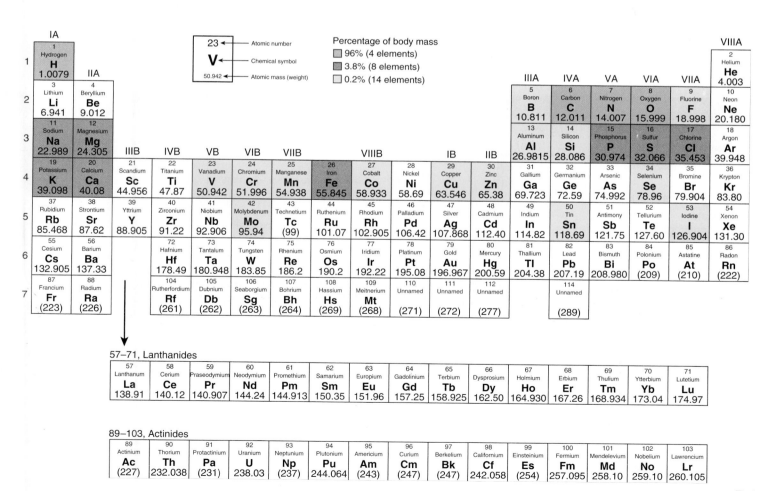

Appendix C
Normal Values For Selected Blood Tests

The system of international (SI) units (Système Internationale d'Unités) is used in most countries and in many medical and scientific journals. Clinical laboratories in the United States, by contrast, usually report values for blood and urine tests in conventional units. The laboratory values in this Appendix give conventional units first, followed by SI equivalents in parentheses. Values listed for various blood tests should be viewed as reference values rather than absolute "normal" values for all well people. Values may vary due to age, gender, diet, and environment of the subject or the equipment, methods, and standards of the lab performing the measurement.

Key To Symbols

g = gram
mg = milligram = 10^{-3} gram
μg = microgram = 10^{-6} gram
U = units
L = liter
dL = deciliter
mL = milliliter
μL = microliter
mEq/L = milliequivalents per liter
mmol/L = millimoles per liter
μmol/L = micromoles per liter
> = greater than; < = less than

Blood Tests

Test (Specimen)	U.S. Reference Values (SI Units)	Values Increase In	Values Decrease In
Aminotransferases (serum)			
Alanine aminotransferase (ALT)	0–35 U/L (same)	Liver disease or liver damage due to toxic drugs.	
Aspartate aminotransferase (AST)	0–35 U/L (same)	Myocardial infarction, liver disease, trauma to skeletal muscles, severe burns.	Beriberi, uncontrolled diabetes mellitus with acidosis, pregnancy.
Ammonia (plasma)	20–120 μg/dL (12–55 μmol/L)	Liver disease, heart failure, emphysema, pneumonia, hemolytic disease of newborn.	Hypertension.
Bilirubin (serum)	Conjugated: <0.5 mg/dL (<5.0 μmol/L)	Conjugated bilirubin: liver dysfunction or gallstones.	
	Unconjugated: 0.2–1.0 mg/dL (18–20 μmol/L) Newborn: 1.0–12.0 mg/dL (<200 μmol/L)	Unconjugated bilirubin: excessive hemolysis of red blood cells.	
Blood urea nitrogen (BUN) (serum)	8–26 mg/dL (2.9–9.3 mmol/L)	Kidney disease, urinary tract obstruction, shock, diabetes, burns, dehydration, myocardial infarction.	Liver failure, malnutrition, overhydration, pregnancy.

Blood Tests

Test (Specimen)	U.S. Reference Values (SI Units)	Values Increase In	Values Decrease In
Carbon dioxide content (bicarbonate + dissolved CO_2) (whole blood)	Arterial: 19–24 mEq/L (19–24 mmol/L) Venous: 22–26 mEq/L (22–26 mmol/L)	Severe diarrhea, severe vomiting, starvation, emphysema, aldosteronism.	Renal failure, diabetic ketoacidosis, shock.
Cholesterol, total (plasma)	<200 mg/dL (<5.2 mmol/L) is desirable	Hypercholesterolemia, uncontrolled diabetes mellitus, hypothyroidism, hypertension, atherosclerosis, nephrosis.	Liver disease, hyperthyroidism, fat malabsorption, pernicious or hemolytic anemia, severe infections.
HDL cholesterol (plasma)	>40 mg/dL (>1.0 mmol/L) is desirable		
LDL cholesterol (plasma)	<130 mg/dL (<3.2 mmol/L) is desirable		
Creatine (serum)	Males: 0.15–0.5 mg/dL (10–40 μmol/L) Females: 0.35–0.9 mg/dL (30–70 μmol/L)	Muscular dystrophy, damage to muscle tissue, electric shock, chronic alcoholism.	
Creatine Kinase (CK), also known as Creatine phosphokinase (CPK) (serum)	0–130 U/L (same)	Myocardial infarction, progressive muscular dystrophy, hypothyroidism, pulmonary edema.	
Creatinine (serum)	0.5–1.2 mg/dL (45–105 μmol/L)	Impaired renal function, urinary tract obstruction, giantism, acromegaly.	Decreased muscle mass, as occurs in muscular dystrophy or myasthenia gravis.
Electrolytes (plasma)	See Table 27.2 on page 1000.		
Gamma-glutamyl transferase (GGT) (serum)	0–30 U/L (same)	Bile duct obstruction, cirrhosis, alcoholism, metastatic liver cancer, congestive heart failure.	
Glucose (plasma)	70–110 mg/dL (3.9–6.1 mmol/L)	Diabetes mellitus, acute stress, hyperthyroidism, chronic liver disease, Cushing's disease.	Addison's disease, hypothyroidism, hyperinsulinism.
Hemoglobin (whole blood)	Males: 14–18 g/100 mL (140–180 g/L) Females: 12–16 g/100 mL (120–160 g/L) Newborns: 14–20 g/100 mL (140–200 g/L)	Polycythemia, congestive heart failure, chronic obstructive pulmonary disease, living at high altitude.	Anemia, severe hemorrhage, cancer, hemolysis, Hodgkin's disease, nutritional deficiency of vitamin B_{12}, systemic lupus erythematosus, kidney disease.
Iron, total (serum)	Males: 80–180 μg/dL (14–32 μmol/L) Females: 60–160 μg/dL (11–29 μmol/L)	Liver disease, hemolytic anemia, iron poisoning.	Iron-deficiency anemia, chronic blood loss, pregnancy (late), chronic heavy menstruation.
Lactic dehydrogenase (LDH) (serum)	71–207 U/L (same)	Myocardial infarction, liver disease, skeletal muscle necrosis, extensive cancer.	
Lipids (serum)		Hyperlipidemia, diabetes mellitus, hypothyroidism.	Fat malabsorption.
Total	400–850 mg/dL (4.0–8.5 g/L)		
Triglycerides	10–190 mg/dL (0.1–1.9 g/L)		
Platelet (thrombocyte) count (whole blood)	150,000–400,000/μL	Cancer, trauma, leukemia, cirrhosis.	Anemias, allergic conditions hemorrhage.
Protein (serum)		Dehydration, shock, chronic infections.	Liver disease, poor protein intake, hemorrhage, diarrhea, malabsorption, chronic renal failure, severe burns.
Total	6–8 g/dL (60–80 g/L)		
Albumin	4–6 g/dL (40–60 g/L)		
Globulin	2.3–3.5 g/dL (23–35 g/L)		

(continues)

Blood Tests

Test (Specimen)	U.S. Reference Values (SI Units)	Values Increase In	Values Decrease In
Red blood cell (erythrocyte) count (whole blood)	Males: 4.5–6.5 million/μL Females: 3.9–5.6 million/μL	Polycythemia, dehydration, living at high altitude.	Hemorrhage, hemolysis, anemias, cancer, overhydration.
Uric acid (urate) (serum)	2.0–7.0 mg/dL (120–420 μmol/L)	Impaired renal function, gout, metastatic cancer, shock, starvation.	
White blood cell (leukocyte) count, total (whole blood)	5,000–10,000/μL (See Table 19.3 on page 646 for relative percentages of different types of WBCs.)	Acute infections, trauma, malignant diseases, cardiovascular diseases. (See also Table 19.2 on page 645.)	Diabetes mellitus, anemia. (See also Table 19.2 on page 645.)

Appendix D
Normal Values For Selected Urine Tests

Urine Tests

Test (Specimen)	U.S. Reference Values (SI Units)	Clinical Implications
Amylase (2 hour)	35–260 Somogyi units/hr (6.5–48.1 units/hr)	Values increase in inflammation of the pancreas (pancreatitis) or salivary glands, obstruction of the pancreatic duct, and perforated peptic ulcer.
Bilirubin* (random)	Negative	Values increase in liver disease and obstructive biliary disease.
Blood* (random)	Negative	Values increase in renal disease, extensive burns, transfusion reactions, and hemolytic anemia.
Calcium (Ca²⁺) (random)	10 mg/dL (2.5 mmol/liter); up to 300 mg/24 hr (7.5 mmol/24 hr)	Amount depends on dietary intake; values increase in hyperparathyroidism, metastatic malignancies, and primary cancer of breasts and lungs; values decrease in hypoparathyroidism and vitamin D deficiency.
Casts (24 hour)		
Epithelial	Occasional	Values increase in nephrosis and heavy metal poisoning.
Granular	Occasional	Values increase in nephritis and pyelonephritis.
Hyaline	Occasional	Values increase in kidney infections.
Red blood cell	Occasional	Values increase in glomerular membrane damage and fever.
White blood cell	Occasional	Values increase in pyelonephritis, kidney stones, and cystitis.
Chloride (Cl⁻) (24 hour)	140–250 mEq/24 hr (140–250 mmol/24 hr)	Amount depends on dietary salt intake; values increase in Addison's disease, dehydration, and starvation; values decrease in pyloric obstruction, diarrhea, and emphysema.
Color (random)	Yellow, straw, amber	Varies with many disease states, hydration, and diet.
Creatinine (24 hour)	Males: 1.0–2.0 g/24 hr (9–18 mmol/24 hr) Females: 0.8–1.8 g/24 hr (7–16 mmol/24 hr)	Values increase in infections; values decrease in muscular atrophy, anemia, and kidney diseases.
Glucose*	Negative	Values increase in diabetes mellitus, brain injury, and myocardial infarction.
Hydroxycortico-steroids (17-hydroxysteroids) (24 hour)	Males: 5–15 mg/24 hr (13–41 μmol/24 hr) Females: 2–13 mg/24 hr (5–36 μmol/24 hr)	Values increase in Cushing's syndrome, burns, and infections; values decrease in Addison's disease.
Ketone bodies* (random)	Negative	Values increase in diabetic acidosis, fever, anorexia, fasting, and starvation.
17-ketosteroids (24 hour)	Males: 8–25 mg/24 hr (28–87 μmol/24 hr) Females: 5–15 mg/24 hr (17–53 μmol/24 hr)	Values decrease in surgery, burns, infections, adrenogenital syndrome, and Cushing's syndrome.
Odor (random)	Aromatic	Becomes acetonelike in diabetic ketosis.
Osmolality (24 hour)	500–1400 mOsm/kg water (500–1400 mmol/kg water)	Values increase in cirrhosis, congestive heart failure (CHF), and high-protein diets; values decrease in aldosteronism, diabetes insipidus, and hypokalemia.
pH* (random)	4.6–8.0	Values increase in urinary tract infections and severe alkalosis; values decrease in acidosis, emphysema, starvation, and dehydration.

Test (Specimen)	U.S. Reference Values (SI Units)	Clinical Implications
Phenylpyruvic acid (random)	Negative	Values increase in phenylketonuria (PKU).
Potassium (K⁺) (24 hour)	40−80 mEq/24 hr (40−80 mmol/24 hr)	Values increase in chronic renal failure, dehydration, starvation, and Cushing's syndrome; values decrease in diarrhea, malabsorption syndrome, and adrenal cortical insufficiency
Protein* (albumin) (random)	Negative	Values increase in nephritis, fever, severe anemias, trauma, and hyperthyroidism.
Sodium (Na⁺) (24 hour)	75−200 mg/24 hr (75−200 mmol/24 hr)	Amount depends on dietary salt intake; values increase in dehydration, starvation, and diabetic acidosis; values decrease in diarrhea, acute renal failure, emphysema, and Cushing's syndrome.
Specific gravity* (random)	1.001−1.035 (same)	Values increase in diabetes mellitus and excessive water loss; values decrease in absence of antidiuretic hormone (ADH) and severe renal damage.
Urea (random)	25−35 g/24 hr (420−580 mmol/24 hr)	Values increase in response to increased protein intake; values decrease in impaired renal function.
Uric acid (24 hour)	0.4−1.0 g/24 hr (1.5−4.0 mmol/24 hr)	Values increase in gout, leukemia, and liver disease; values decrease in kidney disease.
Urobilinogen* (2 hour)	0.3−1.0 Ehrlich units (1.7−6.0 μmol/24 hr)	Values increase in anemias, hepatitis A (infectious), biliary disease, and cirrhosis; values decrease in cholelithiasis and renal insufficiency.
Volume, total (24 hour)	1000−2000 mL/24 hr (1.0−2.0 liters/24 hr)	Varies with many factors.

*Test often performed using a **dipstick,** a plastic strip impregnated with chemicals that is dipped into a urine specimen to detect particular substances. Certain colors indicate the presence or absence of a substance and sometimes give a rough estimate of the amount(s) present.

Appendix E
Answers

Answers to Self-Quiz Questions

Chapter 1
1. cell **2.** catabolism **3.** homeostasis **4.** false **5.** true **6.** true **7.** d
8. b **9.** b **10.** e **11.** e **12.** (a) 3, (b) 4, (c) 6, (d) 5, (e) 1, (f) 2
13. (a) 6, (b) 1, (c) 11, (d) 5, (e) 10, (f) 8, (g) 7, (h) 9, (i) 4, (j) 3, (k) 2
14. (a) 4, (b) 6, (c) 8, (d) 1, (e) 5, (f) 2, (g) 7, (h) 3 **15.** (a) 3, (b) 7,
(c) 1, (d) 5, (e) 6, (f) 4, (g) 8, (h) 2

Chapter 2
1. protons, electrons, neutrons **2.** 8 **3.** atomic number **4.** concentration, temperature **5.** true **6.** false **7.** b **8.** d **9.** a **10.** c **11.** b
12. (a) 1, (b) 2 (c) 1, (d) 4 **13.** a **14.** (a) 8, (b) 1, (c) 7, (d) 2, (e) 5, (f)
3, (g) 4, (h) 6 **15.** (a) 7, (b) 5, (c) 1, (d) 2, (e) 6, (f) 3, (g) 8, (h) 4

Chapter 3
1. plasma membrane, cytoplasm, nucleus **2.** apoptosis, necrosis
3. cytosol **4.** shortening and loss of protective telomeres on
chromosomes, cross-link formation between glucose and proteins, free
radical formation **5.** true **6.** true **7.** a **8.** d **9.** b **10.** a **11.** c
12. (a) 2, (b) 3, (c) 5, (d) 7, (e) 6, (f) 8, (g) 1, (h) 4 **13.** (a) 11, (b) 10,
(c) 1, (d) 2, (e) 5, (f) 9, (g) 12, (h) 7, (i) 6, (j) 3, (k) 4, (l) 8 **14.** (a) 2,
(b) 9, (c) 3, (d) 5, (e) 11, (f) 8, (g) 1, (h) 6, (i) 10, (j) 7, (k) 4 **15.** (a) 3,
(b) 9, (c) 1, (d) 5, (e) 11, (f) 4, (g) 8, (h) 7, (i) 2, (j) 10, (k) 6

Chapter 4
1. tissue **2.** endocrine **3.** cells, ground substance, fibers **4.** true **5.** false
6. a **7.** c **8.** b **9.** c **10.** b **11.** d **12.** (a) 4, (b) 8, (c) 5, (d) 2, (e) 3, (f)
6, (g) 1, (h) 7 **13.** (a) 3, (b) 5, (c) 8, (d) 12, (e) 9, (f) 7, (g) 11, (h) 6, (i) 2,
(j) 4, (k) 10, (l) 1 **14.** (a) 2, (b) 5, (c) 3, (d) 4, (e) 1 **15.** (a) 3, (b) 2, (c) 1

Chapter 5
1. palms, soles, fingertips **2.** strength, extensibility, elasticity
3. ceruminous **4.** melanin, carotene, hemoglobin **5.** a **6.** b **7.** a
8. c **9.** c **10.** e **11.** (a) 3, (b) 4, (c) 1, (d) 2 **12.** (a) 5, (b) 13, (c) 3,
(d) 7, (e) 4, (f) 2, (g) 6, (h) 8, (i) 10, (j) 12, (k) 11, (l) 1, (m) 9
13. (a) 3, (b) 5, (c) 4, (d) 1, (e) 2, (f) 6, (g) 7, (h) 8 **14.** (a) 3, (b) 1,
(c) 2, (d) 5, (e) 4 **15.** (a) 4, (b) 3, (c) 2, (d) 1, inflammatory, migratory, proliferative, maturation

Chapter 6
1. hardness, tensile strength **2.** osteons (Haversian systems), trabeculae **3.** hyaline cartilage, loose fibrous connective tissue membranes
4. true **5.** false **6.** b **7.** b **8.** a **9.** d **10.** a **11.** d **12.** (a) 3, (b) 9,
(c) 8, (d) 1, (e) 5, (f) 4, (g) 6, (h) 7, (i) 2, (j) 10 **13.** (a) 10, (b) 2, (c) 7,
(d) 1, (e) 8, (f) 3, (g) 6, (h) 4, (i) 9, (j) 5 **14.** (a) 9, (b) 7, (c) 6, (d) 1,
(e) 4, (f) 2, (g) 5, (h) 3, (i) 8, (j) 10 **15.** (a) 2, (b) 5, (c) 3, (d) 4, (e) 1

Chapter 7
1. fontanels **2.** thoracic **3.** intervertebral discs **4.** pituitary gland
5. false **6.** false **7.** a **8.** c **9.** b **10.** e **11.** d **12.** (a) 7, (b) 5, (c) 1,
(d) 6, (e) 2, (f) 4, (g) 8, (h) 9, (i) 3, (j) 10, (k) 11, (l) 13, (m) 12
13. (a) 2, (b) 3, (c) 5, (d) 6, (e) 4, (f) 1, (g) 5, (h) 4, (i) 2, (j) 4, (k) 3
14. (a) 4, (b) 7, (c) 6, (d) 5, (e) 3, (f) 1, (g) 2 **15.** (a) 3, (b) 1, (c) 6,
(d) 9, (e) 13, (f) 12, (g) 2, (h) 4, (i) 5, (j) 7, (k) 10, (l) 15, (m) 8, (n)
11, (o) 14

Chapter 8
1. protection, movement **2.** metacarpals **3.** true **4.** true **5.** b **6.** b
7. d **8.** e **9.** e **10.** c **11.** a **12.** b **13.** b **14.** (a) 4, (b) 3, (c) 3,
(d) 6, (e) 7, (f) 1, (g) 3, (h) 2, (i) 5, (j) 9, (k) 8, (l) 2, (m) 4, (n) 6, (o) 7
15. (a) 2, (b) 6, (c) 9, (d) 7, (e) 4, (f) 5, (g) 8, (h) 10, (i) 1, (j) 3

Chapter 9
1. joint, articulation, or arthrosis **2.** synovial fluid **3.** osteoarthritis
4. d **5.** b **6.** a **7.** c **8.** e **9.** e **10.** true **11.** false **12.** (a) 5, (b) 3,
(c) 6, (d) 1, (e) 4, (f) 7, (g) 2 **13.** (a) 4, (b) 3, (c) 6, (d) 2, (e) 5, (f) 1
14. (a) 6, (b) 4, (c) 5, (d) 1, (e) 3, (f) 2 **15.** (a) 6, (b) 7, (c) 9, (d) 4, (e)
8, (f) 3, (g) 10, (h) 2, (i) 11, (j) 1, (k) 5

Chapter 10
1. electrical excitability, contractility, extensibility, elasticity **2.** actin,
myosin **3.** motor unit **4.** false **5.** true **6.** e **7.** a **8.** b **9.** e **10.** d
11. c **12.** (a) 5, (b) 6, (c) 8, (d) 7, (e) 2, (f) 4, (g) 3, (h) 1 **13.** (a) 5, (b)
8, (c) 7, (d) 10, (e) 6, (f) 9, (g) 4, (h) 1, (i) 2, (j) 3 **14.** (a) 2, (b) 3, (c) 1,
(d) 1 and 2 (e) 2 and 3, (f) 2, (g) 1, (h) 3, (i) 1 and 2, (j) 3 **15.** (a) 10,
(b) 2, (c) 4, (d) 3, (e) 6, (f) 5, (g) 1, (h) 7, (i) 9, (j) 8

Chapter 11
1. origin, insertion **2.** buccinator **3.** true **4.** true **5.** d **6.** e **7.** a
8. e **9.** c **10.** a **11.** b **12.** (a) 12, (b) 9, (c) 8, (d) 6, (e) 3, (f) 11, (g)
10, (h) 1, (i) 2, (j) 7, (k) 4, (l) 5 **13.** (a) 3, (b) 1, (c) 2, (d) 1, (e) 2, (f) 3,
(g) 3 **14.** (a) 6, (b) 3, (c) 7, (d) 4, (e) 2, (f) 8, (g) 5, (h) 1 **15.** (a) 10,
(b) 1, (c) 9, (d) 8, (e) 12, (f) 17, (g) 2, (h) 6, (i) 8, (j) 14, (k) 5, (l) 4, (m)
2, (n) 15, (o) 1, (p) 11, (q) 13, (r) 12, (s) 7, (t) 16, (u) 11, (v) 17, (w) 16,
(x) 15, (y) 3

Chapter 12
1. somatic nervous system, autonomic nervous system, enteric nervous
system **2.** polarized **3.** true **4.** false **5.** c **6.** b **7.** a **8.** e **9.** c
10. e **11.** b **12.** a **13.** (a) 6, (b) 11, (c) 1, (d) 2, (e) 9, (f) 13, (g) 4, (h)
8, (i) 7, (j) 12, (k) 5, (l) 3, (m) 10 **14.** (a) 2, (b) 1, (c) 9, (d) 8, (e) 6, (f)
3, (g) 4, (h) 5, (i) 10, (j) 7 **15.** (a) 4, (b) 5, (c) 14, (d) 8, (e) 7, (f) 1,
(g) 2, (h) 10, (i) 13, (j) 6, (k) 3, (l) 12, (m) 9, (n) 11

Chapter 13

1. reflexes **2.** mixed **3.** true **4.** true **5.** d **6.** e **7.** d **8.** a **9.** c **10.** (a) 1, (b) 8, (c) 4, (d) 2, (e) 10, (f) 1, (g) 6, (h) 5, (i) 3, (j) 9, (k) 1, (l) 11, (m) 7, (n) 2 **11.** a **12.** b **13.** e **14.** (a) 10, (b) 8, (c) 9, (d) 1, (e) 2, (f) 5, (g) 7, (h) 6, (i) 11, (j) 4, (k) 3 **15.** (a) 2, (b) 1, (c) 3, (d) 4, (e) 1, (f) 2, (g) 4, (h) 3, (i) 5, (j) 1

Chapter 14

1. choroid plexuses **2.** longitudinal fissue **3.** true **4.** false **5.** c **6.** d **7.** e **8.** d **9.** e **10.** a **11.** d **12.** (a) 3, (b) 5, (c) 6, (d) 8, (e) 11, (f) 10, (g) 7, (h) 9, (i) 1, (j) 4, (k) 2, (l) 12, (m) 1, (n) 8, (o) 5, (p) 7, (q) 12, (r) 10, (s) 9 **13.** (a) 5, (b) 9, (c) 11, (d) 6, (e) 3, (f) 1, (g) 10, (h) 8, (i) 2, (j) 4, (k) 7 **14.** (a) 9, (b) 2, (c) 6, (d) 10, (e) 4, (f) 11, (g) 1, (h) 2, (i) 5, (j) 8, (k) 12, (l) 7, (m) 3, (n) 6 and 8, (o) 13, (p) 7, (q) 1 **15.** (a) 9, (b) 2, (c) 6, (d) 8, (e) 7, (f) 5, (g) 3, (h) 10, (i) 12, (j) 11, (k) 4, (l) 1

Chapter 15

1. sensation, perception **2.** adaptation **3.** true **4.** true **5.** d **6.** b **7.** a **8.** d **9.** e **10.** e **11.** d **12.** d **13.** (a) 9, (b) 8, (c) 4, (d) 7, (e) 10, (f) 2, (g) 3, (h) 1, (i) 5, (j) 6, (k) 11 **14.** (a) 10, (b) 8, (c) 7, (d) 1, (e) 4, (f) 3, (g) 5, (h) 6, (i) 9, (j) 2 **15.** (a) 3, (b) 2, (c) 5, (d) 7, (e) 1, (f) 3, (g) 8, (h) 4, (i) 6

Chapter 16

1. sweet, sour, salty, bitter, umami **2.** static, dynamic **3.** true **4.** false **5.** a **6.** d **7.** c **8.** c **9.** c **10.** c **11.** e **12.** a **13.** (a) 1, (b) 5, (c) 7, (d) 6, (e) 1, (f) 8, (g) 2, (h) 4, (i) 3 **14.** (a) 3, (b) 6, (c) 8, (d) 12, (e) 1, (f) 5, (g) 9, (h) 11, (i) 7, (j) 13, (k) 2, (l) 10, (m) 4 **15.** (a) 2, (b) 11, (c) 14, (d) 13, (e) 3, (f) 10, (g) 6, (h) 12, (i) 4, (j) 5, (k) 9, (l) 1, (m) 7, (n) 8

Chapter 17

1. acetylcholine, epinephrine, or norepinephrine **2.** thoracolumbar, craniosacral **3.** true **4.** true **5.** b **6.** c **7.** d **8.** a **9.** a **10.** c **11.** d **12.** e **13.** (a) 2, (b) 5, (c) 6, (d) 1, (e) 3, (f) 4 **14.** (a) 3, (b) 2, (c) 1, (d) 1, (e) 2, (f) 3, (g) 3 **15.** (a) 2, (b) 1, (c) 1, (d) 2, (e) 1, (f) 1, (g) 2, (h) 2

Chapter 18

1. fight-or-flight response, resistance reaction, exhaustion **2.** hypothalamus **3.** exocrine, endocrine **4.** false **5.** true **6.** b **7.** c **8.** d **9.** a **10.** e **11.** a **12.** c **13.** (a) 8, (b) 2, (c) 7, (d) 1, (e) 12, (f) 18, (g) 5, (h) 17, (i) 20, (j) 15, (k) 3, (l) 16, (m) 19, (n) 6, (o) 13, (p) 11, (q) 4, (r) 10, (s) 14, (t) 9 **14.** (a) 10, (b) 8, (c) 2, (d) 4, (e) 1, (f) 6, (g) 9, (h) 7, (i) 5, (j) 3 **15.** (a) 12, (b) 1, (c) 11, (d) 7, (e) 3, (f) 10, (g) 2, (h) 9, (i) 4, (j) 8, (k) 5, (l) 6

Chapter 19

1. blood plasma, formed elements, red blood cells, white blood cells, platelets **2.** serum **3.** platelets, thrombocytes **4.** true **5.** false **6.** e **7.** a **8.** e **9.** a **10.** b **11.** c **12.** d **13.** (a) 6, (b) 8, (c) 3, (d) 10, (e) 2, (f) 4, (g) 1, (h) 7, (i) 9, (j) 5 **14.** (a) 5, (b) 1, (c) 2, (d) 3, (e) 6, (f) 8 (g) 9, (h) 4, (i) 7 **15.** (a) 4, (b) 6, (c) 8, (d) 1, (e) 7, (f) 3, (g) 5, (h) 2

Chapter 20

1. left ventricle **2.** systole, diastole **3.** stroke volume **4.** false **5.** false **6.** true **7.** a **8.** c **9.** b **10.** a **11.** b **12.** (a) 3, (b) 2, (c) 9, (d) 8, (e) 7, (f) 4, (g) 1, (h) 5, (i) 6 **13.** (a) 4, (b) 10, (c) 6, (d) 5, (e) 3, (f) 9, (g) 2, (h) 8, (i) 1, (j) 7 **14.** (a) 3, (b) 7, (c) 2, (d) 5, (e) 1, (f) 6, (g) 4 and 7 **15.** (a) 3, (b) 6, (c) 1, (d) 5, (e) 2, (f) 4

Chapter 21

1. carotid sinus, aortic **2.** skeletal muscle pump, respiratory pump **3.** true **4.** true **5.** a **6.** c **7.** e **8.** a **9.** d **10.** e **11.** c **12.** (a) 2, (b) 5, (c) 1, (d) 4, (e) 3 **13.** (a) 11, (b) 1, (c) 4, (d) 9, (e) 3, (f) 8, (g) 6, (h) 2, (i) 7, (j) 5, (k) 10, (l) 12 **14.** (a) 2, (b) 6, (c) 4, (d) 1, (e) 3, (f) 5 **15.** (a) 5, (b) 3, (c) 1, (d) 4, (e) 2, (f) 4, (g) 1, (h) 5

Chapter 22

1. skin, mucous membranes; antimicrobial proteins, NK cells, phagocytes **2.** specificity, memory **3.** true **4.** true **5.** e **6.** e **7.** d **8.** a **9.** c **10.** e **11.** b **12.** c **13.** (a) 3, (b) 1, (c) 4, (d) 2, (e) 5 **14.** (a) 2, (b) 3, (c) 4, (d) 7, (e) 1, (f) 6, (g) 5 **15.** (a) 11, (b) 11, (c) 8, (d) 1, (e) 2, (f) 5, (g) 4, (h) 7, (i) 9, (j) 12, (k) 3, (l) 6, (m) 10

Chapter 23

1. pulmonary ventilation, external respiration, internal respiration **2.** less, greater **3.** oxyhemoglogin; dissolved CO_2, carbamino compounds, and bicarbonate ion **4.** $CO_2 + H_2O \rightarrow H_2CO_3 \rightarrow H^+ + HCO_3^-$ **5.** false **6.** true **7.** b **8.** d **9.** a **10.** c **11.** e **12.** (a) 2, (b) 8, (c) 3, (d) 7, (e) 1, (f) 9, (g) 5, (h) 10, (i) 4, (j) 6 **13.** (a) 7, (b) 8, (c) 1, (d) 5, (e) 6, (f) 2, (g) 3, (h) 4 **14.** (a) 3, (b) 8, (c) 5, (d) 2, (e) 7, (f) 1, (g) 4, (h) 6 **15.** (a) 7, (b) 3, (c) 5, (d) 1, (e) 6, (f) 4, (g) 2

Chapter 24

1. mucosa, submucosa, muscularis, serosa **2.** submucosal plexus or plexus of Meissner; myenteric plexus or plexus of Auerbach **3.** monosaccharides; amino acids; monoglycerides, fatty acids; pentoses, phosphates, nitrogenous bases **4.** false **5.** false **6.** a **7.** c **8.** e **9.** b **10.** a **11.** b **12.** d **13.** (a) 3, (b) 10, (c) 5, (d) 8, (e) 6, (f) 1, (g) 4, (h) 9, (i) 11, (j) 2, (k) 7 **14.** (a) 4, (b) 6, (c) 7, (d) 1, (e) 5, (f) 3, (g) 8, (h) 2, (i) 10, (j) 9 **15.** (a) 4, (b) 8, (c) 2, (d) 10, (e) 7, (f) 11, (g) 1, (h) 6, (i) 3, (j) 9, (k) 5

Chapter 25

1. hypothalamus **2.** glucose 6-phosphate, pyruvic acid, acetyl coenzyme A **3.** hormones **4.** false **5.** true **6.** c **7.** d **8.** a **9.** d **10.** a **11.** a **12.** b **13.** (a) 2, (b) 8, (c) 5, (d) 10, (e) 6, (f) 7, (g) 3, (h) 9, (i) 4, (j) 1 **14.** (a) 9, (b) 12, (c) 11, (d) 10, (e) 4, (f) 7, (g) 3, (h) 5, (i) 8, (j) 1, (k) 6, (l) 2 **15.** (a) 8, (b) 5, (c) 1, (d) 10, (e) 6, (f) 2, (g) 4, (h) 9, (i) 3, (j) 7

Chapter 26

1. transitional epithelial **2.** voluntary, involuntary **3.** glomerular filtration, tubular reabsorption, tubular secretion **4.** true **5.** true **6.** c **7.** a **8.** d **9.** a **10.** e **11.** b **12.** c **13.** (a) 8, (b) 2, (c) 10, (d) 5, (e) 3, (f) 1, (g) 7, (h) 4, (i) 9, (j) 6 **14.** (a) 4, (b) 3, (c) 7, (d) 6, (e) 2, (f) 1, (g) 5 **15.** (a) 5, (b) 4, (c) 6, (d) 1, (e) 2, (f) 7, (g) 3

Chapter 27

1. intracellular fluid, extracellular fluid **2.** bicarbonate ion, carbonic acid **3.** weak acid, weak base **4.** protein buffer, carbonic acid-bicarbonate, phosphate **5.** true **6.** true **7.** a **8.** c **9.** b **10.** e **11.** d **12.** a **13.** b **14.** (a) 8, (b) 9, (c) 7, (d) 1, (e) 6, (f) 2, (g) 4, (h) 5, (i) 3 **15.** (a) 8, (b) 7, (c) 5, (d) 6, (e) 1, (f) 3, (g) 4, (h) 2

Chapter 28

1. puberty, menarche, menopause **2.** semen **3.** true **4.** true **5.** e **6.** b **7.** a **8.** c **9.** b **10.** a **11.** e **12.** (a) 7, (b) 2, (c) 1, (d) 4, (e) 8, (f) 5, (g) 3, (h) 6 **13.** (a) 6, (b) 4, (c) 2, (d) 1, (e) 3, (f) 8, (g) 7, (h) 5 **14.** (a) 6, (b) 4, (c) 1, (d) 12, (e) 8, (f) 5, (g) 7, (h) 13, (i) 11, (j) 3, (k) 2, (l) 10, (m) 9 **15.** (a) 10, (b) 12, (c) 1, (d) 5, (e) 2, (f) 4, (g) 6, (h) 11, (i) 8, (j) 3, (k) 9, (l) 7

Chapter 29

1. dilation, expulsion, placental **2.** corpus luteum **3.** mesoderm; ectoderm; endoderm **4.** true **5.** a **6.** b **7.** e **8.** (a) 3, (b) 4, (c) 5, (d) 1, (e) 2, (f) 6, (g) 7, (h) 8, (i) 9 **9.** e **10.** b **11.** (a) 7, (b) 3, (c) 6, (d) 4, (e) 10, (f) 8, (g) 5, (h) 2, (i) 1, (j) 9, (k) 11, (l) 12 **12.** b **13.** (a) 3, (b) 6, (c) 4, (d) 1, (e) 8, (f) 2, (g) 5, (h) 7 **14.** (a) 5, (b) 2, (c) 8, (d) 4, (e) 7, (f) 1, (g) 3, (h) 6 **15.** d

Answers to Critical Thinking Questions

Chapter 1

1. Homeostasis is the relative constancy of the body's internal environment. Body temperature should vary within a narrow range around normal body temperature (38°C or 98.6° F).
2. Bilateral means two sides or both arms in this case. The carpal region is the wrist area.
3. A x ray provides a good image of dense tissue such as bones. An MRI is used for imaging soft tissues, not bones. An MRI cannot be used when metal is present due to the magnetic field used.

Chapter 2

1. Fatty acids are in all lipids including plant oils and the phospholipids that compose cell membranes. Sugars (monosaccharides) are needed for ATP production, are a component of nucleotides, and are the basic units of disaccharides and polysaccharides including starch, glycogen, and cellulose.
2. DNA = deoxyribonucleic acid. DNA is composed of 4 different nucleotides. The order of the nucleotides is unique in every person. The 20 different amino acids form proteins.
3. One pH unit equals a tenfold change in H^+ concentration. Pure water has a pH of 7, which equals 1×10^{-7} moles of H^+ per liter. Blood has a pH of 7.4, which equals 0.4×10^{-7} moles H^+ per liter. Thus, blood has less than half as many H^+ as water.

Chapter 3

1. The tissues are destroyed due to autolysis of the cells caused by the release of acids and digestive enzymes from the lysosomes.
2. Synthesis of mucin by the ribosomes on rough endoplasmic reticulum, to transport vesicle, to entry face of Golgi, to transfer vesicle, to medial cisternae, where protein is modified, to transfer vesicle, to exit face, to secretory vesicle, to plasma membrane, where it undergoes exocytosis.
3. The sense strand is composed of DNA codons transcribed into mRNA. Only the exons will be found in the mRNA. The introns are clipped out.

Chapter 4

1. Many possible adaptations including: more adipose tissue for insulation, thicker bone for support, more red blood cells for oxygen transport, and so on.
2. The surface layer of the skin, the epidermis, is keratinized, stratified squamous epithelium. It is avascular. If the pin were stuck straight in, the vascularized connective tissue would be pierced and the finger would bleed.
3. The gelatin portion of the salad represents the ground substance of the connective tissue matrix. The grapes are cells such as fibroblasts. The shredded carrots and coconut are fibers embedded in the ground substance.

Chapter 5

1. The dust particles are mostly keratinocytes shed from the stratum corneum of the skin.
2. It would be neither wise nor feasible to remove the exocrine glands. Essential exocrine glands include the sudoriferous glands (sweat helps control body temperature), sebaceous glands (sebum lubricates the skin), and ceruminous glands (provide protective lubricant for ear canal).
3. The superficial layer of the epidermis of the skin is the stratum corneum. Its cells are full of intermediate filaments of keratin, kera-

tohyalin, and lipids from lamellar granules, making this layer a water-repellent barrier. The epidermis also contains an abundance of desmosomes.

Chapter 6

1. In greenstick fracture, which occurs only in children due to the greater flexibility of their bones, the bone breaks on one side but bends on the other side, resembling what happens when one tries to break a green (not dry) stick. Lynne Marie probably broke her fall with her extended arm, breaking the radius or ulna.
2. At Aunt Edith's age, the production of several hormones (such as estrogens and human growth hormone) necessary for bone remodeling is likely to be decreased. Her decreased height results, in part, from compression of her vertebrae due to flattening of the discs between them. She may also have osteoporosis and an increased susceptibility to fractures resulting in damage to the vertebrae and loss of height.
3. Exercise causes mechanical stress on bones, but because there is effectively zero gravity in space, the pull of gravity on bones is missing. The lack of stress from gravity results in bone demineralization and weakness.

Chapter 7

1. Fontanels, the soft spots between cranial bones, are fibrous connective tissue membranes that will be replaced by bone as the baby matures. They allow the infant's head to be molded during its passage through the birth canal and allow for brain and skull growth in the infant.
2. Barbara probably broke her coccyx or tailbone. The coccygeal vertebrae usually fuse by age 30. The coccyx points inferiorly in females.
3. Infants are born with a single concave curve in the vertebral column. Adults have four curves in their spinal cord-at the cervical (convex), thoracic (concave), lumbar (convex), and sacral regions. Buying a mattress from this company and sleeping on it would result in potentially severe back problems.

Chapter 8

1. Clawfoot is an abnormal elevation of the medial longitudinal arch of the foot.
2. There are 14 phalanges in each hand: two bones in the thumb and three in each of the other fingers. Farmer White has lost five phalanges on his left hand so he has nine remaining on his left and 14 remaining on his right for a total of 23.
3. Snakes don't have appendages, so they don't need shoulders and hips. Pelvic and pectoral girdles connect the upper and lower limbs (appendages) to the axial skeleton, and thus are considered part of the appendicular skeleton.

Chapter 9

1. Katie's vertebral column, head, thigh, lower leg, and lower arm are all flexed. Her lower arm and shoulder are medially rotated.
2. The elbow is a hinge-type synovial joint. The trochlea of the humerus articulates with the trochlear notch of the ulna. The elbow has monaxial (opening and closing) movement similar to a door.
3. Cartilaginous joints can be composed of hyaline cartilage, such as at the epiphyseal plate, or fibrocartilage, such as the intervertebral disc. Sutures are an example of fibrous joints, which are composed of dense fibrous tissue. Synovial (diarthrotic) joints are held together by an articular capsule, which has an inner synovial membrane layer.

Chapter 10

1. Smooth muscle contains both thick and thin filaments as well as intermediate filaments attached to dense bodies. The smooth muscle fiber contracts like a corkscrew turns; the fiber twists like a helix as it contracts and shortens.
2. Ming's muscles performed isometric contractions. Peng used primarily isotonic concentric contraction to lift, isometric to hold, and then isotonic eccentric contractions to lower the barbell.
3. Mr. Klopfer's students have muscle fatigue in their hands. Fatigue has many causes including an increase in lactic acid and ADP, and a decrease in Ca^{2+}, oxygen, creatine phosphate, and glycogen.

Chapter 11

1. Some of the antagonistic pairs for the regions are: upper arm-biceps brachii vs. triceps brachii; upper leg-quadriceps femoris vs. hamstrings; torso-rectus abdominus vs. erector spinae; lower leg-gastrocnemius vs. tibialis anterior.
2. The fulcrum is the knee joint, the load (resistance) is the weight of the upper body and the package, and the effort is the thigh muscles. This is a third-class lever.
3. Lifting the eyebrow uses the frontal belly of the occipitofrontalis; whistling uses the buccinator and orbicularis oris; closing the eyes uses the orbicularis oculi; shaking the head can use several muscles, including sternocleidomastoid, semispinalis capitis, splenius capitis, and longissimus capitis.

Chapter 12

1. The motor neuron is multipolar in structure with an axon and several dendrites projecting from the cell body. The converging circuit has several neurons that converge to form synapses with one common neuron. The bipolar neuron has one dendrite and an axon projecting from the cell body. In a simple circuit, each presynaptic neuron makes a single synapse with one postsynaptic neuron.
2. Gray matter appears gray in color due to the absence of myelin. It is composed of cell bodies, and unmyelinated axons and dendrites. White matter contains many myelinated axons. Lipofuscin, a yellowing pigment, collects in neurons with age.
3. Smelling coffee and hearing alarm are somatic sensory, stretching and yawning are somatic motor, salivating is autonomic (parasympathetic) motor, stomach rumble is enteric motor.

Chapter 13

1. A withdrawal/flexor reflex is ipsilateral and polysynaptic. The route is pain receptor of sensory neuron → spinal cord (integrating center) interneurons → motor neurons → flexor muscles in the leg. Also present is a crossed extensor reflex, which is contralateral and polysynaptic. The route diverges at the spinal cord: Interneurons cross to the other side → motor neurons → extensor muscles in the opposite leg.
2. The spinal cord is anchored in place by the filum terminale and the denticulate ligaments.
3. The needle will pierce the epidermis, the dermis, and the subcutaneous layer and then go between the vertebrae through the epidural space, the dura mater, the subdural space, the arachnoid mater, and into the CSF in the subarachnoid space. CSF is produced in the brain, and the spinal meninges are continuous with cranial meninges.

Chapter 14

1. Movement of the right arm is controlled by the left hemisphere's primary motor area, located in the precentral gyrus. Speech is controlled by Broca's area in the left hemisphere's frontal lobe just superior to the lateral cerebral sulcus.

2. The brain is enclosed by the cranial bones and meninges. The temporal bone houses the middle ear and inner ear, separating these from the brain.
3. The dentist has injected anesthetic into the inferior alveolar nerve, a branch of the mandibular nerve, which numbs the lower teeth and the lower lip. The tongue is numbed by blocking the lingual nerve. The upper teeth and lip are anesthetized by injecting the superior alveolar nerve, a branch of the maxillary nerve.

Chapter 15

1. Chemoreceptors in the nose detect odors. Proprioceptors detect body position and are involved in equilibrium. The receptors for smell are rapidly adapting (phasic), whereas proprioceptors are slowly adapting (tonic).
2. The tickle receptors are free nerve endings in the foot. The impulses travel along the first-order sensory neurons to the posterior gray horn, then along the second-order neurons, crossing to the other side of the spinal cord and up the anterior spinothalamic tract to the thalamus. The third-order neurons extend from the thalamus to the "foot" region of the somatosensory area of the cerebral cortex.
3. Yoshio's perception of feeling in his amputated foot is called phantom limb sensation. Impulses from the remaining proximal section of the sensory neuron are perceived by the brain as still coming from the amputated foot.

Chapter 16

1. Because smell and taste have ties to the cortex and limbic areas, Brenna may be recalling a memory of this or similar food. Pathway: olfactory receptors (cranial nerve I) → olfactory bulbs → olfactory tracts → lateral olfactory area in temporal lobe of cerebral cortex.
2. Fred may have cataracts, a loss of transparency in the lens of the eye. Cataracts are often associated with age, smoking, and exposure to UV light.
3. Auricle → external auditory meatus → tympanic membrane → maleus → incus → stapes → oval window → perilymph of scala vestibuli and tympani → vestibular membrane → endolymph of cochlear duct → basilar membrane → spiral organ.

Chapter 17

1. Digestion and relaxation are controlled by increased stimulation of the parasympathetic division of the ANS. The salivary glands, pancreas, and liver will show increased secretion; the stomach and intestines will have increased activity; the gallbladder will have increased contractions; heart contractions will have decreased force and rate.
2. Stretch receptors in the colon → autonomic sensory neurons → sacral region of spinal cord (integrating center) → parasympathetic preganglionic neuron → terminal ganglion → parasympathetic postganglionic neuron → smooth muscle in colon, rectum, and sphincter (effectors).
3. Nicotine from the cigarette smoke binds to nicotinic receptors on skeletal muscles, mimicking the effect of acetylcholine and causing increased contractions ("twitches"). Nicotine also binds to nicotinic receptors on cells in the adrenal medulla, stimulating release of epinephrine and norepinephrine, which mimics fight-or-flight responses.

Chapter 18

1. The beta cells are one of the cell types in the pancreatic islets of the pancreas. In Type 1 diabetes, only the beta cells are destroyed; the rest of the pancreas is not affected. A successful transplantation of

the beta cells would enable the recipient to produce the hormone insulin and would cure the diabetes.

2. Amanda has an enlarged thyroid gland or goiter. The goiter is probably due to hypothyroidism, which is causing the weight gain, fatigue, mental dullness, and other symptoms.

3. The two adrenal glands are located superior to the two kidneys. The adrenals are about 4 cm high, 2 cm wide, and 1 cm thick. The outer adrenal cortex is composed of three layers: zona glomerulosa, zona fasciculata, and zona reticularis. The inner adrenal medulla is composed of chromaffin cells.

Chapter 19

1. Blood is composed of liquid blood plasma and the formed elements: red blood cells (erythrocytes), white blood cells (leukocytes), and platelets. Blood plasma contains water, proteins, electrolytes, gases, and nutrients. There are about 5 million RBCs, 5000 WBCs, and 250,000 platelets per mL of blood. Leukocytes include neutrophils, eosinophils, basophils, monocytes, and lymphocytes.

2. To determine the blood type, a drop of each of three different antibody solutions (antisera) are added to three separate blood drops. The solutions are anti-A, anti-B, and anti-Rh. In Josef's test, anti-B caused agglutination (clump formation) when added to his blood, indicating the presence of antigen B on the RBCs. Anti-A and anti-Rh did not cause agglutination, indicating the absence of these antigens.

3. Hemostasis occurred. The stages involved are vascular spasm (if an arteriole was cut), platelet plug formation, and coagulation (clotting).

Chapter 20

1. With a normal resting CO of about 5.25 liters/min, and a heart rate of 55 beats per minute, Arian's stroke volume would be 95 mL/beat. His CO during strenuous exercise is 6 times his resting CO, about 31,500 mL/min.

2. Rheumatic fever is caused by inflammation of the bicuspid and the aortic valves following a streptococcal infection. Antibodies produced by the immune system to destroy the bacteria may also attack and damage the heart valves. Heart problems later in life may be related to this damage.

3. Mr. Perkins is suffering from angina pectoris and has several risk factors for coronary artery disease such as smoking, obesity, and male gender. Cardiac angiography involves the use of a cardiac catheter to inject a radiopaque medium into the heart and its vessels. The angiogram may reveal blockages such as atherosclerotic plaques in his coronary arteries.

Chapter 21

1. After birth, the foramen ovale and ductus arteriosus close to establish the pulmonary circulation. The umbilical artery and veins close because the placenta is no longer functioning. The ductus venosus closes so that the liver is no longer bypassed.

2. Right hand: left ventricle → ascending aorta → arch of aorta → brachiocephalic trunk → right subclavian artery → right axillary artery → right brachial artery → right radial and ulnar arteries → right superficial palmar arch. Left hand: left ventricle → ascending aorta → arch of aorta → left subclavian artery → left axillary artery → left brachial artery → left radial and ulnar arteries → left superficial palmar arch.

3. A vascular (venous) sinus is a vein that lacks smooth muscle in its tunica media. Dense connective tissue replaces the tunica media and tunica externa.

Chapter 22

1. Tariq had a hypersensitivity reaction to an insect sting; he was probably allergic to the venom. He had a localized anaphylactic or type I reaction.

2. The site of the embedded splinter had probably become infected with bacteria. The red streaks are the lymphatic vessels that drain the infected area; the swollen tender bumps are the axillary lymph nodes, which are swollen due to the immune response to the infection.

3. Influenza vaccination introduces a weakened or killed virus (which will not cause the disease) to the body. The immune system recognizes the antigen and mounts a primary immune response. Upon exposure to the same flu virus that was in the vaccine, the body will produce a secondary response, which will usually prevent a case of the flu. This is artificially acquired active immunity.

Chapter 23

1. Normal average male volumes are 500 mL for tidal volume during quiet breathing, 3100 mL for inspiratory reserve volume, and 1200 mL for expiratory reserve volume. Average female volumes are less than average male volumes, because females are generally smaller than males.

2. The bones of the external nose are the frontal bone, the two nasal bones (the location of the fracture), and the maxillae. The external nose is also composed of the septal, lateral nasal, and alar cartilages, as well as skin, muscle, and mucous membrane.

3. The cerebral cortex can temporarily allow voluntary breath holding. P_{CO_2} and H^+ levels in the blood and CSF increase with breath holding, strongly stimulating the inspiratory area, which stimulates inspiration to begin again. Breathing will begin again even if the person loses consciousness.

Chapter 24

1. Katie lost the two upper central permanent incisors. The remaining deciduous teeth include the lower central incisors, an upper and lower pair of lateral incisors, an upper and lower pair of cuspids, and upper and lower pair of first molars, and upper and lower pair of second molars.

2. CCK promotes ejection of bile, secretion of pancreatic juice, and contraction of the pyloric sphincter. It also promotes pancreatic growth and enhances the effects of secretin. CCK acts on the hypothalamus to induce satiety (the feeling of fullness) and should therefore decrease the appetite.

3. The smaller left lobe of the liver is separated from the larger right lobe by the falciform ligament. The left lobe is inferior to the diaphragm in the epigastric region of the abdominopelvic cavity.

Chapter 25

1. Deb is eating a diet high in carbohydrates in order to store maximum amounts of glycogen in her skeletal muscles and liver. This practice is called carbohydrate loading. Glycogenolysis, the breakdown of stored glycogen, supplies the muscles with the glucose needed for ATP production via cellular respiration.

2. Mr. Hernandez was suffering from heat exhaustion caused by loss of fluids and electrolytes. Lack of NaCl causes muscle cramps, nausea and vomiting, dizziness, and fainting. Low blood pressure may also result.

3. Sara's normal growth may be maintained despite day-to-day fluctuations in food intake. Many factors govern food intake, including neurons in the hypothalamus, blood glucose level, amount of adipose tissue, CCK, distention of the GI tract, and body temperature.

Chapter 26

1. Without reabsorption, initially 105–125 mL of filtrate would be lost per minute, assuming normal glomerular filtration rate. Fluid loss from the blood would cause a decrease in blood pressure, and therefore a decrease in GBHP. When GBHP dropped below 45mmHg, filtration would stop (assuming normal CHP and BCOP) because NFP would be zero.

2. The urinary bladder can stretch considerably due to the presence of transitional epithelium, rugae, and three layers of smooth muscle in the detrusor muscle.

3. In females the urethra is about 4 cm long; in males the urethra is 15–20 cm long, including its passage through the penis, urogenital diaphragm, and prostate.

Chapter 27

1. Gary has water intoxication. The Na^+ concentration of his plasma and interstitial fluid is below normal. Water moved by osmosis into the cells, resulting in hypotonic intracellular fluid and water intoxication. Decreased plasma volume, due to water movement into the interstitial fluid, caused hypovolemic shock.

2. The average female body contains about 55% water, versus 60% water in the average male. Due to the influence of male and female hormones, the average female has relatively more subcutaneous fat (which contains very little water) and relatively less muscle and other tissues (which have higher water content) than the average male.

3. Excessive vomiting causes a loss of hydrochloric acid in gastric juice, and intake of antacids increases the amount of alkali in the body fluids, resulting in metabolic alkalosis. Vomiting also causes a loss of fluid and may cause dehydration.

4. (Step 1) pH = 7.30 indicates slight acidosis, which could be caused by elevated P_{CO_2} or lowered HCO_3^-. (Step 2) The HCO_3^- is lower than normal (20 mEq/liter), so (step 3) the cause is metabolic. (Step 4) the P_{CO_2} is lower than normal (32 mmHg), so hyperventilation is providing some compensation. Diagnosis: Henry has partially compensated metabolic acidosis. A possible cause is kidney damage that resulted from interruption of blood flow during the heart attack.

Chapter 28

1. LH (luteinizing hormone) is an anterior pituitary hormone that acts on both the male and female reproductive systems. In males, LH stimulates the Leydig cells of the testes to secrete testosterone.

2. The germinal epithelium, located on the surface of the ovaries, is a misnomer. The oocytes are located in the ovarian cortex, deep to the tunica albuginea. The tunica vaginalis covers the testes, superficial to the tunica albuginea. It forms from the peritoneum during the descent of the testes from the abdomen into the scrotum during fetal development.

3. No. The first polar body, formed by meiosis I, would contain half the homologous chromosome pairs, the other half being in the secondary oocyte. The second polar body, formed by meiosis II, would contain chromosomes identical to the ovum; however, fertilization would be by two different sperm, so the babies would still not be identical.

Chapter 29

1. Being able to taste PCT is an autosomal dominant trait. Because Kendra is a PCT nontaster, her genotype is homozygous recessive for this trait. Her parents must both be heterozygous for the trait. They exhibit the dominant phenotype for PCT tasting and are carriers of the recessive trait.

2. Because Huntington disease (HD) is a trinucleotide repeat disease, in which the number of repeats expand with each succeeding generation, it can occur in a child whose parents did not suffer HD. Also, because it is a dominant gene, one mutant allele of the *HD* gene is sufficient to cause HD.

3. All arteries carry blood away from the heart. The umbilical arteries, located in the fetus, carry deoxygenated blood away from the fetal heart towards the placenta where the blood will become oxygenated as it passes through the placenta.

Glossary

Pronunciation Key

1. The most strongly accented syllable appears in capital letters, for example, bilateral (bī-LAT-er-al) and diagnosis (dī-ag-NŌ-sis).
2. If there is a secondary accent, it is noted by a prime (′), for example, constitution (kon′-sti-TOO-shun) and physiology (fiz′-ē-OL-ō-jē). Any additional secondary accents are also noted by a prime, for example, decarboxylation (dē′-kar-bok′-si-LĀ-shun).
3. Vowels marked by a line above the letter are pronounced with the long sound, as in the following common words:

ā as in *māke*   ō as in *pōle*
ē as in *bē*   ū as in *cute*
ī as in *īvy*

4. Vowels not marked by a line above the letter are pronounced with the short sound, as in the following words:

a as in *above* or *at*   o as in *not*
e as in *bet*   u as in *bud*
i as in *sip*

5. Other vowel sounds are indicated as follows:

oy as in *oil*
oo as in *root*

6. Consonant sounds are pronounced as in the following words:

b as in *bat*   m as in *mother*
ch as in *chair*   n as in *no*
d as in *dog*   p as in *pick*
f as in *father*   r as in *rib*
g as in *get*   s as in *so*
h as in *hat*   t as in *tea*
j as in *jump*   v as in *very*
k as in *can*   w as in *welcome*
ks as in *tax*   z as in *zero*
kw as in *quit*   zh as in *lesion*
l as in *let*

A

Abdomen (ab-DŌ-men *or* AB-dō-men) The area between the diaphragm and pelvis.

Abdominal (ab-DOM-i-nal) **cavity** Superior portion of the abdominopelvic cavity that contains the stomach, spleen, liver, gallbladder, most of the small intestine, and part of the large intestine.

Abdominal thrust maneuver A first-aid procedure for choking. Employs a quick, upward thrust against the diaphragm that forces air out of the lungs with sufficient force to eject any lodged material. Also called the **Heimlich** (HĪM-lik) **maneuver.**

Abdominopelvic (ab-dom′-i-nō-PEL-vic) **cavity** Inferior component of the ventral body cavity that is subdivided into a superior abdominal cavity and an inferior pelvic cavity.

Abduction (ab-DUK-shun) Movement away from the midline of the body.

Abortion (a-BOR-shun) The premature loss (spontaneous) or removal (induced) of the embryo or nonviable fetus; miscarriage due to a failure in the normal process of developing or maturing.

Abscess (AB-ses) A localized collection of pus and liquefied tissue in a cavity.

Absorption (ab-SORP-shun) Intake of fluids or other substances by cells of the skin or mucous membranes; the passage of digested foods from the gastrointestinal tract into blood or lymph.

Absorptive (fed) state Metabolic state during which ingested nutrients are absorbed into the blood or lymph from the gastrointestinal tract.

Accessory duct A duct of the pancreas that empties into the duodenum about 2.5 cm (1 in.) superior to the ampulla of Vater (hepatopancreatic ampulla). Also called the **duct of Santorini** (san′-tō-RĒ-nē).

Accommodation (a-kom-ō-DĀ-shun) An increase in the curvature of the lens of the eye to adjust for near vision.

Acetabulum (as′-e-TAB-ū-lum) The rounded cavity on the external surface of the hip bone that receives the head of the femur.

Acetylcholine (as′-e-til-KŌ-lēn) **(ACh)** A neurotransmitter liberated by many peripheral nervous system neurons and some central nervous system neurons. It is excitatory at neuromuscular junctions but inhibitory at some other synapses (for example, it slows heart rate).

Achalasia (ak′-a-LĀ-zē-a) A condition, caused by malfunction of the myenteric plexus, in which the lower esophageal sphincter fails to relax normally as food approaches. A whole meal may become lodged in the esophagus and enter the stomach very slowly. Distension of the esophagus results in chest pain that is often confused with pain originating from the heart.

Acid (AS-id) A proton donor, or a substance that dissociates into hydrogen ions (H^+) and anions; characterized by an excess of hydrogen ions and a pH less than 7.

Acidosis (as-i-DŌ-sis) A condition in which blood pH is below 7.35. Also known as **acidemia.**

Acini (AS-i-nē) Groups of cells in the pancreas that secrete digestive enzymes.

Acoustic (a-KOOS-tik) Pertaining to sound or the sense of hearing.

Acquired immunodeficiency syndrome (AIDS) A fatal disease caused by the human immunodeficiency virus (HIV). Characterized by a positive HIV-antibody test, low helper T cell count, and certain indicator diseases (for example Kaposi's sarcoma,

Pneumocystis carinii pneumonia, tuberculosis, fungal diseases). Other symptoms include fever or night sweats, coughing, sore throat, fatigue, body aches, weight loss, and enlarged lymph nodes.

Acrosome (AK-rō-sōm) A lysosomelike organelle in the head of a sperm cell containing enzymes that facilitate the penetration of a sperm cell into a secondary oocyte.

Actin (AK-tin) A contractile protein that is part of thin filaments in muscle fibers.

Action potential An electrical signal that propagates along the membrane of a neuron or muscle fiber (cell); a rapid change in membrane potential that involves a depolarization followed by a repolarization. Also called a **nerve action potential** or **nerve impulse** as it relates to a neuron, and a **muscle action potential** as it relates to a muscle fiber.

Activation (ak′-ti-VĀ-shun) **energy** The minimum amount of energy required for a chemical reaction to occur.

Active transport The movement of substances across cell membranes against a concentration gradient, requiring the expenditure of cellular energy (ATP).

Acute (a-KŪT) Having rapid onset, severe symptoms, and a short course; not chronic.

Adaptation (ad′-ap-TĀ-shun) The adjustment of the pupil of the eye to changes in light intensity. The property by which a sensory neuron relays a decreased frequency of action potentials from a receptor, even though the strength of the stimulus remains constant; the decrease in perception of a sensation over time while the stimulus is still present.

Adduction (ad-DUK-shun) Movement toward the midline of the body.

Adenoids (AD-e-noyds) The pharyngeal tonsils.

Adenosine triphosphate (a-DEN-ō-sēn trī-FOS-fāt) **(ATP)** The main energy currency in living cells; used to transfer the chemical energy needed for metabolic reactions. ATP consists of the purine base *adenine* and the five-carbon sugar *ribose,* to which are added, in linear array, three *phosphate* groups.

Adenylate cyclase (a-DEN-i-lāt SĪ-klās) An enzyme that is activated when certain neurotransmitters or hormones bind to their receptors; the enzyme that converts ATP into cyclic AMP, an important second messenger.

Adhesion (ad-HĒ-zhun) Abnormal joining of parts to each other.

Adipocyte (AD-i-pō-sīt) Fat cell, derived from a fibroblast.

Adipose (AD-i-pōz) **tissue** Tissue composed of adipocytes specialized for triglyceride storage and present in the form of soft pads between various organs for support, protection, and insulation.

Adrenal cortex (a-DRĒ-nal KOR-teks) The outer portion of an adrenal gland, divided into three zones; the zona glomerulosa secretes mineralocorticoids, the zona fasciculata secretes glucocorticoids, and the zona reticularis secretes androgens.

Adrenal glands Two glands located superior to each kidney. Also called the **suprarenal** (soo′-pra-RĒ-nal) **glands.**

Adrenal medulla (me-DUL-a) The inner part of an adrenal gland, consisting of cells that secrete epinephrine, norepinephrine, and a small amount of dopamine in response to stimulation by sympathetic preganglionic neurons.

Adrenergic (ad′-ren-ER-jik) **neuron** A neuron that releases epinephrine (adrenaline) or norepinephrine (noradrenaline) as its neurotransmitter.

Adrenocorticotropic (ad-rē′-nō-kor-ti-kō-TRŌP-ik) **hormone (ACTH)** A hormone produced by the anterior pituitary that influences the production and secretion of certain hormones of the adrenal cortex.

Adventitia (ad-ven-TISH-a) The outermost covering of a structure or organ.

Aerobic (air-Ō-bik) Requiring molecular oxygen.

Afferent arteriole (AF-er-ent ar-TĒ-rē-ōl) A blood vessel of a kidney that divides into the capillary network called a glomerulus; there is one afferent arteriole for each glomerulus.

Agglutination (a-gloo′-ti-NĀ-shun) Clumping of microorganisms or blood cells, typically due to an antigen–antibody reaction.

Aggregated lymphatic follicles Clusters of lymph nodules that are most numerous in the ileum. Also called **Peyer's** (PĪ-erz) **patches.**

Albinism (AL-bin-izm) Abnormal, nonpathological, partial, or total absence of pigment in skin, hair, and eyes.

Albumin (al-BŪ-min) The most abundant (60%) and smallest of the plasma proteins; it is the main contributor to blood colloid osmotic pressure (BCOP).

Aldosterone (al-DOS-ter-ōn) A mineralocorticoid produced by the adrenal cortex that promotes sodium and water reabsorption by the kidneys and potassium excretion in urine.

Alkaline (AL-ka-līn) Containing more hydroxide ions (OH^-) than hydrogen ions (H^+); a pH higher than 7.

Alkalosis (al-ka-LŌ-sis) A condition in which blood pH is higher than 7.45. Also known as **alkalemia.**

Allantois (a-LAN-tō-is) A small, vascularized outpouching of the yolk sac that serves as an early site for blood formation and development of the urinary bladder.

Alleles (a-LĒLZ) Alternate forms of a single gene that control the same inherited trait (such as type A blood) and are located at the same position on homologous chromosomes.

Allergen (AL-er-jen) An antigen that evokes a hypersensitivity reaction.

Alpha (AL-fa) **cell** A type of cell in the pancreatic islets (islets of Langerhans) in the pancreas that secretes the hormone glucagon. Also termed an **A cell.**

Alpha receptor A type of receptor for norepinephrine and epinephrine; present on visceral effectors innervated by sympathetic postganglionic neurons.

Alveolar duct Branch of a respiratory bronchiole around which alveoli and alveolar sacs are arranged.

Alveolar macrophage (MAK-rō-fāj) Highly phagocytic cell found in the alveolar walls of the lungs. Also called a **dust cell.**

Alveolar (al-VĒ-ō-lar) **pressure** Air pressure within the lungs. Also called **intrapulmonic pressure.**

Alveolar sac A cluster of alveoli that share a common opening.

Alveolus (al-VĒ-ō-lus) A small hollow or cavity; an air sac in the lungs; milk-secreting portion of a mammary gland. *Plural is* **alveoli** (al-VĒ-ol-ī).

Alzheimer (ALTZ-hī-mer) **disease (AD)** Disabling neurological disorder characterized by dysfunction and death of specific cerebral neurons, resulting in widespread intellectual impairment, personality changes, and fluctuations in alertness.

Amnesia (am-NE-zē-a) A lack or loss of memory.

Amenorrhea (ā-men-ō-RĒ-a) Absence of menstruation.

Amino (a-MĒ-no) **acid** An organic acid, containing an acidic carboxyl group ($-COOH$) and a basic amino group ($-NH_2$); the monomer used to synthesize polypeptides and proteins.

Amnion (AM-nē-on) A thin, protective fetal membrane that develops from the epiblast; holds the fetus suspended in amniotic fluid. Also called the "**bag of waters.**"

Amniotic (am′-nē-OT-ik) **fluid** Fluid in the amniotic cavity, the space between the developing embryo (or fetus) and amnion; the fluid is initially produced as a filtrate from maternal blood and later includes fetal urine. It functions as a shock absorber, helps regulate fetal body temperature, and helps prevent desiccation.

Amphiarthrosis (am′-fē-ar-THRŌ-sis) A slightly movable joint, in which the articulating bony surfaces are separated by fibrous connective tissue or fibrocartilage to which both are attached; types are syndesmosis and symphysis.

Ampulla (am-PUL-la) A saclike dilation of a canal or duct.

Anabolism (a-NAB-ō-lizm) Synthetic, energy-requiring reactions whereby small molecules are built up into larger ones.

Anaerobic (an-ār-Ō-bik) Not requiring oxygen.

Anal (Ā-nal) **canal** The last 2 or 3 cm (1 in.) of the rectum; opens to the exterior through the anus.

Anal column A longitudinal fold in the mucous membrane of the anal canal that contains a network of arteries and veins.

Anal triangle The subdivision of the female or male perineum that contains the anus.

Analgesia (an-al-JĒ-zē-a) Pain relief; absence of the sensation of pain.

Anaphylaxis (an′-a-fi-LAK-sis) A hypersensitivity (allergic) reaction in which IgE antibodies attach to mast cells and basophils, causing them to produce mediators of anaphylaxis (histamine, leukotrienes, kinins, and prostaglandins) that bring about increased blood permeability, increased smooth muscle contraction, and increased mucus production. Examples are hay fever, hives, and anaphylactic shock.

Anaphase (AN-a-fāz) The third stage of mitosis in which the chromatids that have separated at the centromeres move to opposite poles of the cell.

Anastomosis (a-nas-tō-MŌ-sis) An end-to-end union or joining of blood vessels, lymphatic vessels, or nerves.

Anatomical (an′-a-TOM-i-kal) **position** A position of the body universally used in anatomical descriptions in which the body is erect, the head is level, the eyes face forward, the upper limbs are at the sides, the palms face forward, and the feet are flat on the floor.

Anatomic dead space Spaces of the nose, pharynx, larynx, trachea, bronchi, and bronchioles totaling about 150 mL of the 500 mL in a quiet breath (tidal volume); air in the anatomic dead space does not reach the alveoli to participate in gas exchange.

Anatomy (a-NAT-ō-mē) The structure or study of structure of the body and the relation of its parts to each other.

Androgens (AN-drō-jenz) Masculinizing sex hormones produced by the testes in males and the adrenal cortex in both sexes; responsible for libido (sexual desire); the two main androgens are testosterone and dihydrotestosterone.

Anemia (a-NĒ-mē-a) Condition of the blood in which the number of functional red blood cells or their hemoglobin content is below normal.

Anesthesia (an′-es-THĒ-zē-a) A total or partial loss of feeling or sensation; may be general or local.

Aneuploid (an′-ū-PLOYD) A cell that has one or more chromosomes of a set added or deleted.

Aneurysm (AN-ū-rizm) A saclike enlargement of a blood vessel caused by a weakening of its wall.

Angina pectoris (an-JI-na or AN-ji-na PEK-tō-ris) A pain in the chest related to reduced coronary circulation due to coronary artery disease (CAD) or spasms of vascular smooth muscle in coronary arteries.

Angiotensin (an-jē-ō-TEN-sin) Either of two forms of a protein associated with regulation of blood pressure. Angiotensin I is produced by the action of renin on angiotensinogen and is converted by the action of ACE (angiotensin-converting enzyme) into angiotensin II, which stimulates aldosterone secretion by the adrenal cortex, stimulates the sensation of thirst, and causes vasoconstriction with resulting increase in systemic vascular resistance.

Anion (AN-ī-on) A negatively charged ion. An example is the chloride ion (Cl^-).

Ankylosis (ang′-ki-LŌ-sis) Severe or complete loss of movement at a joint as the result of a disease process.

Anoxia (an-OK-sē-a) Deficiency of oxygen.

Antagonist (an-TAG-ō-nist) A muscle that has an action opposite that of the prime mover (agonist) and yields to the movement of the prime mover.

Antagonistic (an-tag-ō-NIST-ik) **effect** A hormonal interaction in which the effect of one hormone on a target cell is opposed by another hormone. For example, calcitonin (CT) lowers blood calcium level, whereas parathyroid hormone (PTH) raises it.

Anterior (an-TĒR-ē-or) Nearer to or at the front of the body. Equivalent to **ventral** in bipeds.

Anterior pituitary Anterior lobe of the pituitary gland. Also called the **adenohypophysis** (ad′-e-nō-hī-POF-i-sis).

Anterior root The structure composed of axons of motor (efferent) neurons that emerges from the anterior aspect of the spinal cord and extends laterally to join a posterior root, forming a spinal nerve. Also called a **ventral root.**

Anterolateral (an′-ter-ō-LAT-er-al) **pathway** Sensory pathway that conveys information related to pain, temperature, crude touch, pressure, tickle, and itch.

Antibody (AN-ti-bod′-ē) A protein produced by plasma cells in response to a specific antigen; the antibody combines with that antigen to neutralize, inhibit, or destroy it. Also called an **immunoglobulin** (im-ū-nō-GLOB-ū-lin) or **Ig.**

Antibody-mediated immunity That component of immunity in which B lymphocytes (B cells) develop into plasma cells that produce antibodies that destroy antigens. Also called **humoral** (HŪ-mor-al) **immunity.**

Anticoagulant (an-tī-cō-AG-ū-lant) A substance that can delay, suppress, or prevent the clotting of blood.

Antidiuretic (an′-ti-dī-ū-RET-ik) Substance that inhibits urine formation.

Antidiuretic hormone (ADH) Hormone produced by neurosecretory cells in the paraventricular and supraoptic nuclei of the hypothalamus that stimulates water reabsorption from kidney tubule cells into the blood and vasoconstriction of arterioles. Also called **vasopressin** (vāz-ō-PRES-in).

Antigen (AN-ti-jen) A substance that has immunogenicity (the ability to provoke an immune response) and reactivity (the ability to react with the antibodies or cells that result from the immune response); contraction of *anti*body *gen*erator. Also termed a **complete antigen.**

Antigen-presenting cell (APC) Special class of migratory cell that processes and presents antigens to T cells during an immune response; APCs include macrophages, B cells, and dendritic cells, which are present in the skin, mucous membranes, and lymph nodes.

Antiporter A transmembrane transporter protein that moves two substances, often Na$^+$ and another substance, in opposite directions across a plasma membrane. Also called a **countertransporter.**

Anulus fibrosus (AN-ū-lus fī-BRŌ-sus) A ring of fibrous tissue and fibrocartilage that encircles the pulpy substance (nucleus pulposus) of an intervertebral disc.

Anuria (an-Ū-rē-a) Absence of urine formation or daily urine output of less than 50 mL.

Anus (Ā-nus) The distal end and outlet of the rectum.

Aorta (ā-OR-ta) The main systemic trunk of the arterial system of the body that emerges from the left ventricle.

Aortic (ā-OR-tik) **body** Cluster of chemoreceptors on or near the arch of the aorta that respond to changes in blood levels of oxygen, carbon dioxide, and hydrogen ions (H$^+$).

Aortic reflex A reflex that helps maintain normal systemic blood pressure; initiated by baroreceptors in the wall of the ascending aorta and arch of the aorta. Nerve impulses from aortic baroreceptors reach the cardiovascular center via sensory axons of the vagus nerves (cranial nerve X).

Aperture (AP-er-chur) An opening or orifice.

Apex (Ā-peks) The pointed end of a conical structure, such as the apex of the heart.

Aphasia (a-FA-zē-a) Loss of ability to express oneself properly through speech or loss of verbal comprehension.

Apnea (AP-nē-a) Temporary cessation of breathing.

Apneustic (ap-NOO-stik) **area** A part of the respiratory center in the pons that sends stimulatory nerve impulses to the inspiratory area that activate and prolong inhalation and inhibit exhalation.

Apocrine (AP-ō-krin) **gland** A type of gland in which the secretory products gather at the free end of the secreting cell and are pinched off, along with some of the cytoplasm, to become the secretion, as in mammary glands.

Aponeurosis (ap'-ō-noo-RŌ-sis) A sheetlike tendon joining one muscle with another or with bone.

Apoptosis (ap'-ō-TŌ-sis *or* ap'-ōp-TŌ-sis) Programmed cell death; a normal type of cell death that removes unneeded cells during embryological development, regulates the number of cells in tissues, and eliminates many potentially dangerous cells such as cancer cells. During apoptosis, the DNA fragments, the nucleus condenses, mitochondria cease to function, and the cytoplasm shrinks, but the plasma membrane remains intact. Phagocytes engulf and digest the apoptotic cells, and an inflammatory response does not occur.

Appositional (ap'-ō-ZISH-o-nal) **growth** Growth due to surface deposition of material, as in the growth in diameter of cartilage and bone. Also called **exogenous** (eks-OJ-e-nus) **growth.**

Aqueous humor (AK-wē-us HŪ-mer) The watery fluid, similar in composition to cerebrospinal fluid, that fills the anterior cavity of the eye.

Arachnoid (a-RAK-noyd) **mater** The middle of the three meninges (coverings) of the brain and spinal cord. Also termed the **arachnoid.**

Arachnoid villus (VIL-us) Berrylike tuft of the arachnoid mater that protrudes into the superior sagittal sinus and through which cerebrospinal fluid is reabsorbed into the bloodstream.

Arbor vitae (AR-bor VĪ-tē) The white matter tracts of the cerebellum, which have a treelike appearance when seen in midsagittal section.

Arch of the aorta The most superior portion of the aorta, lying between the ascending and descending segments of the aorta.

Areola (a-RĒ-ō-la) Any tiny space in a tissue. The pigmented ring around the nipple of the breast.

Arm The part of the upper limb from the shoulder to the elbow.

Arousal (a-ROW-zal) Awakening from sleep, a response due to stimulation of the reticular activating system (RAS).

Arrector pili (a-REK-tor PI-lē) Smooth muscles attached to hairs; contraction pulls the hairs into a vertical position, resulting in "goose bumps."

Arrhythmia (a-RITH-mē-a) An irregular heart rhythm. Also called a **dysrhythmia.**

Arteriole (ar-TĒ-rē-ōl) A small, almost microscopic, artery that delivers blood to a capillary.

Arteriosclerosis (ar-tē-rē-ō-skle-RŌ-sis) Group of diseases characterized by thickening of the walls of arteries and loss of elasticity.

Artery (AR-ter-ē) A blood vessel that carries blood away from the heart.

Arthritis (ar-THRI-tis) Inflammation of a joint.

Arthrology (ar-THROL-ō-jē) The study or description of joints.

Arthroscopy (ar-THROS-co-pē) A procedure for examining the interior of a joint, usually the knee, by inserting an arthroscope into a small incision; used to determine extent of damage, remove torn cartilage, repair cruciate ligaments, and obtain samples for analysis.

Arthrosis (ar-THRŌ-sis) A joint or articulation.

Articular (ar-TIK-ū-lar) **capsule** Sleevelike structure around a synovial joint composed of a fibrous capsule and a synovial membrane.

Articular cartilage (KAR-ti-lij) Hyaline cartilage attached to articular bone surfaces.

Articular disc Fibrocartilage pad between articular surfaces of bones of some synovial joints. Also called a **meniscus** (men-IS-kus).

Articulation (ar-tik'-ū-LĀ-shun) A joint; a point of contact between bones, cartilage and bones, or teeth and bones.

Arytenoid (ar'-i-TĒ-noyd) **cartilages** A pair of small, pyramidal cartilages of the larynx that attach to the vocal folds and intrinsic pharyngeal muscles and can move the vocal folds.

Ascending colon (KŌ-lon) The part of the large intestine that passes superiorly from the cecum to the inferior border of the liver, where it bends at the right colic (hepatic) flexure to become the transverse colon.

Ascites (as-SĪ-tēz) Abnormal accumulation of serous fluid in the peritoneal cavity.

Association areas Large cortical regions on the lateral surfaces of the occipital, parietal, and temporal lobes and on the frontal lobes anterior to the motor areas connected by many motor and sensory axons to other parts of the cortex. The association areas are concerned with motor patterns, memory, concepts of word-hearing and word-seeing, reasoning, will, judgment, and personality traits.

Asthma (AZ-ma) Usually allergic reaction characterized by smooth muscle spasms in bronchi resulting in wheezing and difficult breathing. Also called **bronchial asthma.**

Astigmatism (a-STIG-ma-tizm) An irregularity of the lens or cornea of the eye causing the image to be out of focus and producing faulty vision.

Astrocyte (AS-trō-sīt) A neuroglial cell having a star shape that participates in brain development and the metabolism of neurotransmitters, helps form the blood–brain barrier, helps maintain the proper balance of K$^+$ for generation of nerve impulses, and provides a link between neurons and blood vessels.

Ataxia (a-TAK-sē-a) A lack of muscular coordination, lack of precision.

Atherosclerotic (ath′-er-ō-skle-RO-tic) **plaque** (PLAK) A lesion that results from accumulated cholesterol and smooth muscle fibers (cells) of the tunica media of an artery; may become obstructive.

Atom Unit of matter that makes up a chemical element; consists of a nucleus (containing positively charged protons and uncharged neutrons) and negatively charged electrons that orbit the nucleus.

Atomic mass (weight) Average mass of all stable atoms of an element, reflecting the relative proportion of atoms with different mass numbers.

Atomic number Number of protons in an atom.

Atresia (a-TRĒ-zē-a) Degeneration and reabsorption of an ovarian follicle before it fully matures and ruptures; abnormal closure of a passage, or absence of a normal body opening.

Atrial fibrillation (Ā-trē-al fib-ri-LĀ-shun) Asynchronous contraction of cardiac muscle fibers in the atria that results in the cessation of atrial pumping.

Atrial natriuretic (na′-trē-ū-RET-ik) **peptide (ANP)** Peptide hormone, produced by the atria of the heart in response to stretching, that inhibits aldosterone production and thus lowers blood pressure; causes natriuresis, increased urinary excretion of sodium.

Atrioventricular (AV) (ā′-trē-ō-ven-TRIK-ū-lar) **bundle** The part of the conduction system of the heart that begins at the atrioventricular (AV) node, passes through the cardiac skeleton separating the atria and the ventricles, then extends a short distance down the interventricular septum before splitting into right and left bundle branches. Also called the **bundle of His** (HISS).

Atrioventricular (AV) node The part of the conduction system of the heart made up of a compact mass of conducting cells located in the septum between the two atria.

Atrioventricular (AV) valve A heart valve made up of membranous flaps or cusps that allows blood to flow in one direction only, from an atrium into a ventricle.

Atrium (Ā-trē-um) A superior chamber of the heart.

Atrophy (AT-rō-fē) Wasting away or decrease in size of a part, due to a failure, abnormality of nutrition, or lack of use.

Auditory ossicle (AW-di-tō-rē OS-si-kul) One of the three small bones of the middle ear called the **malleus, incus,** and **stapes.**

Auditory tube The tube that connects the middle ear with the nose and nasopharynx region of the throat. Also called the **Eustachian** (ū-STĀ-shun *or* ū-STĀ-kē-an) **tube** or **pharyngotympanic tube.**

Auscultation (aws-kul-TĀ-shun) Examination by listening to sounds in the body.

Autocrine (AW-tō-krin) Local hormone, such as interleukin-2, that acts on the same cell that secreted it.

Autoimmunity An immunological response against a person's own tissues.

Autolysis (aw-TOL-i-sis) Self-destruction of cells by their own lysosomal digestive enzymes after death or in a pathological process.

Autonomic ganglion (aw′-tō-NOM-ik GANG-lē-on) A cluster of cell bodies of sympathetic or parasympathetic neurons located outside the central nervous system.

Autonomic nervous system (ANS) Visceral sensory (afferent) and visceral motor (efferent) neurons. Autonomic motor neurons, both sympathetic and parasympathetic, conduct nerve impulses from the central nervous system to smooth muscle, cardiac muscle, and glands. So named because this part of the nervous system was thought to be self-governing or spontaneous.

Autonomic plexus (PLEK-sus) A network of sympathetic and parasympathetic axons; examples are the cardiac, celiac, and pelvic plexuses, which are located in the thorax, abdomen, and pelvis, respectively.

Autophagy (aw-TOF-a-jē) Process by which worn-out organelles are digested within lysosomes.

Autopsy (AW-top-sē) The examination of the body after death.

Autoregulation (aw-tō-reg-ū-LĀ-shun) A local, automatic adjustment of blood flow in a given region of the body in response to tissue needs.

Autorhythmic cells Cardiac or smooth muscle fibers that are self-excitable (generate impulses without an external stimulus); act as the heart's pacemaker and conduct the pacing impulse through the conduction system of the heart; self-excitable neurons in the central nervous system, as in the inspiratory area of the brain stem.

Autosome (AW-tō-sōm) Any chromosome other than the X and Y chromosomes (sex chromosomes).

Axilla (ak-SIL-a) The small hollow beneath the arm where it joins the body at the shoulders. Also called the **armpit.**

Axon (AK-son) The usually single, long process of a nerve cell that propagates a nerve impulse toward the axon terminals.

Axon terminal Terminal branch of an axon where synaptic vesicles undergo exocytosis to release neurotransmitter molecules.

Azygos (AZ-ī-gos) An anatomical structure that is not paired; occurring singly.

B

B cell A lymphocyte that can develop into a clone of antibody-producing plasma cells or memory cells when properly stimulated by a specific antigen.

Babinski (ba-BIN-skē) **sign** Extension of the great toe, with or without fanning of the other toes, in response to stimulation of the outer margin of the sole; normal up to 18 months of age and indicative of damage to descending motor pathways such as the corticospinal tracts after that.

Back The posterior part of the body; the dorsum.

Ball-and-socket joint A synovial joint in which the rounded surface of one bone moves within a cup-shaped depression or socket of another bone, as in the shoulder or hip joint. Also called a **spheroid** (SFĒ-royd) **joint.**

Baroreceptor (bar′-ō-re-SEP-tor) Neuron capable of responding to changes in blood, air, or fluid pressure. Also called a **pressoreceptor.**

Basal ganglia (GANG-glē-a) Paired clusters of gray matter deep in each cerebral hemisphere including the globus pallidus, putamen, and caudate nucleus. Together, the caudate nucleus and putamen are known as the **corpus striatum.** Nearby structures that are functionally linked to the basal ganglia are the substantia nigra of the midbrain and the subthalamic nuclei of the diencephalon.

Basal metabolic (BĀ-sal met′-a-BOL-ik) **rate (BMR)** The rate of metabolism measured under standard or basal conditions (awake, at rest, fasting).

Base A nonacid or a proton acceptor, characterized by excess of hydroxide ions (OH^-) and a pH greater than 7. A ring-shaped, nitrogen-containing organic molecule that is one of the components of a nucleotide, namely, adenine, guanine, cytosine, thymine, and uracil; also known as a **nitrogenous base.**

Basement membrane Thin, extracellular layer between epithelium and connective tissue consisting of a basal lamina and a reticular lamina.

Basilar (BĀS-i-lar) **membrane** A membrane in the cochlea of the internal ear that separates the cochlear duct from the scala tympani and on which the spiral organ (organ of Corti) rests.

Basophil (BĀ-sō-fil) A type of white blood cell characterized by a pale nucleus and large granules that stain blue-purple with basic dyes.

Belly The abdomen. The gaster or prominent, fleshy part of a skeletal muscle.

Beta (BĀ-ta) **cell** A type of cell in the pancreatic islets (islets of Langerhans) in the pancreas that secretes the hormone insulin. Also termed a **B cell.**

Beta receptor A type of adrenergic receptor for epinephrine and norepinephrine; found on visceral effectors innervated by sympathetic postganglionic neurons.

Bicuspid (bī-KUS-pid) **valve** Atrioventricular (AV) valve on the left side of the heart. Also called the **mitral valve.**

Bilateral (bī-LAT-er-al) Pertaining to two sides of the body.

Bile (BĪL) A secretion of the liver consisting of water, bile salts, bile pigments, cholesterol, lecithin, and several ions; it emulsifies lipids prior to their digestion.

Bilirubin (bil-ē-ROO-bin) An orange pigment that is one of the end products of hemoglobin breakdown in the hepatocytes and is excreted as a waste material in bile.

Blastocele (BLAS-tō-sēl) The fluid-filled cavity within the blastocyst.

Blastocyst (BLAS-tō-sist) In the development of an embryo, a hollow ball of cells that consists of a blastocele (the internal cavity), trophoblast (outer cells), and inner cell mass.

Blastomere (BLAS-tō-mēr) One of the cells resulting from the cleavage of a fertilized ovum.

Blind spot Area in the retina at the end of the optic nerve (cranial nerve II) in which there are no photoreceptors.

Blood The fluid that circulates through the heart, arteries, capillaries, and veins and that constitutes the chief means of transport within the body.

Blood–brain barrier (BBB) A barrier consisting of specialized brain capillaries and astrocytes that prevents the passage of materials from the blood to the cerebrospinal fluid and brain.

Blood island Isolated mass of mesoderm derived from angioblasts and from which blood vessels develop.

Blood pressure (BP) Force exerted by blood against the walls of blood vessels due to contraction of the heart and influenced by the elasticity of the vessel walls; clinically, a measure of the pressure in arteries during ventricular systole and ventricular diastole. *See also* **mean arterial blood pressure.**

Blood reservoir (REZ-er-vwar) Systemic veins that contain large amounts of blood that can be moved quickly to parts of the body requiring the blood.

Blood–testis barrier (BTB) A barrier formed by Sertoli cells that prevents an immune response against antigens produced by spermatogenic cells by isolating the cells from the blood.

Body cavity A space within the body that contains various internal organs.

Body fluid Body water and its dissolved substances; constitutes about 60% of total body mass.

Bohr (BŌR) **effect** In an acidic environment, oxygen unloads more readily from hemoglobin because when hydrogen ions (H^+) bind to hemoglobin, they alter the structure of hemoglobin, thereby reducing its oxygen-carrying capacity.

Bolus (BŌ-lus) A soft, rounded mass, usually food, that is swallowed.

Bony labyrinth (LAB-i-rinth) A series of cavities within the petrous portion of the temporal bone forming the vestibule, cochlea, and semicircular canals of the inner ear.

Brachial plexus (BRĀ-kē-al PLEK-sus) A network of nerve axons of the ventral rami of spinal nerves C5, C6, C7, C8, and T1. The nerves that emerge from the brachial plexus supply the upper limb.

Bradycardia (brād′-i-KAR-dē-a) A slow resting heart or pulse rate (under 60 beats per minute).

Brain The part of the central nervous system contained within the cranial cavity.

Brain stem The portion of the brain immediately superior to the spinal cord, made up of the medulla oblongata, pons, and midbrain.

Brain waves Electrical signals that can be recorded from the skin of the head due to electrical activity of brain neurons.

Broad ligament A double fold of parietal peritoneum attaching the uterus to the side of the pelvic cavity.

Broca's (BRŌ-kaz) **area** Motor area of the brain in the frontal lobe that translates thoughts into speech. Also called the **motor speech area.**

Bohr effect Shifting of the oxygen–hemoglobin dissociation curve to the right when pH decreases; at any given partial pressure of oxygen, hemoglobin is less saturated with O_2 at lower pH.

Bronchi (BRONG-kē) Branches of the respiratory passageway including primary bronchi (the two divisions of the trachea), secondary or lobar bronchi (divisions of the primary bronchi that are distributed to the lobes of the lung), and tertiary or segmental bronchi (divisions of the secondary bronchi that are distributed to bronchopulmonary segments of the lung). *Singular is* **bronchus.**

Bronchial tree The trachea, bronchi, and their branching structures up to and including the terminal bronchioles.

Bronchiole (BRONG-kē-ōl) Branch of a tertiary bronchus further dividing into terminal bronchioles (distributed to lobules of the lung), which divide into respiratory bronchioles (distributed to alveolar sacs).

Bronchitis (brong-KĪ-tis) Inflammation of the mucous membrane of the bronchial tree; characterized by hypertrophy and hyperplasia of seromucous glands and goblet cells that line the bronchi and which results in a productive cough.

Bronchopulmonary (brong′-kō-PUL-mō-ner-ē) **segment** One of the smaller divisions of a lobe of a lung supplied by its own branches of a bronchus.

Buccal (BUK-al) Pertaining to the cheek or mouth.

Buffer (BUF-er) **system** A pair of chemicals—one a weak acid and the other the salt of the weak acid, which functions as a weak base—that resists changes in pH.

Bulbourethral (bul′-bō-ū-RĒ-thral) **gland** One of a pair of glands located inferior to the prostate on either side of the urethra that secretes an alkaline fluid into the cavernous urethra. Also called a **Cowper's** (KOW-perz) **gland.**

Bulimia (boo-LIM-ē-a *or* boo-LĒ-mē-a) A disorder characterized by overeating at least twice a week followed by purging by self-induced vomiting, strict dieting or fasting, vigorous exercise, or use of laxatives or diuretics. Also called **binge–purge syndrome.**

Bulk flow The movement of large numbers of ions, molecules, or particles in the same direction due to pressure differences (osmotic, hydrostatic, or air pressure).

Bundle branch One of the two branches of the atrioventricular (AV) bundle made up of specialized muscle fibers (cells) that transmit electrical impulses to the ventricles.

Bursa (BUR-sa) A sac or pouch of synovial fluid located at friction points, especially about joints.

Buttocks (BUT-oks) The two fleshy masses on the posterior aspect of the inferior trunk, formed by the gluteal muscles.

C

Calcaneal (kal-KĀ-nē-al) **tendon** The tendon of the soleus, gastrocnemius, and plantaris muscles at the back of the heel. Also called the **Achilles** (a-KIL-ēz) **tendon.**

Calcification (kal-si-fi-KĀ-shun) Deposition of mineral salts, primarily hydroxyapatite, in a framework formed by collagen fibers in which the tissue hardens. Also called **mineralization** (min′-e-ral-i-ZĀ-shun).

Calcitonin (kal-si-TŌ-nin) **(CT)** A hormone produced by the parafollicular cells of the thyroid gland that can lower the amount of blood calcium and phosphates by inhibiting bone resorption (breakdown of bone matrix) and by accelerating uptake of calcium and phosphates into bone matrix.

Calculus (KAL-kū-lus) A stone, or insoluble mass of crystallized salts or other material, formed within the body, as in the gallbladder, kidney, or urinary bladder.

Callus (KAL-lus) A growth of new bone tissue in and around a fractured area, ultimately replaced by mature bone. An acquired, localized thickening.

Calorie (KAL-ō-rē) A unit of heat. A calorie (cal) is the standard unit and is the amount of heat needed to raise the temperature of 1 g of water from 14°C to 15°C. The **kilocalorie (kcal)** or **Calorie** (spelled with an uppercase C), used to express the caloric value of foods and to measure metabolic rate, is equal to 1000 cal.

Calyx (KĀL-iks) Any cuplike division of the kidney pelvis. *Plural is* **calyces** (KĀ-li-sēz).

Canal (ka-NAL) A narrow tube, channel, or passageway.

Canaliculus (kan′-a-LIK-ū-lus) A small channel or canal, as in bones, where they connect lacunae. *Plural is* **canaliculi** (kan′-a-LIK-ū-lī).

Cancellous (KAN-sel-us) Having a reticular or latticework structure, as in spongy tissue of bone.

Capacitation (ka′-pas-i-TĀ-shun) The functional changes that sperm undergo in the female reproductive tract that allow them to fertilize a secondary oocyte.

Capillary (KAP-i-lar′-ē) A microscopic blood vessel located between an arteriole and venule through which materials are exchanged between blood and interstitial fluid.

Carbohydrate (kar′-bō-HĪ-drāt) An organic compound containing carbon, hydrogen, and oxygen in a particular amount and arrangement and composed of monosaccharide subunits; usually has the general formula $(CH_2O)n$.

Carcinogen (kar-SIN-ō-jen) A chemical substance or radiation that causes cancer.

Cardiac (KAR-dē-ak) **arrest** Cessation of an effective heartbeat in which the heart is completely stopped or in ventricular fibrillation.

Cardiac cycle A complete heartbeat consisting of systole (contraction) and diastole (relaxation) of both atria plus systole and diastole of both ventricles.

Cardiac muscle Striated muscle fibers (cells) that form the wall of the heart; stimulated by an intrinsic conduction system and autonomic motor neurons.

Cardiac notch An angular notch in the anterior border of the left lung into which part of the heart fits.

Cardiac output (CO) The volume of blood pumped from one ventricle of the heart (usually measured from the left ventricle) in 1 min; normally about 5.2 liters/min in an adult at rest.

Cardiac reserve The maximum percentage that cardiac output can increase above normal.

Cardinal ligament A ligament of the uterus, extending laterally from the cervix and vagina as a continuation of the broad ligament.

Cardiology (kar-dē-OL-ō-jē) The study of the heart and diseases associated with it.

Cardiovascular (kar-dē-ō-VAS-kū-lar) **center** Groups of neurons scattered within the medulla oblongata that regulate heart rate, force of contraction, and blood vessel diameter.

Carotene (KAR-o-tēn) Antioxidant precursor of vitamin A, which is needed for synthesis of photopigments; yellow-orange pigment present in the stratum corneum of the epidermis. Accounts for the yellowish coloration of skin. Also termed **beta-carotene.**

Carotid (ka-ROT-id) **body** Cluster of chemoreceptors on or near the carotid sinus that respond to changes in blood levels of oxygen, carbon dioxide, and hydrogen ions.

Carotid sinus A dilated region of the internal carotid artery just above the point where it branches from the common carotid artery; it contains baroreceptors that monitor blood pressure.

Carotid sinus reflex A reflex that helps maintain normal blood pressure in the brain. Nerve impulses propagate from the carotid sinus baroreceptors over sensory axons in the glossopharyngeal nerves (cranial nerve IX) to the cardiovascular center in the medulla oblongata.

Carpal bones The eight bones of the wrist. Also called **carpals.**

Carpus (KAR-pus) A collective term for the eight bones of the wrist.

Cartilage (KAR-ti-lij) A type of connective tissue consisting of chondrocytes in lacunae embedded in a dense network of collagen and elastic fibers and a matrix of chondroitin sulfate.

Cartilaginous (kar′-ti-LAJ-i-nus) **joint** A joint without a synovial (joint) cavity where the articulating bones are held tightly together by cartilage, allowing little or no movement.

Cast A small mass of hardened material formed within a cavity in the body and then discharged from the body; can originate in different areas and can be composed of various materials.

Catabolism (ka-TAB-ō-lizm) Chemical reactions that break down complex organic compounds into simple ones, with the net release of energy.

Catalyst (KAT-a-list) A substance that speeds up a chemical reaction without itself being altered; an enzyme.

Cataract (KAT-a-rakt) Loss of transparency of the lens of the eye or its capsule or both.

Cation (KAT-ī-on) A positively charged ion. An example is a sodium ion (Na^+).

Cauda equina (KAW-da ē-KWĪ-na) A tail-like array of roots of spinal nerves at the inferior end of the spinal cord.

Caudal (KAW-dal) Pertaining to any tail-like structure; inferior in position.

Cecum (SĒ-kum) A blind pouch at the proximal end of the large intestine that attaches to the ileum.

Celiac plexus (PLEK-sus) A large mass of autonomic ganglia and axons located at the level of the superior part of the first lumbar vertebra. Also called the **solar plexus.**

Cell The basic structural and functional unit of all organisms; the smallest structure capable of performing all the activities vital to life.

Cell cycle Growth and division of a single cell into two daughter cells; consists of interphase and cell division.

Cell division Process by which a cell reproduces itself that consists of a nuclear division (mitosis) and a cytoplasmic division (cytokinesis); types include somatic and reproductive cell division.

Cell-mediated immunity That component of immunity in which specially sensitized T lymphocytes (T cells) attach to antigens to destroy them. Also called **cellular immunity.**

Cementum (se-MEN-tum) Calcified tissue covering the root of a tooth.

Center of ossification (os'-i-fi-KĀ-shun) An area in the cartilage model of a future bone where the cartilage cells hypertrophy and then secrete enzymes that result in the calcification of their matrix, resulting in the death of the cartilage cells, followed by the invasion of the area by osteoblasts that then lay down bone.

Central canal A microscopic tube running the length of the spinal cord in the gray commissure. A circular channel running longitudinally in the center of an osteon (Haversian system) of mature compact bone, containing blood and lymphatic vessels and nerves. Also called a **Haversian** (ha-VER-shan) **canal.**

Central fovea (FŌ-vē-a) A depression in the center of the macula lutea of the retina, containing cones only and lacking blood vessels; the area of highest visual acuity (sharpness of vision).

Central nervous system (CNS) That portion of the nervous system that consists of the brain and spinal cord.

Centrioles (SEN-trē-ōlz) Paired, cylindrical structures of a centrosome, each consisting of a ring of microtubules and arranged at right angles to each other.

Centromere (SEN-trō-mēr) The constricted portion of a chromosome where the two chromatids are joined; serves as the point of attachment for the microtubules that pull chromatids during anaphase of cell division.

Centrosome (SEN-trō-sōm) A dense network of small protein fibers near the nucleus of a cell, containing a pair of centrioles and pericentriolar material.

Cephalic (se-FAL-ik) Pertaining to the head; superior in position.

Cerebellar peduncle (ser-e-BEL-ar pe-DUNG-kul) A bundle of nerve axons connecting the cerebellum with the brain stem.

Cerebellum (ser-e-BEL-um) The part of the brain lying posterior to the medulla oblongata and pons; governs balance and coordinates skilled movements.

Cerebral aqueduct (SER-ē-bral AK-we-dukt) A channel through the midbrain connecting the third and fourth ventricles and containing cerebrospinal fluid. Also termed the **aqueduct of Sylvius.**

Cerebral arterial circle A ring of arteries forming an anastomosis at the base of the brain between the internal carotid and basilar arteries and arteries supplying the cerebral cortex. Also called the **circle of Willis.**

Cerebral cortex The surface of the cerebral hemispheres, 2–4 mm thick, consisting of gray matter; arranged in six layers of neuronal cell bodies in most areas.

Cerebral peduncle One of a pair of nerve axon bundles located on the anterior surface of the midbrain, conducting nerve impulses between the pons and the cerebral hemispheres.

Cerebrospinal (se-rē'-brō-SPĪ-nal) **fluid (CSF)** A fluid produced by ependymal cells that cover choroid plexuses in the ventricles of the brain; the fluid circulates in the ventricles, the central canal, and the subarachnoid space around the brain and spinal cord.

Cerebrovascular (se rē'-brō-VAS-kū-lar) **accident (CVA)** Destruction of brain tissue (infarction) resulting from obstruction or rupture of blood vessels that supply the brain. Also called a **stroke** or **brain attack.**

Cerebrum (SER-e-brum *or* se-RĒ-brum) The two hemispheres of the forebrain (derived from the telencephalon), making up the largest part of the brain.

Cerumen (se-ROO-men) Waxlike secretion produced by ceruminous glands in the external auditory meatus (ear canal). Also termed **ear wax.**

Ceruminous (se-ROO-mi-nus) **gland** A modified sudoriferous (sweat) gland in the external auditory meatus that secretes cerumen (ear wax).

Cervical ganglion (SER-vi-kul GANG-glē-on) A cluster of cell bodies of postganglionic sympathetic neurons located in the neck, near the vertebral column.

Cervical plexus (PLEK-sus) A network formed by nerve axons from the ventral rami of the first four cervical nerves and receiving gray rami communicates from the superior cervical ganglion.

Cervix (SER-viks) Neck; any constricted portion of an organ, such as the inferior cylindrical part of the uterus.

Chemical bond Force of attraction in a molecule or compound that holds its atoms together. Examples include ionic and covalent bonds.

Chemical element Unit of matter that cannot be decomposed into a simpler substance by ordinary chemical reactions. Examples include hydrogen (H), carbon (C), and oxygen (O).

Chemically gated channel A channel in a membrane that opens and closes in response to a specific chemical stimulus, such as a neurotransmitter, hormone, or specific type of ion.

Chemical reaction The combination or separation of atoms in which chemical bonds are formed or broken and new products with different properties are produced.

Chemiosmosis (kem'-ē-oz-MŌ-sis) Mechanism for ATP generation that links chemical reactions (electrons passing along the electron transport chain) with pumping of H^+ out of the mitochondrial matrix. ATP synthesis occurs as H^+ diffuse back into the mitochondrial matrix through special H^+ channels in the membrane.

Chemoreceptor (kē'-mō-rē-SEP-tor) Sensory receptor that detects the presence of a specific chemical.

Chemotaxis (kē-mō-TAK-sis) Attraction of phagocytes to microbes by a chemical stimulus.

Chiasm (KĪ-azm) A crossing; especially the crossing of axons in the optic nerve (cranial nerve II).

Chief cell The secreting cell of a gastric gland that produces pepsinogen, the precursor of the enzyme pepsin, and the enzyme gastric lipase. Also called a **zymogenic** (zī'-mō-JEN-ik) **cell.** Cell in the parathyroid glands that secretes parathyroid hormone (PTH). Also called a **principal cell.**

Chiropractic (kī-rō-PRAK-tik) A system of treating disease by using one's hands to manipulate body parts, mostly the vertebral column.

Chloride shift Exchange of bicarbonate ions (HCO_3^-) for chloride ions (Cl^-) between red blood cells and plasma; maintains electrical balance inside red blood cells as bicarbonate ions are produced or eliminated during respiration.

Cholecystectomy (kō'-lē-sis-TEK-tō-mē) Surgical removal of the gallbladder.

Cholecystitis (kō'-lē-sis-TĪ-tis) Inflammation of the gallbladder.

Cholesterol (kō-LES-te-rol) Classified as a lipid, the most abundant steroid in animal tissues; located in cell membranes and used for the synthesis of steroid hormones and bile salts.

Cholinergic (kō'-lin-ER-jik) **neuron** A neuron that liberates acetylcholine as its neurotransmitter.

Chondrocyte (KON-drō-sīt) Cell of mature cartilage.

Chondroitin (kon-DROY-tin) **sulfate** An amorphous matrix material found outside connective tissue cells.

Chordae tendineae (KOR-dē TEN-di-nē-ē) Tendonlike, fibrous cords that connect atrioventricular valves of the heart with papillary muscles.

Chorion (KŌ-rē-on) The most superficial fetal membrane that becomes the principal embryonic portion of the placenta; serves a protective and nutritive function.

Chorionic villi (kō-rē-ON-ik VIL-lī) Fingerlike projections of the chorion that grow into the decidua basalis of the endometrium and contain fetal blood vessels.

Choroid (KŌ-royd) One of the vascular coats of the eyeball.

Choroid plexus (PLEK-sus) A network of capillaries located in the roof of each of the four ventricles of the brain; ependymal cells around choroid plexuses produce cerebrospinal fluid.

Chromaffin (KRŌ-maf-in) **cell** Cell that has an affinity for chrome salts, due in part to the presence of the precursors of the neurotransmitter epinephrine; found, among other places, in the adrenal medulla.

Chromatid (KRŌ-ma-tid) One of a pair of identical connected nucleoprotein strands that are joined at the centromere and separate during cell division, each becoming a chromosome of one of the two daughter cells.

Chromatin (KRŌ-ma-tin) The threadlike mass of genetic material, consisting of DNA and histone proteins, that is present in the nucleus of a nondividing or interphase cell.

Chromatolysis (krō-ma-TOL-i-sis) The breakdown of Nissl bodies into finely granular masses in the cell body of a neuron whose axon has been damaged.

Chromosome (KRŌ-mō-sōm) One of the small, threadlike structures in the nucleus of a cell, normally 46 in a human diploid cell, that bears the genetic material; composed of DNA and proteins (histones) that form a delicate chromatin thread during interphase; becomes packaged into compact rodlike structures that are visible under the light microscope during cell division.

Chronic (KRON-ik) Long term or frequently recurring; applied to a disease that is not acute.

Chronic obstructive pulmonary disease (COPD) A disease, such as bronchitis or emphysema, in which there is some degree of obstruction of airways and consequent increase in airway resistance.

Chyle (KĪL) The milky-appearing fluid found in the lacteals of the small intestine after absorption of lipids in food.

Chylomicron (kī-lō-MĪ-kron) Protein-coated spherical structure that contains triglycerides, phospholipids, and cholesterol and is absorbed into the lacteal of a villus in the small intestine.

Chyme (KĪM) The semifluid mixture of partly digested food and digestive secretions found in the stomach and small intestine during digestion of a meal.

Ciliary (SIL-ē-ar′-ē) **body** One of the three parts of the vascular tunic of the eyeball, the others being the choroid and the iris; includes the ciliary muscle and the ciliary processes.

Ciliary ganglion (GANG-glē-on) A very small parasympathetic ganglion whose preganglionic axons come from the oculomotor nerve (cranial nerve III) and whose postganglionic axons carry nerve impulses to the ciliary muscle and the sphincter muscle of the iris.

Cilium (SIL-ē-um) A hair or hairlike process projecting from a cell that may be used to move the entire cell or to move substances along the surface of the cell. *Plural is* **cilia.**

Circadian (ser-KĀ-dē-an) **rhythm** A cycle of active and nonactive periods in organisms determined by internal mechanisms and repeating about every 24 hours.

Circular folds Permanent, deep, transverse folds in the mucosa and submucosa of the small intestine that increase the surface area for absorption. Also called **plicae circulares** (PLĪ-kē SER-kū-lar-ēs).

Circulation time Time required for blood to pass from the right atrium, through pulmonary circulation, back to the left ventricle, through systemic circulation to the foot, and back again to the right atrium; normally about 1 min.

Circumduction (ser′-kum-DUK-shun) A movement at a synovial joint in which the distal end of a bone moves in a circle while the proximal end remains relatively stable.

Cirrhosis (si-RŌ-sis) A liver disorder in which the parenchymal cells are destroyed and replaced by connective tissue.

Cisterna chyli (sis-TER-na-KĪ-lē) The origin of the thoracic duct.

Cleavage The rapid mitotic divisions following the fertilization of a secondary oocyte, resulting in an increased number of progressively smaller cells, called blastomeres.

Climacteric (klī-mak-TER-ik) Cessation of the reproductive function in the female or diminution of testicular activity in the male.

Climax The peak period or moments of greatest intensity during sexual excitement.

Clitoris (KLI-to-ris) An erectile organ of the female, located at the anterior junction of the labia minora, that is homologous to the male penis.

Clone (KLŌN) A population of identical cells.

Clot The end result of a series of biochemical reactions that changes liquid plasma into a gelatinous mass; specifically, the conversion of fibrinogen into a tangle of polymerized fibrin molecules.

Clot retraction (rē-TRAK-shun) The consolidation of a fibrin clot to pull damaged tissue together.

Clotting Process by which a blood clot is formed. Also known as **coagulation** (cō-ag-ū-LĀ-shun).

Coccyx (KOK-six) The fused bones at the inferior end of the vertebral column.

Cochlea (KŌK-lē-a) A winding, cone-shaped tube forming a portion of the inner ear and containing the spiral organ (organ of Corti).

Cochlear duct The membranous cochlea consisting of a spirally arranged tube enclosed in the bony cochlea and lying along its outer wall. Also called the **scala media** (SCA-la MĒ-dē-a).

Coenzyme A nonprotein organic molecule that is associated with and activates an enzyme; many are derived from vitamins. An example is nicotinamide adenine dinucleotide (NAD), derived from the B vitamin niacin.

Coitus (KŌ-i-tus) Sexual intercourse.

Collagen (KOL-a-jen) A protein that is the main organic constituent of connective tissue.

Collateral circulation The alternate route taken by blood through an anastomosis.

Colliculus (ko-LIK-ū-lus) A small elevation.

Colloid (KOL-loyd) The material that accumulates in the center of thyroid follicles, consisting of thyroglobulin and stored thyroid hormones.

Colon The portion of the large intestine consisting of ascending, transverse, descending, and sigmoid portions.

Colony-stimulating factor (CSF) One of a group of molecules that stimulates development of white blood cells. Examples are macrophage CSF and granulocyte CSF.

Colostrum (kō-LOS-trum) A thin, cloudy fluid secreted by the mammary glands a few days prior to or after delivery before true milk is produced.

Column (KOL-um) Group of white matter tracts in the spinal cord.

Common bile duct A tube formed by the union of the common hepatic duct and the cystic duct that empties bile into the duodenum at the hepatopancreatic ampulla (ampulla of Vater).

Compact (dense) bone tissue Bone tissue that contains few spaces between osteons (Haversian systems); forms the external portion of all bones and the bulk of the diaphysis (shaft) of long bones; is found immediately deep to the periosteum and external to spongy bone.

Complement (KOM-ple-ment) A group of at least 20 normally inactive proteins found in plasma that forms a component of nonspecific resistance and immunity by bringing about cytolysis, inflammation, and opsonization.

Compliance The ease with which the lungs and thoracic wall or blood vessels can be expanded.

Compound A substance that can be broken down into two or more other substances by chemical means.

Concha (KONG-ka) A scroll-like bone found in the skull. *Plural is conchae* (KONG-kē).

Concussion (kon-KUSH-un) Traumatic injury to the brain that produces no visible bruising but may result in abrupt, temporary loss of consciousness.

Conduction system A group of autorhythmic cardiac muscle fibers that generates and distributes electrical impulses to stimulate coordinated contraction of the heart chambers; includes the sinoatrial (SA) node, the atrioventricular (AV) node, the atrioventricular (AV) bundle, the right and left bundle branches, and the Purkinje fibers.

Conductivity (kon'-duk-TIV-i-tē) The ability of a cell to propagate (conduct) action potentials along its plasma membrane; characteristic of neurons and muscle fibers (cells).

Condyloid (KON-di-loyd) **joint** A synovial joint structured so that an oval-shaped condyle of one bone fits into an elliptical cavity of another bone, permitting side-to-side and back-and-forth movements, such as the joint at the wrist between the radius and carpals. Also called an **ellipsoidal** (ē-lip-SOYD-al) **joint.**

Cone (KŌN) The type of photoreceptor in the retina that is specialized for highly acute color vision in bright light.

Congenital (kon-JEN-i-tal) Present at the time of birth.

Conjunctiva (kon'-junk-TĪ-va) The delicate membrane covering the eyeball and lining the eyes.

Connective tissue One of the most abundant of the four basic tissue types in the body, performing the functions of binding and supporting; consists of relatively few cells in a generous matrix (the ground substance and fibers between the cells).

Consciousness (KON-shus-nes) A state of wakefulness in which an individual is fully alert, aware, and oriented, partly as a result of feedback between the cerebral cortex and reticular activating system.

Continuous conduction (kon-DUK-shun) Propagation of an action potential (nerve impulse) in a step-by-step depolarization of each adjacent area of an axon membrane.

Contraception (kon'-tra-SEP-shun) The prevention of fertilization or impregnation without destroying fertility.

Contractility (kon'-trak-TIL-i-tē) The ability of cells or parts of cells to actively generate force to undergo shortening for movements. Muscle fibers (cells) exhibit a high degree of contractility.

Contralateral (kon'-tra-LAT-er-al) On the opposite side; affecting the opposite side of the body.

Control center The component of a feedback system, such as the brain, that determines the point at which a controlled condition, such as body temperature, is maintained.

Conus medullaris (KŌ-nus med-ū-LAR-is) The tapered portion of the spinal cord inferior to the lumbar enlargement.

Convergence (con-VER-jens) A synaptic arrangement in which the synaptic end bulbs of several presynaptic neurons terminate on one postsynaptic neuron. The medial movement of the two eyeballs so that both are directed toward a near object being viewed in order to produce a single image.

Convulsion (con-VUL-shun) Violent, involuntary contractions or spasms of an entire group of muscles.

Cornea (KOR-nē-a) The nonvascular, transparent fibrous coat through which the iris of the eye can be seen.

Corona radiata The innermost layer of granulosa cells that is firmly attached to the zona pellucida around a secondary oocyte.

Coronary artery disease (CAD) A condition such as atherosclerosis that causes narrowing of coronary arteries so that blood flow to the heart is reduced. The result is **coronary heart disease (CHD),** in which the heart muscle receives inadequate blood flow due to an interruption of its blood supply.

Coronary circulation The pathway followed by the blood from the ascending aorta through the blood vessels supplying the heart and returning to the right atrium. Also called **cardiac circulation.**

Coronary sinus (SĪ-nus) A wide venous channel on the posterior surface of the heart that collects the blood from the coronary circulation and returns it to the right atrium.

Corpus (KOR-pus) The principal part of any organ; any mass or body.

Corpus albicans (KOR-pus AL-bi-kanz) A white fibrous patch in the ovary that forms after the corpus luteum regresses.

Corpus callosum (kal-LŌ-sum) The great commissure of the brain between the cerebral hemispheres.

Corpus luteum (LOO-tē-um) A yellowish body in the ovary formed when a follicle has discharged its secondary oocyte; secretes estrogens, progesterone, relaxin, and inhibin.

Corpuscle of touch The sensory receptor for the sensation of touch; found in the dermal papillae, especially in palms and soles. Also called a **Meissner** (MĪZ-ner) **corpuscle.**

Cortex (KOR-teks) An outer layer of an organ. The convoluted layer of gray matter covering each cerebral hemisphere.

Costal (KOS-tal) Pertaining to a rib.

Costal cartilage (KAR-ti-lij) Hyaline cartilage that attaches a rib to the sternum.

Countercurrent mechanism One mechanism involved in the ability of the kidneys to produce hypertonic urine.

Cramp A spasmodic, usually painful contraction of a muscle.

Cranial (KRĀ-ne-al) **cavity** A subdivision of the dorsal body cavity formed by the cranial bones and containing the brain.

Cranial nerve One of 12 pairs of nerves that leave the brain; pass through foramina in the skull; and supply sensory and motor neurons to the head, neck, part of the trunk, and viscera of the thorax and abdomen. Each is designated by a Roman numeral and a name.

Craniosacral (krā-nē-ō-SĀ-kral) **outflow** The axons of parasympathetic preganglionic neurons, which have their cell bodies located in nuclei in the brain stem and in the lateral gray matter of the sacral portion of the spinal cord.

Cranium (KRĀ-nē-um) The skeleton of the skull that protects the brain and the organs of sight, hearing, and balance; includes the frontal, parietal, temporal, occipital, sphenoid, and ethmoid bones.

Creatine phosphate (KRĒ-a-tin FOS-fāt) Molecule in striated muscle fibers that contains high-energy phosphate bonds; used to generate ATP rapidly from ADP by transfer of a phosphate group. Also called **phosphocreatine** (fos′-fō-KRĒ-a-tin).

Crenation (krē-NĀ-shun) The shrinkage of red blood cells into knobbed, starry forms when they are placed in a hypertonic solution.

Crista (KRIS-ta) A crest or ridged structure. A small elevation in the ampulla of each semicircular duct that contains receptors for dynamic equilibrium.

Crossed extensor reflex A reflex in which extension of the joints in one limb occurs together with contraction of the flexor muscles of the opposite limb.

Crossing-over The exchange of a portion of one chromatid with another during meiosis. It permits an exchange of genes among chromatids and is one factor that results in genetic variation of progeny.

Crus (KRUS) **of penis** Separated, tapered portion of the corpora cavernosa penis. *Plural is* **crura** (KROO-ra).

Cryptorchidism (krip-TOR-ki-dizm) The condition of undescended testes.

Cuneate (KŪ-nē-āt) **nucleus** A group of neurons in the inferior part of the medulla oblongata in which axons of the cuneate fasciculus terminate.

Cupula (KUP-ū-la) A mass of gelatinous material covering the hair cells of a crista; a sensory receptor in the ampulla of a semicircular canal stimulated when the head moves.

Cushing's syndrome Condition caused by a hypersecretion of glucocorticoids characterized by spindly legs, "moon face," "buffalo hump," pendulous abdomen, flushed facial skin, poor wound healing, hyperglycemia, osteoporosis, hypertension, and increased susceptibility to disease.

Cutaneous (kū-TĀ-nē-us) Pertaining to the skin.

Cyanosis (sī-a-NŌ-sis) A blue or dark purple discoloration, most easily seen in nail beds and mucous membranes, that results from an increased concentration of deoxygenated (reduced) hemoglobin (more than 5 gm/dL).

Cyclic AMP (cyclic adenosine-3′, 5′-monophosphate) Molecule formed from ATP by the action of the enzyme adenylate cyclase; serves as second messenger for some hormones and neurotransmitters.

Cyst (SIST) A sac with a distinct connective tissue wall, containing a fluid or other material.

Cystic (SIS-tik) **duct** The duct that carries bile from the gallbladder to the common bile duct.

Cystitis (sis-TĪ-tis) Inflammation of the urinary bladder.

Cytosol (SĪ-tō-sol) Fluid located within cells. Also called **intracellular** (in′-tra-SEL-ū-lar) **fluid (ICF).**

Cytolysis (sī-TOL-i-sis) The rupture of living cells in which the contents leak out.

Cytochrome (SĪ-tō-krōm) A protein with an iron-containing group (heme) capable of alternating between a reduced form (Fe^{2+}) and an oxidized form (Fe^{3+}).

Cytokines (SĪ-to-kīns) Small protein hormones produced by lymphocytes, fibroblasts, endothelial cells, and antigen-presenting cells that act as autocrine or paracrine substances to stimulate or inhibit cell growth and differentiation, regulate immune responses, or aid nonspecific defenses.

Cytokinesis (sī-tō-ki-NĒ-sis) Distribution of the cytoplasm into two separate cells during cell division; coordinated with nuclear division (mitosis).

Cytoplasm (SĪ-tō-plazm) Cytosol plus all organelles except the nucleus.

Cytoskeleton Complex internal structure of cytoplasm consisting of microfilaments, microtubules, and intermediate filaments.

Cytosol (SĪ-tō-sol) Semifluid portion of cytoplasm in which organelles and inclusions are suspended and solutes are dissolved. Also called **intracellular fluid.**

D

Dartos (DAR-tōs) The contractile tissue deep to the skin of the scrotum.

Decibel (DES-i-bel) **(dB)** A unit for expressing the relative intensity (loudness) of sound.

Decidua (dē-SID-ū-a) That portion of the endometrium of the uterus (all but the deepest layer) that is modified during pregnancy and shed after childbirth.

Deciduous (dē-SID-ū-us) Falling off or being shed seasonally or at a particular stage of development. In the body, referring to the first set of teeth.

Decussation (dē-ku-SĀ-shun) A crossing-over to the opposite (contralateral) side; an example is the crossing of 90% of the axons in the large motor tracts to opposite sides in the medullary pyramids.

Deep Away from the surface of the body or an organ.

Deep fascia (FASH-ē-a) A sheet of connective tissue wrapped around a muscle to hold it in place.

Deep inguinal (IN-gwi-nal) **ring** A slitlike opening in the aponeurosis of the transversus abdominis muscle that represents the origin of the inguinal canal.

Defecation (def-e-KĀ-shun) The discharge of feces from the rectum.

Deglutition (dē-gloo-TISH-un) The act of swallowing.

Dehydration (dē-hī-DRĀ-shun) Excessive loss of water from the body or its parts.

Delta cell A cell in the pancreatic islets (islets of Langerhans) in the pancreas that secretes somatostatin. Also termed a **D cell.**

Demineralization (de-min′-er-al-i-ZĀ-shun) Loss of calcium and phosphorus from bones.

Denaturation (de-nā-chur-Ā-shun) Disruption of the tertiary structure of a protein by heat, changes in pH, or other physical or chemical methods, in which the protein loses its physical properties and biological activity.

Dendrite (DEN-drīt) A neuronal process that carries electrical signals, usually graded potentials, toward the cell body.

Dendritic (den-DRIT-ik) **cell** One type of antigen-presenting cell with long branchlike projections that commonly is present in mucosal linings such as the vagina, in the skin (Langerhans cells in the epidermis), and in lymph nodes (follicular dendritic cells).

Dental caries (KA-rēz) Gradual demineralization of the enamel and dentin of a tooth that may invade the pulp and alveolar bone. Also called **tooth decay.**

Denticulate (den-TIK-ū-lāt) Finely toothed or serrated; characterized by a series of small, pointed projections.

Dentin (DEN-tin) The bony tissues of a tooth enclosing the pulp cavity.

Dentition (den-TI-shun) The eruption of teeth. The number, shape, and arrangement of teeth.

Deoxyribonucleic (dē-ok′-sē-rī′-bo-noo-KLĒ-ik) **acid (DNA)** A nucleic acid constructed of nucleotides consisting of one of four bases (adenine, cytosine, guanine, or thymine), deoxyribose, and a phosphate group; encoded in the nucleotides is genetic information.

Depolarization (dē-pō-lar-i-ZĀ-shun) A reduction of voltage across a plasma membrane; expressed as a change toward less negative (more positive) voltages on the interior surface of the plasma membrane.

Depression (de-PRESH-un) Movement in which a part of the body moves inferiorly.

Dermal papilla (pa-PILL-a) Fingerlike projection of the papillary region of the dermis that may contain blood capillaries or corpuscles of touch (Meissner corpuscles).

Dermatology (der-ma-TOL-ō-jē) The medical specialty dealing with diseases of the skin.

Dermatome (DER-ma-tōm) The cutaneous area developed from one embryonic spinal cord segment and receiving most of its sensory innervation from one spinal nerve. An instrument for incising the skin or cutting thin transplants of skin.

Dermis (DER-mis) A layer of dense irregular connective tissue lying deep to the epidermis.

Descending colon (KŌ-lon) The part of the large intestine descending from the left colic (splenic) flexure to the level of the left iliac crest.

Detritus (de-TRĪ-tus) Particulate matter produced by or remaining after the wearing away or disintegration of a substance or tissue; scales, crusts, or loosened skin.

Detrusor (de-TROO-ser) **muscle** Smooth muscle that forms the wall of the urinary bladder.

Developmental biology The study of development from the fertilized egg to the adult form.

Diagnosis (dī-ag-NŌ-sis) Distinguishing one disease from another or determining the nature of a disease from signs and symptoms by inspection, palpation, laboratory tests, and other means.

Dialysis (dī-AL-i-sis) The removal of waste products from blood by diffusion through a selectively permeable membrane.

Diaphragm (DĪ-a-fram) Any partition that separates one area from another, especially the dome-shaped skeletal muscle between the thoracic and abdominal cavities. Also a dome-shaped device that is placed over the cervix, usually with a spermicide, to prevent conception.

Diaphysis (dī-AF-i-sis) The shaft of a long bone.

Diarrhea (dī-a-RE-a) Frequent defecation of liquid feces caused by increased motility of the intestines.

Diarthrosis (dī-ar-THRŌ-sis) A freely movable joint; types are gliding, hinge, pivot, condyloid, saddle, and ball-and-socket.

Diastole (dī-AS-tō-lē) In the cardiac cycle, the phase of relaxation or dilation of the heart muscle, especially of the ventricles.

Diastolic (dī-as-TOL-ik) **blood pressure** The force exerted by blood on arterial walls during ventricular relaxation; the lowest blood pressure measured in the large arteries, normally about 80 mmHg in a young adult.

Diencephalon (dī′-en-SEF-a-lon) A part of the brain consisting of the thalamus, hypothalamus, epithalamus, and subthalamus.

Diffusion (dif-Ū-zhun) A passive process in which there is a net or greater movement of molecules or ions from a region of high concentration to a region of low concentration until equilibrium is reached.

Digestion (dī-JES-chun) The mechanical and chemical breakdown of food to simple molecules that can be absorbed and used by body cells.

Dilate (DĪ-lāt) To expand or swell.

Diploid (DIP-loyd) Having the number of chromosomes characteristically found in the somatic cells of an organism; having two haploid sets of chromosomes, one each from the mother and father. Symbolized 2n.

Direct motor pathways Collections of upper motor neurons with cell bodies in the motor cortex that project axons into the spinal cord, where they synapse with lower motor neurons or interneurons in the anterior horns. Also called the **pyramidal pathways.**

Disease Any change from a state of health.

Dislocation (dis-lō-KĀ-shun) Displacement of a bone from a joint with tearing of ligaments, tendons, and articular capsules. Also called **luxation** (luks-Ā-shun).

Dissect (di-SEKT) To separate tissues and parts of a cadaver or an organ for anatomical study.

Distal (DIS-tal) Farther from the attachment of a limb to the trunk; farther from the point of origin or attachment.

Diuretic (dī-ū-RET-ik) A chemical that increases urine volume by decreasing reabsorption of water, usually by inhibiting sodium reabsorption.

Divergence (dī-VER-jens) A synaptic arrangement in which the synaptic end bulbs of one presynaptic neuron terminate on several postsynaptic neurons.

Diverticulum (dī-ver-TIK-ū-lum) A sac or pouch in the wall of a canal or organ, especially in the colon.

Dominant allele An allele that overrides the influence of an alternate allele on the homologous chromosome; the allele that is expressed.

Dorsal body cavity Cavity near the dorsal (posterior) surface of the body that consists of a cranial cavity and vertebral canal.

Dorsal ramus (RĀ-mus) A branch of a spinal nerve containing motor and sensory axons supplying the muscles, skin, and bones of the posterior part of the head, neck, and trunk.

Dorsiflexion (dor′-si-FLEK-shun) Bending the foot in the direction of the dorsum (upper surface).

Down-regulation Phenomenon in which there is a decrease in the number of receptors in response to an excess of a hormone or neurotransmitter.

Ductus arteriosus (DUK-tus ar-tē-rē-Ō-sus) A small vessel connecting the pulmonary trunk with the aorta; found only in the fetus.

Ductus (vas) deferens (DEF-er-ens) The duct that carries sperm from the epididymis to the ejaculatory duct. Also called the **seminal duct.**

Ductus epididymis (ep′-i-DID-i-mis) A tightly coiled tube inside the epididymis, distinguished into a head, body, and tail, in which sperm undergo maturation.

Ductus venosus (ve-NŌ-sus) A small vessel in the fetus that helps the circulation bypass the liver.

Duodenal (doo-ō-DĒ-nal) **gland** Gland in the submucosa of the duodenum that secretes an alkaline mucus to protect the lining of the small intestine from the action of enzymes and to help neutralize the acid in chyme. Also called **Brunner's** (BRUN-erz) **gland.**

Duodenum (doo′-ō-DĒ-num *or* doo-OD-e-num) The first 25 cm (10 in.) of the small intestine, which connects the stomach and the ileum.

Dura mater (DOO-ra MĀ-ter) The outermost of the three meninges (coverings) of the brain and spinal cord.

Dynamic equilibrium (ē-kwi-LIB-rē-um) The maintenance of body position, mainly the head, in response to sudden movements such as rotation.

Dysfunction (dis-FUNK-shun) Absence of completely normal function.

Dysmenorrhea (dis′-men-ō-RĒ-a) Painful menstruation.

Dysplasia (dis-PLĀ-zē-a) Change in the size, shape, and organization of cells due to chronic irritation or inflammation; may either revert to normal if stress is removed or progress to neoplasia.

Dyspnea (DISP-nē-a) Shortness of breath.

E

Eardrum A thin, semitransparent partition of fibrous connective tissue between the external auditory meatus and the middle ear. Also called the **tympanic membrane.**

Ectoderm The primary germ layer that gives rise to the nervous system and the epidermis of skin and its derivatives.

Ectopic (ek-TOP-ik) Out of the normal location, as in ectopic pregnancy.

Edema (e-DĒ-ma) An abnormal accumulation of interstitial fluid.

Effector (e-FEK-tor) An organ of the body, either a muscle or a gland, that is innervated by somatic or autonomic motor neurons.

Efferent arteriole (EF-er-ent ar-TĒ-rē-ōl) A vessel of the renal vascular system that carries blood from a glomerulus to a peritubular capillary.

Efferent (EF-er-ent) **ducts** A series of coiled tubes that transport sperm from the rete testis to the epididymis.

Eicosanoids (ī-KŌ-sa-noyds) Local hormones derived from a 20-carbon fatty acid (arachidonic acid); two important types are prostaglandins and leukotrienes.

Ejaculation (e-jak-ū-LĀ-shun) The reflex ejection or expulsion of semen from the penis.

Ejaculatory (e-JAK-ū-la-tō-rē) **duct** A tube that transports sperm from the ductus (vas) deferens to the prostatic urethra.

Elasticity (e-las-TIS-i-tē) The ability of tissue to return to its original shape after contraction or extension.

Electrocardiogram (e-lek′-trō-KAR-dē-ō-gram) **(ECG** or **EKG)** A recording of the electrical changes that accompany the cardiac cycle that can be detected at the surface of the body; may be resting, stress, or ambulatory.

Electroencephalogram (e-lek′-trō-en-SEF-a-lō-gram) **(EEG)** A recording of the electrical activity of the brain from the scalp surface; used to diagnose certain diseases (such as epilepsy), furnish information regarding sleep and wakefulness, and confirm brain death.

Electrolyte (ē-LEK-trō-līt) Any compound that separates into ions when dissolved in water and that conducts electricity.

Electromyography (e-lek′-trō-mī-OG-ra-fē) Evaluation of the electrical activity of resting and contracting muscle to ascertain causes of muscular weakness, paralysis, involuntary twitching, and abnormal levels of muscle enzymes; also used as part of biofeedback studies.

Electron transport chain A sequence of electron carrier molecules on the inner mitochondrial membrane that undergo oxidation and reduction as they pump hydrogen ions (H$^+$) through the membrane. ATP synthesis then occurs as H$^+$ diffuse back into the mitochondrial matrix through special H$^+$ channels. *See also* **Chemiosmosis.**

Elevation (el-e-VĀ-shun) Movement in which a part of the body moves superiorly.

Embolism (EM-bō-lizm) Obstruction or closure of a vessel by an embolus.

Embolus (EM-bō-lus) A blood clot, bubble of air or fat from broken bones, mass of bacteria, or other debris or foreign material transported by the blood.

Embryo (EM-brē-ō) The young of any organism in an early stage of development; in humans, the developing organism from fertilization to the end of the eighth week of development.

Embryology (em′-brē-OL-ō-jē) The study of development from the fertilized egg to the end of the eighth week of development.

Emesis (EM-e-sis) Vomiting.

Emigration (em′-e-GRĀ-shun) Process whereby white blood cells (WBCs) leave the bloodstream by rolling along the endothelium, sticking to it, and squeezing between the endothelial cells. Adhesion molecules help WBCs stick to the endothelium. Also known as **migration** or **extravasation.**

Emission (ē-MISH-un) Propulsion of sperm into the urethra due to peristaltic contractions of the ducts of the testes, epididymides, and ductus (vas) deferens as a result of sympathetic stimulation.

Emmetropia (em′-e-TRŌ-pē-a) Normal vision in which light rays are focused exactly on the retina.

Emphysema (em′-fi′-SĒ-ma) A lung disorder in which alveolar walls disintegrate, producing abnormally large air spaces and loss of elasticity in the lungs; typically caused by exposure to cigarette smoke.

Emulsification (ē-mul′-si-fi-KĀ-shun) The dispersion of large lipid globules into smaller, uniformly distributed particles in the presence of bile.

Enamel (e-NAM-el) The hard, white substance covering the crown of a tooth.

End-diastolic (dī-as-TO-lik) **volume (EDV)** The volume of blood, about 130 mL, remaining in a ventricle at the end of its diastole (relaxation).

Endergonic (end′-er-GON-ik) **reaction** Type of chemical reaction in which the energy released as new bonds form is less than the energy needed to break apart old bonds; an energy-requiring reaction.

Endocardium (en-dō-KAR-dē-um) The layer of the heart wall, composed of endothelium and smooth muscle, that lines the inside of the heart and covers the valves and tendons that hold the valves open.

Endochondral ossification (en′-dō-KON-dral os′-i-fi-KĀ-shun) The replacement of cartilage by bone. Also called **intracartilaginous** (in′-tra-kar′-ti-LAJ-i-nus) **ossification.**

Endocrine (EN-dō-krin) **gland** A gland that secretes hormones into interstitial fluid and then the blood; a ductless gland.

Endocrinology (en′-dō-kri-NOL-ō-jē) The science concerned with the structure and functions of endocrine glands and the diagnosis and treatment of disorders of the endocrine system.

Endocytosis (en′-dō-sī-TŌ-sis) The uptake into a cell of large molecules and particles in which a segment of plasma membrane surrounds the substance, encloses it, and brings it in; includes phagocytosis, pinocytosis, and receptor-mediated endocytosis.

Endoderm (EN-dō-derm) A primary germ layer of the developing embryo; gives rise to the gastrointestinal tract, urinary bladder, urethra, and respiratory tract.

Endodontics (en′-dō-DON-tiks) The branch of dentistry concerned with the prevention, diagnosis, and treatment of diseases that affect the pulp, root, periodontal ligament, and alveolar bone.

Endogenous (en-DOJ-e-nus) Growing from or beginning within the organism.

Endolymph (EN-dō-limf′) The fluid within the membranous labyrinth of the internal ear.

Endometrium (en′-dō-MĒ-trē-um) The mucous membrane lining the uterus.

Endomysium (en′-dō-MĪZ-ē-um) Invagination of the perimysium separating each individual muscle fiber (cell).

Endoneurium (en′-dō-NOO-rē-um) Connective tissue wrapping around individual nerve axons (cells).

Endoplasmic reticulum (en′-do-PLAZ-mik re-TIK-ū-lum) **(ER)** A network of channels running through the cytoplasm of a cell that serves in intracellular transportation, support, storage, synthesis, and packaging of molecules. Portions of ER where ribosomes are attached to the outer surface are called **rough ER;** portions that have no ribosomes are called **smooth ER.**

Endorphin (en-DOR-fin) A neuropeptide in the central nervous system that acts as a painkiller.

Endosteum (en-DOS-tē-um) The membrane that lines the medullary (marrow) cavity of bones, consisting of osteogenic cells and scattered osteoclasts.

Endothelium (en′-dō-THĒ-lē-um) The layer of simple squamous epithelium that lines the cavities of the heart, blood vessels, and lymphatic vessels.

End-systolic (sis-TO-lik) **volume (ESV)** The volume of blood, about 60 mL, remaining in a ventricle after its systole (contraction).

Energy The capacity to do work.

Enkephalin (en-KEF-a-lin) A peptide found in the central nervous system that acts as a painkiller.

Enteric (EN-ter-ik) **nervous system** The part of the nervous system that is embedded in the submucosa and muscularis of the gastrointestinal (GI) tract; governs motility and secretions of the GI tract.

Enteroendocrine (en-ter-ō-EN-dō-krin) **cell** A cell of the mucosa of the gastrointestinal tract that secretes a hormone that governs function of the GI tract; hormones secreted include gastrin, cholecystokinin, glucose-dependent insulinotropic peptide (GIP), and secretin.

Enterogastric (en-te-rō-GAS-trik) **reflex** A reflex that inhibits gastric secretion; initiated by food in the small intestine.

Enzyme (EN-zīm) A substance that accelerates chemical reactions; an organic catalyst, usually a protein.

Eosinophil (ē′-ō-SIN-ō-fil) A type of white blood cell characterized by granules that stain red or pink with acid dyes.

Ependymal (e-PEN-de-mal) **cells** Neuroglial cells that cover choroid plexuses and produce cerebrospinal fluid (CSF); they also line the ventricles of the brain and probably assist in the circulation of CSF.

Epicardium (ep′-i-KAR-dē-um) The thin outer layer of the heart wall, composed of serous tissue and mesothelium. Also called the **visceral pericardium.**

Epidemiology (ep′-i-dē-mē-OL-ō-jē) Study of the occurrence and distribution of diseases and disorders in human populations.

Epidermis (ep-i-DERM-is) The superficial, thinner layer of skin, composed of keratinized stratified squamous epithelium.

Epididymis (ep′-i-DID-i-mis) A comma-shaped organ that lies along the posterior border of the testis and contains the ductus epididymis, in which sperm undergo maturation. *Plural is* **epididymides** (ep′-i-DID-i-mi-dēz).

Epidural (ep′-i-DOO-ral) **space** A space between the spinal dura mater and the vertebral canal, containing areolar connective tissue and a plexus of veins.

Epiglottis (ep′-i-GLOT-is) A large, leaf-shaped piece of cartilage lying on top of the larynx, attached to the thyroid cartilage and its unattached portion is free to move up and down to cover the glottis (vocal folds and rima glottidis) during swallowing.

Epimysium (ep′-i-MĪZ-ē-um) Fibrous connective tissue around muscles.

Epinephrine (ep-ē-NEF-rin) Hormone secreted by the adrenal medulla that produces actions similar to those that result from sympathetic stimulation. Also called **adrenaline** (a-DREN-a-lin).

Epineurium (ep′-i-NOO-rē-um) The superficial connective tissue covering around an entire nerve.

Epiphyseal (ep′-i-FIZ-ē-al) **line** The remnant of the epiphyseal plate in the metaphysis of a long bone.

Epiphyseal (ep′-i-FIZ-ē-al) **plate** The hyaline cartilage plate in the metaphysis of a long bone; site of lengthwise growth of long bones.

Epiphysis (ē-PIF-i-sis) The end of a long bone, usually larger in diameter than the shaft (diaphysis).

Episiotomy (e-piz′-ē-OT-ō-mē) A cut made with surgical scissors to avoid tearing of the perineum at the end of the second stage of labor.

Epistaxis (ep′-i-STAK-sis) Loss of blood from the nose due to trauma, infection, allergy, neoplasm, and bleeding disorders. Also called **nosebleed.**

Epithelial (ep′-i-THĒ-lē-al) **tissue** The tissue that forms innermost and outermost surfaces of body structures and forms glands.

Eponychium (ep′-o-NIK-ē-um) Narrow band of stratum corneum at the proximal border of a nail that extends from the margin of the nail wall. Also called the **cuticle.**

Erectile dysfunction Failure to maintain an erection long enough for sexual intercourse. Also known as **impotence** (IM-pō-tens).

Erection (ē-REK-shun) The enlarged and stiff state of the penis or clitoris resulting from the engorgement of the spongy erectile tissue with blood.

Eructation (e-ruk′-TĀ-shun) The forceful expulsion of gas from the stomach. Also called **belching.**

Erythema (er′-i-THĒ-ma) Skin redness usually caused by dilation of the capillaries.

Erythrocyte (e-RITH-rō-sīt) A mature red blood cell.

Erythropoiesis (e-rith′-rō-poy-Ē-sis) The process by which red blood cells are formed.

Erythropoietin (e-rith′-rō-POY-e-tin) A hormone released by the juxtaglomerular cells of the kidneys that stimulates red blood cell production.

Esophagus (e-SOF-a-gus) The hollow muscular tube that connects the pharynx and the stomach.

Essential amino acids Those 10 amino acids that cannot be synthesized by the human body at an adequate rate to meet its needs and therefore must be obtained from the diet.

Estrogens (ES-tro-jenz) Feminizing sex hormones produced by the ovaries; govern development of oocytes, maintenance of female reproductive structures, and appearance of secondary sex characteristics; also affect fluid and electrolyte balance, and protein anabolism. Examples are β-estradiol, estrone, and estriol.

Etiology (ē′-tē-OL-ō-jē) The study of the causes of disease, including theories of the origin and organisms (if any) involved.

Eupnea (ŪP-nē-a) Normal quiet breathing.

Eversion (ē-VER-zhun) The movement of the sole laterally at the ankle joint or of an atrioventricular valve into an atrium during ventricular contraction.

Excitability (ek-sīt′-a-BIL-i-tē) The ability of muscle fibers to receive and respond to stimuli; the ability of neurons to respond to stimuli and generate nerve impulses.

Excitatory postsynaptic potential (EPSP) A small depolarization of a postsynaptic membrane when it is stimulated by an excitatory neurotransmitter. An EPSP is both graded (variable size) and localized (decreases in size away from the point of excitation).

Excrement (EKS-kre-ment) Material eliminated from the body as waste, especially fecal matter.

Excretion (eks-KRĒ-shun) The process of eliminating waste products from the body; also the products excreted.

Exergonic (eks'-er-GON-ik) **reaction** Type of chemical reaction in which the energy released as new bonds form is greater than the energy needed to break apart old bonds; an energy-releasing reaction.

Exocrine (EK-sō-krin) **gland** A gland that secretes its products into ducts that carry the secretions into body cavities, into the lumen of an organ, or to the outer surface of the body.

Exocytosis (ex'-ō-sī-TŌ-sis) A process in which membrane-enclosed secretory vesicles form inside the cell, fuse with the plasma membrane, and release their contents into the interstitial fluid; achieves secretion of materials from a cell.

Exogenous (ex-SOJ-e-nus) Originating outside an organ or part.

Exhalation (eks-ha-LĀ-shun) Breathing out; expelling air from the lungs into the atmosphere. Also called **expiration.**

Exon (EX-on) A region of DNA that codes for synthesis of a protein.

Expiratory (eks-PĪ-ra-tō-rē) **reserve volume** The volume of air in excess of tidal volume that can be exhaled forcibly; about 1200 mL.

Extensibility (ek-sten'-si-BIL-i-tē) The ability of muscle tissue to stretch when it is pulled.

Extension (ek-STEN-shun) An increase in the angle between two bones; restoring a body part to its anatomical position after flexion.

External Located on or near the surface.

External auditory (AW-di-tōr-ē) **canal** or **meatus** (mē-Ā-tus) A curved tube in the temporal bone that leads to the middle ear.

External ear The outer ear, consisting of the pinna, external auditory canal, and tympanic membrane (eardrum).

External nares (NĀ-rez) The external nostrils, or the openings into the nasal cavity on the exterior of the body.

External respiration The exchange of respiratory gases between the lungs and blood. Also called **pulmonary respiration.**

Exteroceptor (eks'-ter-ō-SEP-tor) A sensory receptor adapted for the reception of stimuli from outside the body.

Extracellular fluid (ECF) Fluid outside body cells, such as interstitial fluid and plasma.

Extravasation (eks-trav-a-SĀ-shun) The escape of fluid, especially blood, lymph, or serum, from a vessel into the tissues.

Extrinsic (ek-STRIN-sik) Of external origin.

Extrinsic pathway (of blood clotting) Sequence of reactions leading to blood clotting that is initiated by the release of tissue factor (TF), also known as thromboplastin, that leaks into the blood from damaged cells outside the blood vessels.

Exudate (EKS-oo-dāt) Escaping fluid or semifluid material that oozes from a space and that may contain serum, pus, and cellular debris.

Eyebrow The hairy ridge superior to the eye.

F

Face The anterior aspect of the head.

Facilitated diffusion (fa-SIL-i-tā-ted dif-Ū-zhun) Diffusion in which a substance not soluble by itself in lipids diffuses across a selectively permeable membrane with the help of a transporter protein.

Falciform ligament (FAL-si-form LIG-a-ment) A sheet of parietal peritoneum between the two principal lobes of the liver. The ligamentum teres, or remnant of the umbilical vein, lies within its fold.

Falx cerebelli (FALKS ser'-e-BEL-lē) A small triangular process of the dura mater attached to the occipital bone in the posterior cranial fossa and projecting inward between the two cerebellar hemispheres.

Falx cerebri (FALKS SER-e-brē) A fold of the dura mater extending deep into the longitudinal fissure between the two cerebral hemispheres.

Fascia (FASH-ē-a) A fibrous membrane covering, supporting, and separating muscles.

Fascicle (FAS-i-kul) A small bundle or cluster, especially of nerve or muscle fibers (cells). Also called a **fasciculus** (fa-SIK-ū-lus). *Plural is* **fasciculi** (fa-SIK-yoo-lī).

Fasciculation (fa-sik'-ū-LĀ-shun) Abnormal, spontaneous twitch of all skeletal muscle fibers in one motor unit that is visible at the skin surface; not associated with movement of the affected muscle; present in progressive diseases of motor neurons, for example, poliomyelitis.

Fauces (FAW-sēz) The opening from the mouth into the pharynx.

F cell A cell in the pancreatic islets (islets of Langerhans) that secretes pancreatic polypeptide.

Feces (FĒ-sēz) Material discharged from the rectum and made up of bacteria, excretions, and food residue. Also called **stool.**

Feedback system A sequence of events in which information about the status of a situation is continually reported (fed back) to a control center.

Female reproductive cycle General term for the ovarian and uterine cycles, the hormonal changes that accompany them, and cyclic changes in the breasts and cervix; includes changes in the endometrium of a nonpregnant female that prepares the lining of the uterus to receive a fertilized ovum. Less correctly termed the **menstrual cycle.**

Fertilization (fer'-ti-li-ZĀ-shun) Penetration of a secondary oocyte by a sperm cell, meiotic division of secondary oocyte to form an ovum, and subsequent union of the nuclei of the gametes.

Fetal circulation The cardiovascular system of the fetus, including the placenta and special blood vessels involved in the exchange of materials between fetus and mother.

Fetus (FĒ-tus) In humans, the developing organism *in utero* from the beginning of the third month to birth.

Fever An elevation in body temperature above the normal temperature of 37°C (98.6°F) due to a resetting of the hypothalamic thermostat.

Fibrillation (fi-bri-LĀ-shun) Abnormal, spontaneous twitch of a single skeletal muscle fiber (cell) that can be detected with electromyography but is not visible at the skin surface; not associated with movement of the affected muscle; present in certain disorders of motor neurons, for example, amyotrophic lateral sclerosis (ALS). With reference to cardiac muscle, *see* **Atrial fibrillation** and **Ventricular fibrillation.**

Fibrin (FĪ-brin) An insoluble protein that is essential to blood clotting; formed from fibrinogen by the action of thrombin.

Fibrinogen (fī-BRIN-ō-jen) A clotting factor in blood plasma that by the action of thrombin is converted to fibrin.

Fibrinolysis (fī-bri-NOL-i-sis) Dissolution of a blood clot by the action of a proteolytic enzyme, such as plasmin (fibrinolysin), that dissolves fibrin threads and inactivates fibrinogen and other blood-clotting factors.

Fibroblast (FĪ-brō-blast) A large, flat cell that secretes most of the matrix (extracellular) material of areolar and dense connective tissues.

Fibrous (FĪ-brus) **joint** A joint that allows little or no movement, such as a suture or a syndesmosis.

Fibrous tunic (TOO-nik) The superficial coat of the eyeball, made up of the posterior sclera and the anterior cornea.

Fight-or-flight response The effects produced upon stimulation of the sympathetic division of the autonomic nervous system.

Filiform papilla (FIL-i-form pa-PIL-a) One of the conical projections that are distributed in parallel rows over the anterior two-thirds of the tongue and lack taste buds.

Filtration (fil-TRĀ-shun) The flow of a liquid through a filter (or membrane that acts like a filter) due to a hydrostatic pressure; occurs in capillaries due to blood pressure.

Filtration fraction The percentage of plasma entering the kidneys that becomes glomerular filtrate.

Filtration membrane Site of blood filtration in nephrons of the kidneys, consisting of the endothelium and basement membrane of the glomerulus and the epithelium of the visceral layer of the glomerular (Bowman's) capsule.

Filum terminale (FĪ-lum ter-mi-NAL-ē) Non-nervous fibrous tissue of the spinal cord that extends inferiorly from the conus medullaris to the coccyx.

Fimbriae (FIM-brē-ē) Fingerlike structures, especially the lateral ends of the uterine (Fallopian) tubes.

Fissure (FISH-ur) A groove, fold, or slit that may be normal or abnormal.

Fistula (FIS-tū-la) An abnormal passage between two organs or between an organ cavity and the outside.

Fixator A muscle that stabilizes the origin of the prime mover so that the prime mover can act more efficiently.

Fixed macrophage (MAK-rō-fāj) Stationary phagocytic cell found in the liver, lungs, brain, spleen, lymph nodes, subcutaneous tissue, and red bone marrow. Also called a **histiocyte** (HIS-tē-ō-sīt).

Flaccid (FLAS-sid) Relaxed, flabby, or soft; lacking muscle tone.

Flagellum (fla-JEL-um) A hairlike, motile process on the extremity of a bacterium, protozoan, or sperm cell. *Plural is* **flagella** (fla-JEL-a).

Flatus (FLĀ-tus) Gas in the stomach or intestines, commonly used to denote expulsion of gas through the anus.

Flexion (FLEK-shun) Movement in which there is a decrease in the angle between two bones.

Flexor reflex A protective reflex in which flexor muscles are stimulated while extensor muscles are inhibited.

Follicle (FOL-i-kul) A small secretory sac or cavity; the group of cells that contains a developing oocyte in the ovaries.

Follicle-stimulating hormone (FSH) Hormone secreted by the anterior pituitary that initiates development of ova and stimulates the ovaries to secrete estrogens in females, and initiates sperm production in males.

Fontanel (fon′-ta-NEL) A fibrous connective tissue membrane-filled space where bone formation is not yet complete, especially between the cranial bones of an infant's skull.

Foot The terminal part of the lower limb, from the ankle to the toes.

Foramen (fō-RĀ-men) A passage or opening; a communication between two cavities of an organ, or a hole in a bone for passage of vessels or nerves. *Plural is* **foramina** (fō-RAM-i-na).

Foramen ovale (fō-RĀ-men-ō-VAL-ē) An opening in the fetal heart in the septum between the right and left atria. A hole in the greater wing of the sphenoid bone that transmits the mandibular branch of the trigeminal nerve (cranial nerve V).

Forearm (FOR-arm) The part of the upper limb between the elbow and the wrist.

Fornix (FOR-niks) An arch or fold; a tract in the brain made up of association fibers, connecting the hippocampus with the mammillary bodies; a recess around the cervix of the uterus where it protrudes into the vagina.

Fossa (FOS-a) A furrow or shallow depression.

Fourth ventricle (VEN-tri-kul) A cavity filled with cerebrospinal fluid within the brain lying between the cerebellum and the medulla oblongata and pons.

Fracture (FRAK-choor) Any break in a bone.

Frenulum (FREN-ū-lum) A small fold of mucous membrane that connects two parts and limits movement.

Frontal plane A plane at a right angle to a midsagittal plane that divides the body or organs into anterior and posterior portions. Also called a **coronal** (kō-RŌ-nal) **plane.**

Functional residual (re-ZID-ū-al) **capacity** The sum of residual volume plus expiratory reserve volume; about 2400 mL.

Fundus (FUN-dus) The part of a hollow organ farthest from the opening.

Fungiform papilla (FUN-ji-form pa-PIL-a) A mushroomlike elevation on the upper surface of the tongue appearing as a red dot; most contain taste buds.

G

Gallbladder A small pouch, located inferior to the liver, that stores bile and empties by means of the cystic duct.

Gallstone A solid mass, usually containing cholesterol, in the gallbladder or a bile-containing duct; formed anywhere between bile canaliculi in the liver and the hepatopancreatic ampulla (ampulla of Vater), where bile enters the duodenum. Also called a **biliary calculus.**

Gamete (GAM-ēt) A male or female reproductive cell; a sperm cell or secondary oocyte.

Ganglion (GANG-glē-on) Usually, a group of neuronal cell bodies lying outside the central nervous system (CNS). *Plural is* **ganglia** (GANG-glē-a).

Gastric (GAS-trik) **glands** Glands in the mucosa of the stomach composed of cells that empty their secretions into narrow channels called gastric pits. Types of cells are chief cells (secrete pepsinogen), parietal cells (secrete hydrochloric acid and intrinsic factor), surface mucous and mucous neck cells (secrete mucus), and G cells (secrete gastrin).

Gastroenterology (gas′-trō-en′-ter-OL-ō-jē) The medical specialty that deals with the structure, function, diagnosis, and treatment of diseases of the stomach and intestines.

Gastrointestinal (gas-trō-in-TES-ti-nal) **(GI) tract** A continuous tube running through the ventral body cavity extending from the mouth to the anus. Also called the **alimentary** (al′-i-MEN-tar-ē) **canal.**

Gastrulation (gas′-troo-LĀ-shun) The migration of groups of cells from the epiblast that transform a bilaminar embryonic disc into a trilaminar embryonic disc that consists of the three primary germ layers; transformation of the blastula into the gastrula.

Gene (JĒN) Biological unit of heredity; a segment of DNA located in a definite position on a particular chromosome; a sequence of DNA that codes for a particular mRNA, rRNA, or tRNA.

Generator potential The graded depolarization that results in a change in the resting membrane potential in a receptor (specialized neuronal ending); may trigger a nerve action potential (nerve impulse) if depolarization reaches threshold.

Genetic engineering The manufacture and manipulation of genetic material.

Genetics The study of genes and heredity.

Genitalia (jen′-i-TĀ-lē-a) Reproductive organs.

Genome (JĒ-nōm) The complete set of genes of an organism.

Genotype (JĒ-nō-tīp) The genetic makeup of an individual; the combination of alleles present at one or more chromosomal locations, as distinguished from the appearance, or phenotype, that results from those alleles.

Geriatrics (jer′-ē-AT-riks) The branch of medicine devoted to the medical problems and care of elderly persons.

Gestation (jes-TĀ-shun) The period of development from fertilization to birth.

Gingivae (jin-JI-vē) Gums. They cover the alveolar processes of the mandible and maxilla and extend slightly into each socket.

Gland Specialized epithelial cell or cells that secrete substances; may be exocrine or endocrine.

Glans penis (glanz PĒ-nis) The slightly enlarged region at the distal end of the penis.

Glaucoma (glaw-KŌ-ma) An eye disorder in which there is increased intraocular pressure due to an excess of aqueous humor.

Gliding joint A synovial joint having articulating surfaces that are usually flat, permitting only side-to-side and back-and-forth movements, as between carpal bones, tarsal bones, and the scapula and clavicle. Also called an **arthrodial** (ar-THRŌ-dē-al) **joint.**

Glomerular (glō-MER-ū-lar) **capsule** A double-walled globe at the proximal end of a nephron that encloses the glomerular capillaries. Also called **Bowman's** (BŌ-manz) **capsule.**

Glomerular filtrate (glō-MER-ū-lar FIL-trāt) The fluid produced when blood is filtered by the filtration membrane in the glomeruli of the kidneys.

Glomerular filtration The first step in urine formation in which substances in blood pass through the filtration membrane and the filtrate enters the proximal convoluted tubule of a nephron.

Glomerular filtration rate (GFR) The total volume of fluid that enters all the glomerular (Bowman's) capsules of the kidneys in 1 min; about 100–125 mL/min.

Glomerulus (glō-MER-ū-lus) A rounded mass of nerves or blood vessels, especially the microscopic tuft of capillaries that is surrounded by the glomerular (Bowman's) capsule of each kidney tubule. *Plural is* **glomeruli.**

Glottis (GLOT-is) The vocal folds (true vocal cords) in the larynx plus the space between them (rima glottidis).

Glucagon (GLOO-ka-gon) A hormone produced by the alpha cells of the pancreatic islets (islets of Langerhans) that increases blood glucose level.

Glucocorticoids (gloo-kō-KOR-ti-koyds) Hormones secreted by the cortex of the adrenal gland, especially cortisol, that influence glucose metabolism.

Gluconeogenesis (gloo′-kō-nē-ō-JEN-e-sis) The synthesis of glucose from certain amino acids or lactic acid.

Glucose (GLOO-kōs) A hexose (six-carbon sugar), $C_6H_{12}O_6$, that is a major energy source for the production of ATP by body cells.

Glucosuria (gloo′-kō-SOO-rē-a) The presence of glucose in the urine; may be temporary or pathological. Also called **glycosuria.**

Glycogen (GLĪ-kō-jen) A highly branched polymer of glucose containing thousands of subunits; functions as a compact store of glucose molecules in liver and muscle fibers (cells).

Glycogenesis (glī′-kō-JEN-e-sis) The chemical reactions by which many molecules of glucose are used to synthesize glycogen.

Glycogenolysis (glī-kō-je-NOL-i-sis) The breakdown of glycogen into glucose.

Glycolysis (glī-KOL-i-sis) Series of chemical reactions in the cytosol of a cell in which a molecule of glucose is split into two molecules of pyruvic acid with net production of two ATPs.

Goblet cell A goblet-shaped unicellular gland that secretes mucus; present in epithelium of the airways and intestines.

Goiter (GOY-ter) An enlarged thyroid gland.

Golgi (GOL-jē) **complex** An organelle in the cytoplasm of cells consisting of four to six flattened sacs (cisternae), stacked on one another, with expanded areas at their ends; functions in processing, sorting, packaging, and delivering proteins and lipids to the plasma membrane, lysosomes, and secretory vesicles.

Gomphosis (gom-FŌ-sis) A fibrous joint in which a cone-shaped peg fits into a socket.

Gonad (GŌ-nad) A gland that produces gametes and hormones; the ovary in the female and the testis in the male.

Gonadotropic hormone Anterior pituitary hormone that affects the gonads.

Gracile (GRAS-īl) **nucleus** A group of nerve cells in the inferior part of the medulla oblongata in which axons of the gracile fasciculus terminate.

Gray commissure (KOM-i-shur) A narrow strip of gray matter connecting the two lateral gray masses within the spinal cord.

Gray matter Areas in the central nervous system and ganglia containing neuronal cell bodies, dendrites, unmyelinated axons, axon terminals, and neuroglia; Nissl bodies impart a gray color and there is little or no myelin in gray matter.

Gray ramus communicans (RĀ-mus kō-MŪ-ni-kans) A short nerve containing axons of sympathetic postganglionic neurons; the cell bodies of the neurons are in a sympathetic chain ganglion, and the unmyelinated axons extend via the gray ramus to a spinal nerve and then to the periphery to supply smooth muscle in blood vessels, arrector pili muscles, and sweat glands. *Plural is* **rami communicantes** (RĀ-mē kō-mū-ni-KAN-tēz).

Greater omentum (ō-MEN-tum) A large fold in the serosa of the stomach that hangs down like an apron anterior to the intestines.

Greater vestibular (ves-TIB-ū-lar) **glands** A pair of glands on either side of the vaginal orifice that open by a duct into the space between the hymen and the labia minora. Also called **Bartholin's** (BAR-to-linz) **glands.**

Groin (GROYN) The depression between the thigh and the trunk; the inguinal region.

Gross anatomy The branch of anatomy that deals with structures that can be studied without using a microscope. Also called **macroscopic anatomy.**

Growth An increase in size due to an increase in (1) the number of cells, (2) the size of existing cells as internal components increase in size, or (3) the size of intercellular substances.

Gustatory (GUS-ta-tō′-rē) Pertaining to taste.

Gynecology (gī′-ne-KOL-ō-jē) The branch of medicine dealing with the study and treatment of disorders of the female reproductive system.

Gynecomastia (gīn′-e-kō-MAS-tē-a) Excessive growth (benign) of the male mammary glands due to secretion of estrogens by an adrenal gland tumor (feminizing adenoma).

Gyrus (JĪ-rus) One of the folds of the cerebral cortex of the brain. *Plural is* **gyri** (JĪ-rī). Also called a **convolution.**

H

Hair A threadlike structure produced by hair follicles that develops in the dermis. Also called a **pilus** (PĪ-lus).

Hair follicle (FOL-li-kul) Structure, composed of epithelium and surrounding the root of a hair, from which hair develops.

Hair root plexus (PLEK-sus) A network of dendrites arranged around the root of a hair as free or naked nerve endings that are stimulated when a hair shaft is moved.

Haldane effect Less carbon dioxide binds to hemoglobin when the partial pressure of oxygen is higher.

Hand The terminal portion of an upper limb, including the carpus, metacarpus, and phalanges.

Haploid (HAP-loyd) Having half the number of chromosomes characteristically found in the somatic cells of an organism; characteristic of mature gametes. Symbolized n.

Hard palate (PAL-at) The anterior portion of the roof of the mouth, formed by the maxillae and palatine bones and lined by mucous membrane.

Haustra (HAWS-tra) A series of pouches that characterize the colon; caused by tonic contractions of the teniae coli. *Singular is* **haustrum.**

Head The superior part of a human, cephalic to the neck. The superior or proximal part of a structure.

Heart A hollow muscular organ lying slightly to the left of the midline of the chest that pumps the blood through the cardiovascular system.

Heart block An arrhythmia (dysrhythmia) of the heart in which the atria and ventricles contract independently because of a blocking of electrical impulses through the heart at some point in the conduction system.

Heart murmur (MER-mer) An abnormal sound that consists of a flow noise that is heard before, between, or after the normal heart sounds, or that may mask normal heart sounds.

Heat exhaustion Condition characterized by cool, clammy skin, profuse perspiration, and fluid and electrolyte (especially sodium and chloride) loss that results in muscle cramps, dizziness, vomiting, and fainting. Also called **heat prostration.**

Heat stroke Condition produced when the body cannot easily lose heat and characterized by reduced perspiration and elevated body temperature. Also called **sunstroke.**

Hematocrit (he-MAT-ō-krit) **(Hct)** The percentage of blood made up of red blood cells. Usually measured by centrifuging a blood sample in a graduated tube and then reading the volume of red blood cells and dividing it by the total volume of blood in the sample.

Hematology (hē′-ma-TOL-ō-jē) The study of blood.

Hemiplegia (hem-i-PLĒ-jē-a) Paralysis of the upper limb, trunk, and lower limb on one side of the body.

Hemodynamics (hē-mō-dī-NA-miks) The study of factors and forces that govern the flow of blood through blood vessels.

Hemoglobin (hē′-mō-GLŌ-bin) **(Hb)** A substance in red blood cells consisting of the protein globin and the iron-containing red pigment heme that transports most of the oxygen and some carbon dioxide in blood.

Hemolysis (hē-MOL-i-sis) The escape of hemoglobin from the interior of a red blood cell into the surrounding medium; results from disruption of the cell membrane by toxins or drugs, freezing or thawing, or hypotonic solutions.

Hemolytic disease of the newborn A hemolytic anemia of a newborn child that results from the destruction of the infant's erythrocytes (red blood cells) by antibodies produced by the mother; usually the antibodies are due to an Rh blood type incompatibility. Also called **erythroblastosis fetalis** (e-rith′-rō-blas-TŌ-sis fe-TAL-is).

Hemophilia (hē′-mō-FIL-ē-a) A hereditary blood disorder where there is a deficient production of certain factors involved in blood clotting, resulting in excessive bleeding into joints, deep tissues, and elsewhere.

Hemopoiesis (hē-mō-poy-Ē-sis) Blood cell production, which occurs in red bone marrow after birth. Also called **hematopoiesis** (hem′-a-tō-poy-Ē-sis).

Hemorrhage (HEM-or-rij) Bleeding; the escape of blood from blood vessels, especially when the loss is profuse.

Hemorrhoids (HEM-ō-royds) Dilated or varicosed blood vessels (usually veins) in the anal region. Also called **piles.**

Hemostasis (hē-MOS-tā-sis) The stoppage of bleeding.

Heparin (HEP-a-rin) An anticoagulant given to slow the conversion of prothrombin to thrombin, thus reducing the risk of blood clot formation; found in basophils, mast cells, and various other tissues, especially the liver and lungs.

Hepatic (he-PAT-ik) Refers to the liver.

Hepatic duct A duct that receives bile from the bile capillaries. Small hepatic ducts merge to form the larger right and left hepatic ducts that unite to leave the liver as the common hepatic duct.

Hepatic portal circulation The flow of blood from the gastrointestinal organs to the liver before returning to the heart.

Hepatocyte (he-PAT-ō-cyte) A liver cell.

Hepatopancreatic (hep′-a-tō-pan′-krē-A-tik) **ampulla** A small, raised area in the duodenum where the combined common bile duct and main pancreatic duct empty into the duodenum. Also called the **ampulla of Vater** (VA-ter).

Hernia (HER-nē-a) The protrusion or projection of an organ or part of an organ through a membrane or cavity wall, usually the abdominal cavity.

Herniated (HER-nē-Ā′-ted) **disc** A rupture of an intervertebral disc so that the nucleus pulposus protrudes into the vertebral cavity. Also called a **slipped disc.**

Heterozygous (he-ter-ō-ZĪ-gus) Possessing different alleles on homologous chromosomes for a particular hereditary trait.

Hiatus (hī-Ā-tus) An opening; a foramen.

Hilus (HĪ-lus) An area, depression, or pit where blood vessels and nerves enter or leave an organ. Also called a **hilum.**

Hinge joint A synovial joint in which a convex surface of one bone fits into a concave surface of another bone, such as the elbow, knee, ankle, and interphalangeal joints. Also called a **ginglymus** (JIN-gli-mus) **joint.**

Hirsutism (HER-soot-izm) An excessive growth of hair in females and children, with a distribution similar to that in adult males, due to the conversion of vellus hairs into large terminal hairs in response to higher-than-normal levels of androgens.

Histamine (HISS-ta-mēn) Substance found in many cells, especially mast cells, basophils, and platelets, released when the cells are injured; results in vasodilation, increased permeability of blood vessels, and constriction of bronchioles.

Histology (hiss-TOL-ō-jē) Microscopic study of the structure of tissues.

Holocrine (HŌL-ō-krin) **gland** A type of gland in which entire secretory cells, along with their accumulated secretions, make up the secretory product of the gland, as in the sebaceous (oil) glands.

Homeostasis (hō′-mē-ō-STĀ-sis) The condition in which the body's internal environment remains relatively constant, within physiological limits.

Homologous chromosomes Two chromosomes that belong to a pair. Also called **homologues.**

Homozygous (hō-mō-ZĪ-gus) Possessing the same alleles on homologous chromosomes for a particular hereditary trait.

Hormone (HOR-mōn) A secretion of endocrine cells that alters the physiological activity of target cells of the body.

Horn An area of gray matter (anterior, lateral, or posterior) in the spinal cord.

Human chorionic gonadotropin (kō-rē-ON-ik gō-nad-ō-TRŌ-pin) **(hCG)** A hormone produced by the developing placenta that maintains the corpus luteum.

Human chorionic somatomammotropin (sō-mat-ō-mam-ō-TRŌ-pin) **(hCS)** Hormone produced by the chorion of the placenta that stimulates breast tissue for lactation, enhances body growth, and regulates metabolism. Also called **human placental lactogen (hPL).**

Human growth hormone (hGH) Hormone secreted by the anterior pituitary that stimulates growth of body tissues, especially skeletal and muscular tissues. Also known as **somatotropin** and **somatotropic hormone (STH).**

Hyaluronic (hī′-a-loo-RON-ik) **acid** A viscous, amorphous extracellular material that binds cells together, lubricates joints, and maintains the shape of the eyeballs.

Hyaluronidase (hī′-a-loo-RON-i-dās) An enzyme that breaks down hyaluronic acid, increasing the permeability of connective tissues by dissolving the substances that hold body cells together.

Hydride ion (HĪ-drīd Ī-on) A hydrogen nucleus with two orbiting electrons (H^-), in contrast to the more common **hydrogen ion,** which has no orbiting electrons (H^+).

Hymen (HĪ-men) A thin fold of vascularized mucous membrane at the vaginal orifice.

Hypercalcemia (hī′-per-kal-SĒ-mē-a) An excess of calcium in the blood.

Hypercapnia (hī′-per-KAP-nē-a) An abnormal increase in the amount of carbon dioxide in the blood.

Hyperextension (hī′-per-ek-STEN-shun) Continuation of extension beyond the anatomical position, as in bending the head backward.

Hyperglycemia (hī′-per-glī-SĒ-mē-a) An elevated blood glucose level.

Hyperkalemia (hī′-per-kā-LĒ-mē-a) An excess of potassium ions in the blood.

Hypermagnesemia (hī′-per-mag′-ne-SĒ-mē-a) An excess of magnesium ions in the blood.

Hypermetropia (hī′-per-mē-TRŌ-pē-a) A condition in which visual images are focused behind the retina, with resulting defective vision of near objects; farsightedness.

Hyperphosphatemia (hī-per-fos′-fa-TĒ-mē-a) An abnormally high level of phosphates in the blood.

Hyperplasia (hī′-per-PLĀ-zē-a) An abnormal increase in the number of normal cells in a tissue or organ, increasing its size.

Hyperpolarization (hī′-per-PŌL-a-ri-zā′-shun) Increase in the internal negativity across a cell membrane, thus increasing the voltage and moving it farther away from the threshold value.

Hypersecretion (hī′-per-se-KRĒ-shun) Overactivity of glands resulting in excessive secretion.

Hypersensitivity (hī′-per-sen-si-TI-vi-tē) Overreaction to an allergen that results in pathological changes in tissues. Also called **allergy.**

Hypertension (hī′-per-TEN-shun) High blood pressure.

Hyperthermia (hī′-per-THERM-ē-a) An elevated body temperature.

Hypertonia (hī′-per-TŌ-nē-a) Increased muscle tone that is expressed as spasticity or rigidity.

Hypertonic (hī′-per-TON-ik) Solution that causes cells to shrink due to loss of water by osmosis.

Hypertrophy (hī-PER-trō-fē) An excessive enlargement or overgrowth of tissue without cell division.

Hyperventilation (hī′-per-ven-ti-LĀ-shun) A rate of respiration higher than that required to maintain a normal partial pressure of carbon dioxide in the blood.

Hypocalcemia (hī′-pō-kal-SĒ-mē-a) A below-normal level of calcium in the blood.

Hypochloremia (hī′-pō-klō-RĒ-mē-a) Deficiency of chloride ions in the blood.

Hypoglycemia (hī′-pō-glī-SĒ-mē-a) An abnormally low concentration of glucose in the blood; can result from excess insulin (injected or secreted).

Hypokalemia (hī′-pō-ka-LĒ-mē-a) Deficiency of potassium ions in the blood.

Hypomagnesemia (hī′-pō-mag′-ne-SĒ-mē-a) Deficiency of magnesium ions in the blood.

Hyponatremia (hī′-pō-na-TRĒ-mē-a) Deficiency of sodium ions in the blood.

Hyponychium (hī′-pō-NIK-ē-um) Free edge of the fingernail.

Hypophosphatemia (hī-pō-fos′-fa-TĒ-mē-a) An abnormally low level of phosphates in the blood.

Hypophyseal fossa (hī′-pō-FIZ-ē-al FOS-a) A depression on the superior surface of the sphenoid bone that houses the pituitary gland.

Hypophyseal (hī′-pō-FIZ-ē-al) **pouch** An outgrowth of ectoderm from the roof of the mouth from which the anterior pituitary develops.

Hypophysis (hī-POF-i-sis) Pituitary gland.

Hyposecretion (hī′-pō-se-KRĒ-shun) Underactivity of glands resulting in diminished secretion.

Hypothalamohypophyseal (hī-pō-tha-lam′-ō-hī-pō-FIZ-ē-al) **tract** A bundle of axons containing secretory vesicles filled with oxytocin or antidiuretic hormone that extend from the hypothalamus to the posterior pituitary.

Hypothalamus (hī′-pō-THAL-a-mus) A portion of the diencephalon, lying beneath the thalamus and forming the floor and part of the wall of the third ventricle.

Hypothermia (hī′-pō-THER-mē-a) Lowering of body temperature below 35°C (95°F); in surgical procedures, it refers to deliberate cooling of the body to slow down metabolism and reduce oxygen needs of tissues.

Hypotonia (hī′-pō-TŌ-nē-a) Decreased or lost muscle tone in which muscles appear flaccid.

Hypotonic (hī′-pō-TON-ik) Solution that causes cells to swell and perhaps rupture due to gain of water by osmosis.

Hypoventilation (hī-pō-ven-ti-LĀ-shun) A rate of respiration lower than that required to maintain a normal partial pressure of carbon dioxide in plasma.

Hypovolemic (hī-pō-vō-LĒ-mik) **shock** A type of shock characterized by decreased blood volume; may be caused by acute hemorrhage or excessive loss of other body fluids, for example, by vomiting, diarrhea, or excessive sweating.

Hypoxia (hī-POKS-ē-a) Lack of adequate oxygen at the tissue level.

Hysterectomy (hiss-te-REK-tō-mē) The surgical removal of the uterus.

I

Ileocecal (il-ē-ō-SĒ-kal) **sphincter** A fold of mucous membrane that guards the opening from the ileum into the large intestine. Also called the **ileocecal valve.**

Ileum (IL-ē-um) The terminal part of the small intestine.

Immunity (im-Ū-ni-tē) The state of being resistant to injury, particularly by poisons, foreign proteins, and invading pathogens.

Immunogenicity (im′-ū-nō-je-NIS-i-tē) Ability of an antigen to provoke an immune response.

Immunoglobulin (im-ū-nō-GLOB-ū-lin) **(Ig)** An antibody synthesized by plasma cells derived from B lymphocytes in response to the introduction of an antigen. Immunoglobulins are divided into five kinds (IgG, IgM, IgA, IgD, IgE).

Immunology (im′-ū-NOL-ō-jē) The study of the responses of the body when challenged by antigens.

Implantation (im-plan-TĀ-shun) The insertion of a tissue or a part into the body. The attachment of the blastocyst to the stratum basalis of the endometrium about 6 days after fertilization.

Incontinence (in-KON-ti-nens) Inability to retain urine, semen, or feces through loss of sphincter control.

Indirect motor pathways Motor tracts that convey information from the brain down the spinal cord for automatic movements, coordination of body movements with visual stimuli, skeletal muscle tone and posture, and balance. Also known as **extrapyramidal pathways.**

Infarction (in-FARK-shun) A localized area of necrotic tissue, produced by inadequate oxygenation of the tissue.

Infection (in-FEK-shun) Invasion and multiplication of microorganisms in body tissues, which may be inapparent or characterized by cellular injury.

Inferior (in-FĒR-ē-or) Away from the head or toward the lower part of a structure. Also called **caudad** (KAW-dad).

Inferior vena cava (VĒ-na CĀ-va) **(IVC)** Large vein that collects blood from parts of the body inferior to the heart and returns it to the right atrium.

Infertility Inability to conceive or to cause conception. Also called **sterility.**

Inflammation (in′-fla-MĀ-shun) Localized, protective response to tissue injury designed to destroy, dilute, or wall off the infecting agent or injured tissue; characterized by redness, pain, heat, swelling, and sometimes loss of function.

Inflation reflex Reflex that prevents overinflation of the lungs. Also called **Hering–Breuer reflex.**

Infundibulum (in′-fun-DIB-ū-lum) The stalklike structure that attaches the pituitary gland to the hypothalamus of the brain. The funnel-shaped, open, distal end of the uterine (Fallopian) tube.

Ingestion (in-JES-chun) The taking in of food, liquids, or drugs, by mouth.

Inguinal (IN-gwi-nal) Pertaining to the groin.

Inguinal canal An oblique passageway in the anterior abdominal wall just superior and parallel to the medial half of the inguinal ligament that transmits the spermatic cord and ilioinguinal nerve in the male and round ligament of the uterus and ilioinguinal nerve in the female.

Inhalation (in-ha-LĀ-shun) The act of drawing air into the lungs. Also termed **inspiration.**

Inheritance The acquisition of body traits by transmission of genetic information from parents to offspring.

Inhibin A hormone secreted by the gonads that inhibits release of follicle-stimulating hormone (FSH) by the anterior pituitary.

Inhibiting hormone Hormone secreted by the hypothalamus that can suppress secretion of hormones by the anterior pituitary.

Inhibitory postsynaptic potential (IPSP) A small hyperpolarization in a neuron; caused by an inhibitory neurotransmitter at a synapse.

Inner cell mass A region of cells of a blastocyst that differentiates into the three primary germ layers—ectoderm, mesoderm, and endoderm—from which all tissues and organs develop; also called an **embryoblast.**

Inorganic (in′-or-GAN-ik) **compound** Compound that usually lacks carbon, usually is small, and often contains ionic bonds. Examples include water and many acids, bases, and salts.

Insertion (in-SER-shun) The attachment of a muscle tendon to a movable bone or the end opposite the origin.

Inspiratory (in-SPĪ-ra-tor-ē) **capacity** Total inspiratory capacity of the lungs; the total of tidal volume plus inspiratory reserve volume; averages 3600 mL.

Inspiratory (in-SPĪ-ra-tor-ē) **reserve volume** Additional inspired air over and above tidal volume; averages 3100 mL.

Insula (IN-soo-la) A triangular area of the cerebral cortex that lies deep within the lateral cerebral fissue, under the parietal, frontal, and temporal lobes.

Insulin (IN-soo-lin) A hormone produced by the beta cells of a pancreatic islet (islet of Langerhans) that decreases the blood glucose level.

Insulinlike growth factor (IGF) Small protein, produced by the liver and other tissues in response to stimulation by human growth hormone (hGH), that mediates most of the effects of human growth hormone. Previously called **somatomedin** (sō′-ma-tō-MĒ-din).

Integrins (IN-te-grinz) A family of transmembrane glycoproteins in plasma membranes that function in cell adhesion; they are present in hemidesmosomes, which anchor cells to a basement membrane, and they mediate adhesion of neutrophils to endothelial cells during emigration.

Integumentary (in-teg′-ū-MEN-tar-e) Relating to the skin.

Intercalated (in-TER-ka-lāt-ed) **disc** An irregular transverse thickening of sarcolemma that contains desmosomes, which hold cardiac muscle fibers (cells) together, and gap junctions, which aid in conduction of muscle action potentials from one fiber to the next.

Intercostal (in′-ter-KOS-tal) **nerve** A nerve supplying a muscle located between the ribs.

Interferons (in′-ter-FĒR-ons) **(IFNs)** Antiviral proteins produced by virus-infected host cells; induce uninfected host cells to synthesize proteins that inhibit viral replication and enhance phagocytic activity of macrophages; types include alpha interferon, beta interferon, and gamma interferon.

Intermediate Between two structures, one of which is medial and one of which is lateral.

Intermediate filament Protein filament, ranging from 8 to 12 nm in diameter, that may provide structural reinforcement, hold organelles in place, and give shape to a cell.

Internal Away from the surface of the body.

Internal capsule A large tract of projection fibers lateral to the thalamus that is the major connection between the cerebral cortex and the brain stem and spinal cord; contains axons of sensory neurons carrying auditory, visual, and somatic sensory signals to the cerebral cortex plus axons of motor neurons descending from the cerebral cortex to the thalamus, subthalamus, brain stem, and spinal cord.

Internal ear The inner ear or labyrinth, lying inside the temporal bone, containing the organs of hearing and balance.

Internal nares (NĀ-rez) The two openings posterior to the nasal cavities opening into the nasopharynx. Also called the **choanae** (kō-Ā-nē).

Internal respiration The exchange of respiratory gases between blood and body cells. Also called **tissue respiration.**

Interneurons (in′-ter-NOO-ronz) Neurons whose axons extend only for a short distance and contact nearby neurons in the brain, spinal cord, or a ganglion; they comprise the vast majority of neurons in the body.

Interoceptor (in′-ter-ō-SEP-tor) Sensory receptor located in blood vessels and viscera that provides information about the body's internal environment.

Interphase (IN-ter-fāz) The period of the cell cycle between cell divisions, consisting of the G_1-(gap or growth) phase, when the cell is engaged in growth, metabolism, and production of substances required for division; S-(synthesis) phase, during which chromosomes are replicated; and G_2-phase.

Interstitial (in′-ter-STISH-al) **fluid** The portion of extracellular fluid that fills the microscopic spaces between the cells of tissues; the internal environment of the body. Also called **intercellular** or **tissue fluid.**

Interstitial growth Growth from within, as in the growth of cartilage. Also called **endogenous** (en-DOJ-e-nus) **growth.**

Interventricular (in′-ter-ven-TRIK-ū-lar) **foramen** A narrow, oval opening through which the lateral ventricles of the brain communicate with the third ventricle. Also called the **foramen of Monro.**

Intervertebral (in′-ter-VER-te-bral) **disc** A pad of fibrocartilage located between the bodies of two vertebrae.

Intestinal gland A gland that opens onto the surface of the intestinal mucosa and secretes digestive enzymes. Also called a **crypt of Lieberkühn** (LĒ-ber-kūn).

Intrafusal (in′-tra-FŪ-sal) **fibers** Three to ten specialized muscle fibers (cells), partially enclosed in a spindle-shaped connective tissue capsule, that make up a muscle spindle.

Intramembranous ossification (in′-tra-MEM-bra-nus os′-i′-fi-KĀ-shun) The method of bone formation in which the bone is formed directly in membranous tissue.

Intraocular (in′-tra-OK-ū-lar) **pressure (IOP)** Pressure in the eyeball, produced mainly by aqueous humor.

Intrapleural pressure Air pressure between the two pleurae of the lungs, usually subatmospheric. Also called **intrathoracic pressure.**

Intrinsic (in-TRIN-sik) Of internal origin.

Intrinsic pathway (of blood clotting) Sequence of reactions leading to blood clotting that is initiated by damage to blood vessel endothelium or platelets; activators of this pathway are contained within blood itself or are in direct contact with blood.

Intrinsic factor (IF) A glycoprotein, synthesized and secreted by the parietal cells of the gastric mucosa, that facilitates vitamin B_{12} absorption in the small intestine.

Intron (IN-tron) A region of DNA that does not code for the synthesis of a protein.

In utero (Ū-ter-ō) Within the uterus.

Invagination (in-vaj′-i-NĀ-shun) The pushing of the wall of a cavity into the cavity itself.

Inversion (in-VER-zhun) The movement of the sole medially at the ankle joint.

In vitro (VĒ-trō) Literally, in glass; outside the living body and in an artificial environment such as a laboratory test tube.

In vivo (VĒ-vō) In the living body.

Ion (Ī-on) Any charged particle or group of particles; usually formed when a substance, such as a salt, dissolves and dissociates.

Ionization (ī′-on-i-ZĀ-shun) Separation of inorganic acids, bases, and salts into ions when dissolved in water. Also called **dissociation.**

Ipsilateral (ip′-si-LAT-er-al) On the same side, affecting the same side of the body.

Iris The colored portion of the vascular tunic of the eyeball seen through the cornea that contains circular and radial smooth muscle; the hole in the center of the iris is the pupil.

Irritable bowel syndrome (IBS) Disease of the entire gastrointestinal tract in which a person reacts to stress by developing symptoms (such as cramping and abdominal pain) associated with alternating patterns of diarrhea and constipation. Excessive amounts of mucus may appear in feces, and other symptoms include flatulence, nausea, and loss of appetite. Also known as **irritable colon** or **spastic colitis.**

Ischemia (is-KĒ-mē-a) A lack of sufficient blood to a body part due to obstruction or constriction of a blood vessel.

Isoantibody A specific antibody in blood plasma that reacts with specific isoantigens and causes the clumping of bacteria, red blood cells, or particles. Also called an **agglutinin.**

Isoantigen A genetically determined antigen located on the surface of red blood cells; basis for the ABO grouping and Rh system of blood classification. Also called an **agglutinogen.**

Isometric contraction A muscle contraction in which tension on the muscle increases, but there is only minimal muscle shortening so that no visible movement is produced.

Isotonic (Ī′-sō-TON-ik) Having equal tension or tone. A solution having the same concentration of impermeable solutes as cytosol.

Isotonic (i′-so-TON-ik) **contraction** Contraction in which the tension remains the same; occurs when a constant load is moved through the range of motions possible at a joint.

Isotopes (Ī-sō-tōps′) Chemical elements that have the same number of protons but different numbers of neutrons. Radioactive isotopes change into other elements with the emission of alpha or beta particles or gamma rays.

Isovolumetric (ī-sō-vol-ū′-MET-rik) **contraction** The period of time, about 0.05 sec, between the start of ventricular systole and the opening of the semilunar valves; the ventricles contract and ventricular pressure rises rapidly, but ventricular volume does not change.

Isovolumetric relaxation The period of time, about 0.05 sec, between the closing of the semilunar valves and the opening of the atrioventricular (AV) valves; there is a drastic decrease in ventricular pressure without a change in ventricular volume.

Isthmus (IS-mus) A narrow strip of tissue or narrow passage connecting two larger parts.

J

Jaundice (JAWN-dis) A condition characterized by yellowness of the skin, the white of the eyes, mucous membranes, and body fluids because of a buildup of bilirubin.

Jejunum (je-JOO-num) The middle part of the small intestine.

Joint kinesthetic (kin′-es-THET-ik) **receptor** A proprioceptive receptor located in a joint, stimulated by joint movement.

Juxtaglomerular (juks-ta-glō-MER-ū-lar) **apparatus (JGA)** Consists of the macula densa (cells of the distal convoluted tubule adjacent to the afferent and efferent arteriole) and juxtaglomerular cells (modified cells of the afferent and sometimes efferent arteriole); secretes renin when blood pressure starts to fall.

K

Keratin (KER-a-tin) An insoluble protein found in the hair, nails, and other keratinized tissues of the epidermis.

Keratinocyte (ke-RAT-in′-ō-sīt) The most numerous of the epidermal cells; produces keratin.

Ketone (KĒ-tōn) **bodies** Substances produced primarily during excessive triglyceride catabolism, such as acetone, acetoacetic acid, and β-hydroxybutyric acid.

Ketosis (kē-TŌ-sis) Abnormal condition marked by excessive production of ketone bodies.

Kidney (KID-nē) One of the paired reddish organs located in the lumbar region that regulates the composition, volume, and pressure of blood and produces urine.

Kidney stone A solid mass, usually consisting of calcium oxalate, uric acid, or calcium phosphate crystals, that may form in any portion of the urinary tract. Also called a **renal calculus.**

Kinesiology (ki-nē′-sē-OL-ō-jē) The study of the movement of body parts.

Kinesthesia (kin-es-THĒ-zē-a) The perception of the extent and direction of movement of body parts; this sense is possible due to nerve impulses generated by proprioceptors.

Korotkoff (kō-ROT-kof) **sounds** The various sounds that are heard while taking blood pressure.

Krebs cycle A series of biochemical reactions that occurs in the matrix of mitochondria in which electrons are transferred to coenzymes and carbon dioxide is formed. The electrons carried by the coenzymes then enter the electron transport chain, which generates a large quantity of ATP. Also called the **citric acid cycle** or **tricarboxylic acid (TCA) cycle.**

Kyphosis (kī-FŌ-sis) An exaggeration of the thoracic curve of the vertebral column, resulting in a "round-shouldered" appearance. Also called **hunchback.**

L

Labia majora (LĀ-bē-a ma-JŌ-ra) Two longitudinal folds of skin extending downward and backward from the mons pubis of the female.

Labia minora (min-OR-a) Two small folds of mucous membrane lying medial to the labia majora of the female.

Labial frenulum (LĀ-bē-al FREN-ū-lum) A medial fold of mucous membrane between the inner surface of the lip and the gums.

Labium (LĀ-bē-um) A lip. A liplike structure. *Plural is* **labia** (LĀ-bē-a).

Labor The process of giving birth in which a fetus is expelled from the uterus through the vagina.

Lacrimal canal A duct, one on each eyelid, beginning at the punctum at the medial margin of an eyelid and conveying tears medially into the nasolacrimal sac.

Lacrimal gland Secretory cells, located at the superior anterolateral portion of each orbit, that secrete tears into excretory ducts that open onto the surface of the conjunctiva.

Lacrimal sac The superior expanded portion of the nasolacrimal duct that receives the tears from a lacrimal canal.

Lactation (lak-TĀ-shun) The secretion and ejection of milk by the mammary glands.

Lacteal (LAK-tē-al) One of many lymphatic vessels in villi of the intestines that absorb triglycerides and other lipids from digested food.

Lacuna (la-KOO-na) A small, hollow space, such as that found in bones in which the osteocytes lie. *Plural is* **lacunae** (la-KOO-nē).

Lambdoid (lam-DOYD) **suture** The joint in the skull between the parietal bones and the occipital bone; sometimes contains sutural (Wormian) bones.

Lamellae (la-MEL-ē) Concentric rings of hard, calcified matrix found in compact bone.

Lamellated corpuscle Oval-shaped pressure receptor located in the dermis or subcutaneous tissue and consisting of concentric layers of connective tissue wrapped around the dendrites of a sensory neuron. Also called a **Pacinian** (pa-SIN-ē-an) **corpuscle.**

Lamina (LAM-i-na) A thin, flat layer or membrane, as the flattened part of either side of the arch of a vertebra. *Plural is* **laminae** (LAM-i-nē).

Lamina propria (PRŌ-prē-a) The connective tissue layer of a mucosa.

Langerhans (LANG-er-hans) **cell** Epidermal dendritic cell that functions as an antigen-presenting cell (APC) during an immune response.

Lanugo (la-NOO-gō) Fine downy hairs that cover the fetus.

Large intestine The portion of the gastrointestinal tract extending from the ileum of the small intestine to the anus, divided structurally into the cecum, colon, rectum, and anal canal.

Laryngopharynx (la-rin′-gō-FAR-inks) The inferior portion of the pharynx, extending downward from the level of the hyoid bone that divides posteriorly into the esophagus and anteriorly into the larynx. Also called the **hypopharynx.**

Laryngotracheal (la-rin′-gō-TRĀ-ke-al) **bud** An outgrowth of endoderm of the foregut from which the respiratory system develops.

Larynx (LAR-inks) The voice box, a short passageway that connects the pharynx with the trachea.

Lateral (LAT-er-al) Farther from the midline of the body or a structure.

Lateral ventricle (VEN-tri-kul) A cavity within a cerebral hemisphere that communicates with the lateral ventricle in the other cerebral hemisphere and with the third ventricle by way of the interventricular foramen.

Leg The part of the lower limb between the knee and the ankle.

Lens A transparent organ constructed of proteins (crystallins) lying posterior to the pupil and iris of the eyeball and anterior to the vitreous body.

Lesion (LĒ-zhun) Any localized, abnormal change in a body tissue.

Lesser omentum (ō-MEN-tum) A fold of the peritoneum that extends from the liver to the lesser curvature of the stomach and the first part of the duodenum.

Lesser vestibular (ves-TIB-ū-lar) **gland** One of the paired mucus-secreting glands with ducts that open on either side of the urethral orifice in the vestibule of the female.

Leukemia (loo-KĒ-mē-a) A malignant disease of the blood-forming tissues characterized by either uncontrolled production and accumulation of immature leukocytes in which many cells fail to reach maturity (acute) or an accumulation of mature leukocytes in the blood because they do not die at the end of their normal life span (chronic).

Leukocyte (LOO-kō-sīt) A white blood cell.

Leukocytosis (loo′-kō-sī-TŌ-sis) An increase in the number of white blood cells, above 10,000 per μL, characteristic of many infections and other disorders.

Leukopenia (loo-kō-PĒ-nē-a) A decrease in the number of white blood cells below 5000 cells per μL.

Leukotriene (loo′-kō-TRĪ-ēn) A type of eicosanoid produced by basophils and mast cells; acts as a local hormone; produces increased vascular permeability and acts as a chemotactic agent for phagocytes in tissue inflammation.

Leydig (LĪ-dig) **cell** A type of cell that secretes testosterone; located in the connective tissue between seminiferous tubules in a mature testis. Also known as **interstitial cell of Leydig** or **interstitial endocrinocyte.**

Libido (li-BĒ-dō) Sexual desire.

Ligament (LIG-a-ment) Dense regular connective tissue that attaches bone to bone.

Ligand (LĪ-gand) A chemical substance that binds to a specific receptor.

Limbic system A part of the forebrain, sometimes termed the visceral brain, concerned with various aspects of emotion and behavior; includes the limbic lobe, dentate gyrus, amygdala, septal nuclei, mammillary bodies, anterior thalamic nucleus, olfactory bulbs, and bundles of myelinated axons.

Lingual frenulum (LIN-gwal FREN-ū-lum) A fold of mucous membrane that connects the tongue to the floor of the mouth.

Lipase An enzyme that splits fatty acids from triglycerides and phospholipids.

Lipid (LIP-id) An organic compound composed of carbon, hydrogen, and oxygen that is usually insoluble in water, but soluble in alcohol, ether, and chloroform; examples include triglycerides (fats and oils), phospholipids, steroids, and eicosanoids.

Lipid bilayer Arrangement of phospholipid, glycolipid, and cholesterol molecules in two parallel sheets in which the hydrophilic "heads" face outward and the hydrophobic "tails" face inward; found in cellular membranes.

Lipogenesis (li-pō-GEN-e-sis) The synthesis of triglycerides.

Lipolysis (lip-OL-i-sis) The splitting of fatty acids from a triglyceride or phospholipid.

Lipoprotein (lip′-ō-PRŌ-tēn) One of several types of particles containing lipids (cholesterol and triglycerides) and proteins that make it water soluble for transport in the blood; high levels of **low-density lipoproteins (LDLs)** are associated with increased risk of atherosclerosis, whereas high levels of **high-density lipoproteins (HDLs)** are associated with decreased risk of atherosclerosis.

Liver Large organ under the diaphragm that occupies most of the right hypochondriac region and part of the epigastric region. Functionally, it produces bile and synthesizes most plasma proteins; interconverts nutrients; detoxifies substances; stores glycogen, iron, and vitamins; carries on phagocytosis of worn-out blood cells and bacteria; and helps synthesize the active form of vitamin D.

Long-term potentiation (LTP) Prolonged, enhanced synaptic transmission that occurs at certain synapses within the hippocampus of the brain; believed to underlie some aspects of memory.

Lordosis (lor-DŌ-sis) An exaggeration of the lumbar curve of the vertebral column. Also called **swayback.**

Lower limb The appendage attached at the pelvic (hip) girdle, consisting of the thigh, knee, leg, ankle, foot, and toes. Also called **lower extremity.**

Lumbar (LUM-bar) Region of the back and side between the ribs and pelvis; loin.

Lumbar plexus (PLEK-sus) A network formed by the anterior (ventral) branches of spinal nerves L1 through L4.

Lumen (LOO-men) The space within an artery, vein, intestine, renal tubule, or other tubular structure.

Lungs Main organs of respiration that lie on either side of the heart in the thoracic cavity.

Lunula (LOO-noo-la) The moon-shaped white area at the base of a nail.

Luteinizing (LOO-tē-in′-īz-ing) **hormone (LH)** A hormone secreted by the anterior pituitary that stimulates ovulation, stimulates progesterone secretion by the corpus luteum, and readies the mammary glands for milk secretion in females; stimulates testosterone secretion by the testes in males.

Lymph (LIMF) Fluid confined in lymphatic vessels and flowing through the lymphatic system until it is returned to the blood.

Lymph node An oval or bean-shaped structure located along lymphatic vessels.

Lymphatic (lim-FAT-ik) **capillary** Closed-ended microscopic lymphatic vessel that begins in spaces between cells and converges with other lymphatic capillaries to form lymphatic vessels.

Lymphatic tissue A specialized form of reticular tissue that contains large numbers of lymphocytes.

Lymphatic vessel A large vessel that collects lymph from lymphatic capillaries and converges with other lymphatic vessels to form the thoracic and right lymphatic ducts.

Lymphocyte (LIM-fō-sīt) A type of white blood cell that helps carry out cell-mediated and antibody-mediated immune responses; found in blood and in lymphatic tissues.

Lysosome (LĪ-sō-sōm) An organelle in the cytoplasm of a cell, enclosed by a single membrane and containing powerful digestive enzymes.

Lysozyme (LĪ-sō-zīm) A bactericidal enzyme found in tears, saliva, and perspiration.

M

Macrophage (MAK-rō-fāj) Phagocytic cell derived from a monocyte; may be fixed or wandering.

Macula (MAK-ū-la) A discolored spot or a colored area. A small, thickened region on the wall of the utricle and saccule that contains receptors for static equilibrium.

Macula lutea (MAK-ū-la LOO-tē-a) The yellow spot in the center of the retina.

Major histocompatibility (MHC) antigens Surface proteins on white blood cells and other nucleated cells that are unique for each person (except for identical siblings); used to type tissues and help prevent rejection of transplanted tissues. Also known as **human leukocyte antigens (HLA).**

Malignant (ma-LIG-nant) Referring to diseases that tend to become worse and cause death, especially the invasion and spreading of cancer.

Mammary (MAM-ar-ē) **gland** Modified sudoriferous (sweat) gland of the female that produces milk for the nourishment of the young.

Mammillary (MAM-i-ler-ē) **bodies** Two small rounded bodies on the inferior aspect of the hypothalamus that are involved in reflexes related to the sense of smell.

Marrow (MAR-ō) Soft, spongelike material in the cavities of bone. Red bone marrow produces blood cells; yellow bone marrow contains adipose tissue that stores triglycerides.

Mast cell A cell found in areolar connective tissue that releases histamine, a dilator of small blood vessels, during inflammation.

Mastication (mas′-ti-KĀ-shun) Chewing.

Matrix (MĀ-triks) The ground substance and fibers between cells in a connective tissue.

Matter Anything that occupies space and has mass.

Mature follicle A large, fluid-filled follicle containing a secondary oocyte and surrounding granulosa cells that secrete estrogens. Also called a **Graafian** (GRAF-ē-an) **follicle.**

Maximal oxygen uptake Maximum rate of oxygen consumption during aerobic catabolism of pyruvic acid that is determined by age, sex, body size, and aerobic training.

Mean arterial blood pressure (MABP) The average force of blood pressure exerted against the walls of arteries; approximately equal to diastolic pressure plus one-third of pulse pressure; for example, 93 mmHg when systolic pressure is 120 mmHg and diastolic pressure is 80 mmHg.

Meatus (mē-Ā-tus) A passage or opening, especially the external portion of a canal.

Mechanoreceptor (me-KAN-ō-rē-sep-tor) Sensory receptor that detects mechanical deformation of the receptor itself or adjacent cells; stimuli so detected include those related to touch, pressure, vibration, proprioception, hearing, equilibrium, and blood pressure.

Medial (MĒ-dē-al) Nearer the midline of the body or a structure.

Medial lemniscus (lem-NIS-kus) A white matter tract that originates in the gracile and cuneate nuclei of the medulla oblongata and extends to the thalamus on the same side; sensory axons in this tract conduct nerve impulses for the sensations of proprioception, fine touch, vibration, hearing, and equilibrium.

Median aperture (AP-er-choor) One of the three openings in the roof of the fourth ventricle through which cerebrospinal fluid enters the subarachnoid space of the brain and cord. Also called the **foramen of Magendie.**

Median plane A vertical plane dividing the body into right and left halves. Situated in the middle.

Mediastinum (mē′-dē-as-TĪ-num) The broad, median partition between the pleurae of the lungs, that extends from the sternum to the vertebral column in the thoracic cavity.

Medulla (me-DUL-la) An inner layer of an organ, such as the medulla of the kidneys.

Medulla oblongata (me-DUL-la ob′-long-GA-ta) The most inferior part of the brain stem. Also termed the **medulla.**

Medullary (MED-ū-lar′-ē) **cavity** The space within the diaphysis of a bone that contains yellow bone marrow. Also called the **marrow cavity.**

Medullary rhythmicity (rith-MIS-i-tē) **area** The neurons of the respiratory center in the medulla oblongata that control the basic rhythm of respiration.

Meiosis (mē-Ō-sis) A type of cell division that occurs during production of gametes, involving two successive nuclear divisions that result in daughter cells with the haploid *(n)* number of chromosomes.

Melanin (MEL-a-nin) A dark black, brown, or yellow pigment found in some parts of the body such as the skin, hair, and pigmented layer of the retina.

Melanocyte (MEL-a-nō-sīt′) A pigmented cell, located between or beneath cells of the deepest layer of the epidermis, that synthesizes melanin.

Melanocyte-stimulating hormone (MSH) A hormone secreted by the anterior pituitary that stimulates the dispersion of melanin granules in melanocytes in amphibians; continued administration produces darkening of skin in humans.

Melatonin (mel-a-TON-in) A hormone secreted by the pineal gland that helps set the timing of the body's biological clock.

Membrane A thin, flexible sheet of tissue composed of an epithelial layer and an underlying connective tissue layer, as in an epithelial membrane, or of areolar connective tissue only, as in a synovial membrane.

Membranous labyrinth (mem-BRA-nus LAB-i-rinth) The part of the labyrinth of the internal ear that is located inside the bony labyrinth and separated from it by the perilymph; made up of the semicircular ducts, the saccule and utricle, and the cochlear duct.

Menarche (me-NAR-kē) The first menses (menstrual flow) and beginning of ovarian and uterine cycles.

Meninges (me-NIN-jēz) Three membranes covering the brain and spinal cord, called the dura mater, arachnoid mater, and pia mater. *Singular is* **meninx** (MEN-inks).

Menopause (MEN-ō-pawz) The termination of the menstrual cycles.

Menstruation (men′-stroo-Ā-shun) Periodic discharge of blood, tissue fluid, mucus, and epithelial cells that usually lasts for 5 days; caused by a sudden reduction in estrogens and progesterone. Also called the **menstrual phase** or **menses.**

Merkel (MER-kel) **cell** Type of cell in the epidermis of hairless skin that makes contact with a tactile (Merkel) disc, which functions in touch.

Merocrine (MER-ō-krin) **gland** Gland made up of secretory cells that remain intact throughout the process of formation and discharge of the secretory product, as in the salivary and pancreatic glands.

Mesenchyme (MEZ-en-kīm) An embryonic connective tissue from which all other connective tissues arise.

Mesentery (MEZ-en-ter′-ē) A fold of peritoneum attaching the small intestine to the posterior abdominal wall.

Mesocolon (mez′-ō-KŌ-lon) A fold of peritoneum attaching the colon to the posterior abdominal wall.

Mesoderm The middle primary germ layer that gives rise to connective tissues, blood and blood vessels, and muscles.

Mesothelium (mez′-ō-THĒ-lē-um) The layer of simple squamous epithelium that lines serous membranes.

Mesovarium (mez′-ō-VAR-ē-um) A short fold of peritoneum that attaches an ovary to the broad ligament of the uterus.

Metabolism (me-TAB-ō-lizm) All the biochemical reactions that occur within an organism, including the synthetic (anabolic) reactions and decomposition (catabolic) reactions.

Metacarpus (met′-a-KAR-pus) A collective term for the five bones that make up the palm.

Metaphase (MET-a-phāz) The second stage of mitosis, in which chromatid pairs line up on the metaphase plate of the cell.

Metaphysis (me-TAF-i-sis) Region of a long bone between the diaphysis and epiphysis that contains the epiphyseal plate in a growing bone.

Metarteriole (met′-ar-TĒ-rē-ōl) A blood vessel that emerges from an arteriole, traverses a capillary network, and empties into a venule.

Metastasis (me-TAS-ta-sis) The spread of cancer to surrounding tissues (local) or to other body sites (distant).

Metatarsus (met′-a-TAR-sus) A collective term for the five bones located in the foot between the tarsals and the phalanges.

Micelle (mī-SEL) A spherical aggregate of bile salts that dissolves fatty acids and monoglycerides so that they can be absorbed into small intestinal epithelial cells.

Microfilament (mī-krō-FIL-a-ment) Rodlike protein filament about 6 nm in diameter; constitutes contractile units in muscle fibers (cells) and provides support, shape, and movement in nonmuscle cells.

Microglia (mī-krō-GLĒ-a) Neuroglial cells that carry on phagocytosis.

Microtubule (mī-krō-TOO-būl′) Cylindrical protein filament, ranging in diameter from 18 to 30 nm, consisting of the protein tubulin; provides support, structure, and transportation.

Microvilli (mī′-krō-VIL-ē) Microscopic, fingerlike projections of the plasma membranes of cells that increase surface area for absorption, especially in the small intestine and proximal convoluted tubules of the kidneys.

Micturition (mik′-choo-RISH-un) The act of expelling urine from the urinary bladder. Also called **urination** (yoo-ri-NĀ-shun).

Midbrain The part of the brain between the pons and the diencephalon. Also called the **mesencephalon** (mes′-en-SEF-a-lon).

Middle ear A small, epithelial-lined cavity hollowed out of the temporal bone, separated from the external ear by the eardrum and from the internal ear by a thin bony partition containing the oval and round windows; extending across the middle ear are the three auditory ossicles. Also called the **tympanic** (tim-PAN-ik) **cavity.**

Midline An imaginary vertical line that divides the body into equal left and right sides.

Midsagittal plane A vertical plane through the midline of the body that divides the body or organs into *equal* right and left sides. Also called a **median plane.**

Milk ejection reflex Contraction of alveolar cells to force milk into ducts of mammary glands, stimulated by oxytocin (TO), which is released from the posterior pituitary in response to suckling action. Also called the **milk letdown reflex.**

Mineral Inorganic, homogeneous solid substance that may perform a function vital to life; examples include calcium and phosphorus.

Mineralocorticoids (min′-er-al-ō-KOR-ti-koyds) A group of hormones of the adrenal cortex that help regulate sodium and potassium balance.

Minimal volume The volume of air in the lungs even after the thoracic cavity has been opened, forcing out some of the residual volume.

Minute ventilation (MV) Total volume of air inhaled and exhaled per minute; about 6000 mL.

Miosis Constriction of the pupil of the eye.

Mitochondrion (mī′-tō-KON-drē-on) A double-membraned organelle that plays a central role in the production of ATP; known as the "powerhouse" of the cell.

Mitosis (mī-TŌ-sis) The orderly division of the nucleus of a cell that ensures that each new daughter nucleus has the same number and kind of chromosomes as the original parent nucleus. The process includes the replication of chromosomes and the distribution of the two sets of chromosomes into two separate and equal nuclei.

Mitotic spindle Collective term for a football-shaped assembly of microtubules (nonkinetochore, kinetochore, and aster) that is responsible for the movement of chromosomes during cell division.

Modality (mō-DAL-i-tē) Any of the specific sensory entities, such as vision, smell, taste, or touch.

Modiolus (mō-DĪ-ō′-lus) The central pillar or column of the cochlea.

Mole The weight, in grams, of the combined atomic weights of the atoms that make up a molecule of a substance.

Molecule (MOL-e-kūl) The chemical combination of two or more atoms covalently bonded together.

Monocyte (MON-ō-sīt′) The largest type of white blood cell, characterized by agranular cytoplasm.

Monounsaturated fat A fatty acid that contains one double covalent bond between its carbon atoms; it is not completely saturated with hydrogen atoms. Plentiful in triglycerides of olive and peanut oils.

Mons pubis (monz PŪ-bis) The rounded, fatty prominence over the pubic symphysis, covered by coarse pubic hair.

Morula (MOR-ū-la) A solid sphere of cells produced by successive cleavages of a fertilized ovum about four days after fertilization.

Motor end plate Region of the sarcolemma of a muscle fiber (cell) that includes acetylcholine (ACh) receptors, which bind ACh released by synaptic end bulbs of somatic motor neurons.

Motor neurons (NOO-ronz) Neurons that conduct impulses from the brain toward the spinal cord or out of the brain and spinal cord into cranial or spinal nerves to effectors that may be either muscles or glands. Also called **efferent neurons.**

Motor unit A motor neuron together with the muscle fibers (cells) it stimulates.

Mucin (MŪ-sin) A protein found in mucus.

Mucosa-associated lymphatic tissue (MALT) Lymphatic nodules scattered throughout the lamina propria (connective tissue) of mucous membranes lining the gastrointestinal tract, respiratory airways, urinary tract, and reproductive tract.

Mucous (MŪ-kus) **cell** A unicellular gland that secretes mucus. Two types are mucous neck cells and surface mucous cells in the stomach.

Mucous membrane A membrane that lines a body cavity that opens to the exterior. Also called the **mucosa** (mū-KŌ-sa).

Mucus The thick fluid secretion of goblet cells, mucous cells, mucous glands, and mucous membranes.

Muscarinic (mus′-ka-RIN-ik) **receptor** Receptor for the neurotransmitter acetylcholine found on all effectors innervated by parasympathetic postganglionic axons and on sweat glands innervated by cholinergic sympathetic postganglionic axons; so named because muscarine activates these receptors but does not activate nicotinic receptors for acetylcholine.

Muscle An organ composed of one of three types of muscle tissue (skeletal, cardiac, or smooth), specialized for contraction to produce voluntary or involuntary movement of parts of the body.

Muscle action potential A stimulating impulse that propagates along the sarcolemma and transverse tubules; in skeletal muscle, it is generated by acetylcholine, which increases the permeability of the sarcolemma to cations, especially sodium ions (Na^+).

Muscle fatigue (fa-TĒG) Inability of a muscle to maintain its strength of contraction or tension; may be related to insufficient oxygen, depletion of glycogen, and/or lactic acid buildup.

Muscle spindle An encapsulated proprioceptor in a skeletal muscle, consisting of specialized intrafusal muscle fibers and nerve endings; stimulated by changes in length or tension of muscle fibers.

Muscle tissue A tissue specialized to produce motion in response to muscle action potentials by its qualities of contractility, extensibility, elasticity, and excitability; types include skeletal, cardiac, and smooth.

Muscle tone A sustained, partial contraction of portions of a skeletal or smooth muscle in response to activation of stretch receptors or a baseline level of action potentials in the innervating motor neurons.

Muscular dystrophies (DIS-trō-fēz′) Inherited muscle-destroying diseases, characterized by degeneration of muscle fibers (cells), which causes progressive atrophy of the skeletal muscle.

Muscularis (MUS-kū-la′-ris) A muscular layer (coat or tunic) of an organ.

Muscularis mucosae (mū-KŌ-sē) A thin layer of smooth muscle fibers that underlie the lamina propria of the mucosa of the gastrointestinal tract.

Mutation (mū-TĀ-shun) Any change in the sequence of bases in a DNA molecule resulting in a permanent alteration in some inheritable trait.

Myasthenia (mī-as-THĒ-nē-a) **gravis** Weakness and fatigue of skeletal muscles caused by antibodies directed against acetylcholine receptors.

Myelin (MĪ-e-lin) **sheath** Multilayered lipid and protein covering, formed by Schwann cells and oligodendrocytes, around axons of many peripheral and central nervous system neurons.

Myenteric plexus A network of autonomic axons and postganglionic cell bodies located in the muscularis of the gastrointestinal tract. Also called the **plexus of Auerbach** (OW-er-bak).

Myocardial infarction (mī′-ō-KAR-dē-al in-FARK-shun) **(MI)** Gross necrosis of myocardial tissue due to interrupted blood supply. Also called a **heart attack.**

Myocardium (mī′-ō-KAR-dē-um) The middle layer of the heart wall, made up of cardiac muscle tissue, lying between the epicardium and the endocardium and constituting the bulk of the heart.

Myofibril (mī-ō-FĪ-bril) A threadlike structure, extending longitudinally through a muscle fiber (cell) consisting mainly of thick filaments (myosin) and thin filaments (actin, troponin, and tropomyosin).

Myoglobin (mī-ō-GLŌ-bin) The oxygen-binding, iron-containing protein present in the sarcoplasm of muscle fibers (cells); contributes the red color to muscle.

Myogram (MĪ-ō-gram) The record or tracing produced by a myograph, an apparatus that measures and records the force of muscular contractions.

Myology (mī-OL-ō-jē) The study of muscles.

Myometrium (mī′-ō-MĒ-trē-um) The smooth muscle layer of the uterus.

Myopathy (mī-OP-a-thē) Any abnormal condition or disease of muscle tissue.

Myopia (mī-Ō-pē-a) Defect in vision in which objects can be seen distinctly only when very close to the eyes; nearsightedness.

Myosin (MĪ-ō-sin) The contractile protein that makes up the thick filaments of muscle fibers.

Myotome (MĪ-ō-tōm) A group of muscles innervated by the motor neurons of a single spinal segment. In an embryo, the portion of a somite that develops into some skeletal muscles.

N

Nail A hard plate, composed largely of keratin, that develops from the epidermis of the skin to form a protective covering on the dorsal surface of the distal phalanges of the fingers and toes.

Nail matrix (MĀ-triks) The part of the nail beneath the body and root from which the nail is produced.

Nasal (NĀ-zal) **cavity** A mucosa-lined cavity on either side of the nasal septum that opens onto the face at the external nares and into the nasopharynx at the internal nares.

Nasal septum (SEP-tum) A vertical partition composed of bone (perpendicular plate of ethmoid and vomer) and cartilage, covered with a mucous membrane, separating the nasal cavity into left and right sides.

Nasolacrimal (nā′-zō-LAK-ri-mal) **duct** A canal that transports the lacrimal secretion (tears) from the nasolacrimal sac into the nose.

Nasopharynx (nā′-zō-FAR-inks) The superior portion of the pharynx, lying posterior to the nose and extending inferiorly to the soft palate.

Neck The part of the body connecting the head and the trunk. A constricted portion of an organ such as the neck of the femur or uterus.

Necrosis (ne-KRŌ-sis) A pathological type of cell death that results from disease, injury, or lack of blood supply in which many adjacent cells swell, burst, and spill their contents into the interstitial fluid, triggering an inflammatory response.

Negative feedback The principle governing most control systems; a mechanism of response in which a stimulus initiates actions that reverse or reduce the stimulus.

Neonatal (nē-ō-NĀ-tal) Pertaining to the first four weeks after birth.

Neoplasm (NĒ-ō-plazm) A new growth that may be benign or malignant.

Nephron (NEF-ron) The functional unit of the kidney.

Nerve A cordlike bundle of neuronal axons and/or dendrites and associated connective tissue coursing together outside the central nervous system.

Nerve fiber General term for any process (axon or dendrite) projecting from the cell body of a neuron.

Nerve impulse A wave of depolarization and repolarization that self-propagates along the plasma membrane of a neuron; also called a **nerve action potential.**

Nervous tissue Tissue containing neurons that initiate and conduct nerve impulses to coordinate homeostasis, and neuroglia that provide support and nourishment to neurons.

Net filtration pressure (NFP) Net pressure that promotes fluid outflow at the arterial end of a capillary, and fluid inflow at the venous end of a capillary; net pressure that promotes glomerular filtration in the kidneys.

Neural plate A thickening of ectoderm, induced by the notochord, that forms early in the third week of development and represents the beginning of the development of the nervous system.

Neuralgia (noo-RAL-jē-a) Attacks of pain along the entire course or branch of a peripheral sensory nerve.

Neuritis (noo-RĪ-tis) Inflammation of one or more nerves.

Neuroglia (noo-RŌG-lē-a) Cells of the nervous system that perform various supportive functions. The neuroglia of the central nervous system are the astrocytes, oligodendrocytes, microglia, and ependymal cells; neuroglia of the peripheral nervous system include Schwann cells and satellite cells. Also called **glial** (GLĒ-al) **cells.**

Neurohypophyseal (noo′-rō-hī′-pō-FIZ-ē-al) **bud** An outgrowth of ectoderm located on the floor of the hypothalamus that gives rise to the posterior pituitary.

Neurolemma (noo-rō-LEM-ma) The peripheral, nucleated cytoplasmic layer of the Schwann cell. Also called **sheath of Schwann** (SCHVON).

Neurology (noo-ROL-ō-jē) The study of the normal functioning and disorders of the nervous system.

Neuromuscular (noo-rō-MUS-kū-lar) **junction** A synapse between the axon terminals of a motor neuron and the sarcolemma of a muscle fiber (cell).

Neuron (NOO-ron) A nerve cell, consisting of a cell body, dendrites, and an axon.

Neuropeptide (noo-rō-PEP-tīd) Chain of three to about 40 amino acids that occurs naturally in the nervous system, and that acts primarily to modulate the response of or to a neurotransmitter. Examples are enkephalins and endorphins.

Neurosecretory (noo-rō-SEC-re-tō-rē) **cell** A neuron that secretes a hypothalamic releasing hormone or inhibiting hormone into blood capillaries of the hypothalmus; a neuron that secretes oxytocin or antidiuretic hormone into blood capillaries of the posterior pituitary.

Neurotransmitter One of a variety of molecules within axon terminals that are released into the synaptic cleft in response to a nerve impulse, and that change the membrane potential of the postsynaptic neuron.

Neutrophil (NOO-trō-fil) A type of white blood cell characterized by granules that stain pale lilac with a combination of acidic and basic dyes.

Nicotinic (nik'-ō-TIN-ik) **receptor** Receptor for the neurotransmitter acetylcholine found on both sympathetic and parasympathetic postganglionic neurons and on skeletal muscle in the motor end plate; so named because nicotine activates these receptors but does not activate muscarinic receptors for acetylcholine.

Nipple A pigmented, wrinkled projection on the surface of the breast that is the location of the openings of the lactiferous ducts for milk release.

Nociceptor (nō'-sē-SEP-tor) A free (naked) nerve ending that detects painful stimuli.

Node of Ranvier (ron-vē-Ā) A space, along a myelinated axon, between the individual Schwann cells that form the myelin sheath and the neurolemma. Also called a **neurofibral node.**

Nondisjunction (non'-dis-JUNGK-shun) Failure of sister chromatids to separate properly during anaphase of mitosis (or meiosis II) or failure of homologous chromosomes to separate properly during meiosis I in which chromatids or chromosomes pass into the same daughter cell; the result is too many copies of that chromosome in the daughter cell or gamete.

Norepinephrine (nor'-ep-ē-NEF-rin) **(NE)** A hormone secreted by the adrenal medulla that produces actions similar to those that result from sympathetic stimulation. Also called **noradrenaline** (nor-a-DREN-a-lin).

Notochord (NŌ-tō-cord) A flexible rod of mesodermal tissue that lies where the future vertebral column will develop and plays a role in induction.

Nuclear medicine The branch of medicine concerned with the use of radioisotopes in the diagnosis and therapy of disease.

Nucleic (noo-KLĒ-ic) **acid** An organic compound that is a long polymer of nucleotides, with each nucleotide containing a pentose sugar, a phosphate group, and one of four possible nitrogenous bases (adenine, cytosine, guanine, and thymine or uracil).

Nucleolus (noo-KLĒ-ō-lus) Spherical body within a cell nucleus composed of protein, DNA, and RNA that is the site of the assembly of small and large ribosomal subunits.

Nucleosome (NOO-klē-ō-sōm) Structural subunit of a chromosome consisting of histones and DNA.

Nucleus (NOO-klē-us) A spherical or oval organelle of a cell that contains the hereditary factors of the cell, called genes. A cluster of unmyelinated nerve cell bodies in the central nervous system. The central part of an atom made up of protons and neutrons.

Nucleus pulposus (pul-PŌ-sus) A soft, pulpy, highly elastic substance in the center of an intervertebral disc; a remnant of the notochord.

Nutrient A chemical substance in food that provides energy, forms new body components, or assists in various body functions.

O

Obesity (ō-BĒS-i-tē) Body weight more than 20% above a desirable standard due to excessive accumulation of fat.

Oblique (ō-BLĒK) **plane** A plane that passes through the body or an organ at an angle between the transverse plane and either the midsagittal, parasagittal, or frontal plane.

Obstetrics (ob-STET-riks) The specialized branch of medicine that deals with pregnancy, labor, and the period of time immediately after delivery (about 6 weeks).

Olfactory (ōl-FAK-tō-rē) Pertaining to smell.

Olfactory bulb A mass of gray matter containing cell bodies of neurons that form synapses with neurons of the olfactory nerve (cranial nerve I), lying inferior to the frontal lobe of the cerebrum on either side of the crista galli of the ethmoid bone.

Olfactory receptor A bipolar neuron with its cell body lying between supporting cells located in the mucous membrane lining the superior portion of each nasal cavity; transduces odors into neural signals.

Olfactory tract A bundle of axons that extends from the olfactory bulb posteriorly to olfactory regions of the cerebral cortex.

Oligodendrocyte (ol'-i-gō-DEN-drō-sīt) A neuroglial cell that supports neurons and produces a myelin sheath around axons of neurons of the central nervous system.

Oligospermia (ol'-i-gō-SPER-mē-a) A deficiency of sperm cells in the semen.

Olive A prominent oval mass on each lateral surface of the superior part of the medulla oblongata.

Oncogenes (ONG-kō-jēnz) Cancer-causing genes; they derive from normal genes, termed **proto-oncogenes,** that encode proteins involved in cell growth or cell regulation but have the ability to transform a normal cell into a cancerous cell when they are mutated or inappropriately activated. One example is *p53*.

Oncology (ong-KOL-ō-jē) The study of tumors.

Oogenesis (ō'-ō-JEN-e-sis) Formation and development of female gametes (oocytes).

Oophorectomy (ō'-of-ō-REK-tō-me) Surgical removal of the ovaries.

Ophthalmic (of-THAL-mik) Pertaining to the eye.

Ophthalmologist (of'-thal-MOL-ō-jist) A physician who specializes in the diagnosis and treatment of eye disorders using drugs, surgery, and corrective lenses.

Ophthalmology (of'-thal-MOL-ō-jē) The study of the structure, function, and diseases of the eye.

Opsin (OP-sin) The glycoprotein portion of a photopigment.

Opsonization (op-sō-ni-ZĀ-shun) The action of some antibodies that renders bacteria and other foreign cells more susceptible to phagocytosis.

Optic (OP-tik) Refers to the eye, vision, or properties of light.

Optic chiasm (KĪ-azm) A crossing point of the optic nerves (cranial nerve II), anterior to the pituitary gland. Also called **optic chiasma.**

Optic disc A small area of the retina containing openings through which the axons of the ganglion cells emerge as the optic nerve (cranial nerve II). Also called the **blind spot.**

Optician (op-TISH-an) A technician who fits, adjusts, and dispenses corrective lenses on prescription of an ophthalmologist or optometrist.

Optic tract A bundle of axons that carry nerve impulses from the retina of the eye between the optic chiasm and the thalamus.

Optometrist (op-TOM-e-trist) Specialist with a doctorate degree in optometry who is licensed to examine and test the eyes and treat visual defects by prescribing corrective lenses.

Ora serrata (Ō-ra ser-RĀ-ta) The irregular margin of the retina lying internal and slightly posterior to the junction of the choroid and ciliary body.

Orbit (OR-bit) The bony, pyramidal-shaped cavity of the skull that holds the eyeball.

Organ A structure composed of two or more different kinds of tissues with a specific function and usually a recognizable shape.

Organelle (or-gan-EL) A permanent structure within a cell with characteristic morphology that is specialized to serve a specific function in cellular activities.

Organic (or-GAN-ik) **compound** Compound that always contains carbon in which the atoms are held together by covalent bonds. Examples include carbohydrates, lipids, proteins, and nucleic acids (DNA and RNA).

Organism (OR-ga-nizm) A total living form; one individual.

Orgasm (OR-gazm) Sensory and motor events involved in ejaculation for the male and involuntary contraction of the perineal muscles in the female at the climax of sexual intercourse.

Orifice (OR-i-fis) Any aperture or opening.

Origin (OR-i-jin) The attachment of a muscle tendon to a stationary bone or the end opposite the insertion.

Oropharynx (or′-ō-FAR-inks) The intermediate portion of the pharynx, lying posterior to the mouth and extending from the soft palate to the hyoid bone.

Orthopedics (or′-thō-PĒ-diks) The branch of medicine that deals with the preservation and restoration of the skeletal system, articulations, and associated structures.

Osmoreceptor (oz′-mō-re-CEP-tor) Receptor in the hypothalamus that is sensitive to changes in blood osmolarity and, in response to high osmolarity (low water concentration), stimulates synthesis and release of antidiuretic hormone (ADH).

Osmosis (os-MŌ-sis) The net movement of water molecules through a selectively permeable membrane from an area of higher water concentration to an area of lower water concentration until equilibrium is reached.

Osmotic pressure The pressure required to prevent the movement of pure water into a solution containing solutes when the solutions are separated by a selectively permeable membrane.

Osseous (OS-ē-us) Bony.

Ossicle (OS-si-kul) One of the small bones of the middle ear (malleus, incus, stapes).

Ossification (os′-i-fi-KĀ-shun) Formation of bone. Also called **osteogenesis.**

Osteoblast (OS-tē-ō-blast) Cell formed from an osteogenic cell that participates in bone formation by secreting some organic components and inorganic salts.

Osteoclast (OS-tē-ō-clast′) A large, multinuclear cell that resorbs (destroys) bone matrix.

Osteocyte (OS-tē-ō-sīt′) A mature bone cell that maintains the daily activities of bone tissue.

Osteogenic (os′-tē-ō-prō-JEN-i-tor) **cell** Stem cell derived from mesenchyme that has mitotic potential and the ability to differentiate into an osteoblast.

Osteogenic (os′-tē-ō-JEN-ik) **layer** The inner layer of the periosteum that contains cells responsible for forming new bone during growth and repair.

Osteology (os′-tē-OL-ō-jē) The study of bones.

Osteon (OS-tē-on) The basic unit of structure in adult compact bone, consisting of a central (Haversian) canal with its concentrically arranged lamellae, lacunae, osteocytes, and canaliculi. Also called a **Haversian** (ha-VER-shan) **system.**

Osteoporosis (os′-tē-ō-pō-RŌ-sis) Age-related disorder characterized by decreased bone mass and increased susceptibility to fractures, often as a result of decreased levels of estrogens.

Otic (Ō-tik) Pertaining to the ear.

Otolith (Ō-tō-lith) A particle of calcium carbonate embedded in the otolithic membrane that functions in maintaining static equilibrium.

Otolithic (ō-tō-LITH-ik) **membrane** Thick, gelatinous, glycoprotein layer located directly over hair cells of the macula in the saccule and utricle of the internal ear.

Otorhinolaryngology (ō′-tō-rī-nō-lar′-in-GOL-ō-jē) The branch of medicine that deals with the diagnosis and treatment of diseases of the ears, nose, and throat.

Oval window A small, membrane-covered opening between the middle ear and inner ear into which the footplate of the stapes fits.

Ovarian (ō-VAR-ē-an) **cycle** A monthly series of events in the ovary associated with the maturation of a secondary oocyte.

Ovarian follicle (FOL-i-kul) A general name for oocytes (immature ova) in any stage of development, along with their surrounding epithelial cells.

Ovarian ligament (LIG-a-ment) A rounded cord of connective tissue that attaches the ovary to the uterus.

Ovary (Ō-var-ē) Female gonad that produces oocytes and the estrogens, progesterone, inhibin, and relaxin hormones.

Ovulation (ov-ū-LĀ-shun) The rupture of a mature ovarian (Graafian) follicle with discharge of a secondary oocyte into the pelvic cavity.

Ovum (Ō-vum) The female reproductive or germ cell; an egg cell; arises through completion of meiosis in a secondary oocyte after penetration by a sperm.

Oxidation (ok-si-DĀ-shun) The removal of electrons from a molecule or, less commonly, the addition of oxygen to a molecule that results in a decrease in the energy content of the molecule. The oxidation of glucose in the body is called **cellular respiration.**

Oxyhemoglobin (ok′-sē-HĒ-mō-glō-bin) (**Hb−O$_2$**) Hemoglobin combined with oxygen.

Oxytocin (ok′-sē-TŌ-sin) (**OT**) A hormone secreted by neurosecretory cells in the paraventricular and supraoptic nuclei of the hypothalamus that stimulates contraction of smooth muscle in the pregnant uterus and myoepithelial cells around the ducts of mammary glands.

P

P wave The deflection wave of an electrocardiogram that signifies atrial depolarization.

Palate (PAL-at) The horizontal structure separating the oral and the nasal cavities; the roof of the mouth.

Palpate (PAL-pāt) To examine by touch; to feel.

Pancreas (PAN-krē-as) A soft, oblong organ lying along the greater curvature of the stomach and connected by a duct to the duodenum. It is both an exocrine gland (secreting pancreatic juice) and an endocrine gland (secreting insulin, glucagon, somatostatin, and pancreatic polypeptide).

Pancreatic (pan′-krē-AT-ik) **duct** A single large tube that unites with the common bile duct from the liver and gallbladder and drains pancreatic juice into the duodenum at the hepatopancreatic ampulla (ampulla of Vater). Also called the **duct of Wirsung.**

Pancreatic islet A cluster of endocrine gland cells in the pancreas that secretes insulin, glucagon, somatostatin, and pancreatic polypeptide. Also called an **islet of Langerhans** (LANG-er-hanz).

Pancreatic polypeptide Hormone secreted by the F cells of pancreatic islets (islets of Langerhans) that regulates release of pancreatic digestive enzymes.

Papanicolaou (pap′-a-NIK-ō-la-oo) **test** A cytological staining test for the detection and diagnosis of premalignant and malignant conditions of the female genital tract. Cells scraped from the epithelium of the cervix of the uterus are examined microscopically. Also called a **Pap test** or **Pap smear.**

Papilla (pa-PIL-a) A small nipple-shaped projection or elevation.

Paracrine (PAR-a-krin) Local hormone, such as histamine, that acts on neighboring cells without entering the bloodstream.

Paralysis (pa-RAL-a-sis) Loss or impairment of motor function due to a lesion of nervous or muscular origin.

Paranasal sinus (par′-a-NĀ-zal SĪ-nus) A mucus-lined air cavity in a skull bone that communicates with the nasal cavity. Paranasal sinuses are located in the frontal, maxillary, ethmoid, and sphenoid bones.

Paraplegia (par-a-PLĒ-jē-a) Paralysis of both lower limbs.

Parasagittal plane (par-a-SAJ-i-tal) A vertical plane that does not pass through the midline and that divides the body or organs into *unequal* left and right portions.

Parasympathetic (par′-a-sim-pa-THET-ik) **division** One of the two subdivisions of the autonomic nervous system, having cell bodies of preganglionic neurons in nuclei in the brain stem and in the lateral gray horn of the sacral portion of the spinal cord; primarily concerned with activities that conserve and restore body energy.

Parathyroid (par′-a-THĪ-royd) **gland** One of usually four small endocrine glands embedded in the posterior surfaces of the lateral lobes of the thyroid gland.

Parathyroid hormone (PTH) A hormone secreted by the chief (principal) cells of the parathyroid glands that increases blood calcium level and decreases blood phosphate level.

Paraurethral (par′-a-ū-RĒ-thral) **gland** Gland embedded in the wall of the urethra whose duct opens on either side of the urethral orifice and secretes mucus. Also called **Skene's** (SKĒNZ) **gland.**

Parenchyma (par-EN-ki-ma) The functional parts of any organ, as opposed to tissue that forms its stroma or framework.

Parietal (pa-RĪ-e-tal) Pertaining to or forming the outer wall of a body cavity.

Parietal cell A type of secretory cell in gastric glands that produces hydrochloric acid and intrinsic factor. Also called an **oxyntic cell.**

Parkinson disease (PD) Progressive degeneration of the basal ganglia and substantia nigra of the cerebrum resulting in decreased production of dopamine (DA) that leads to tremor, slowing of voluntary movements, and muscle weakness.

Parotid (pa-ROT-id) **gland** One of the paired salivary glands located inferior and anterior to the ears and connected to the oral cavity via a duct (Stensen's) that opens into the inside of the cheek opposite the maxillary (upper) second molar tooth.

Parturition (par′-too-RISH-un) Act of giving birth to young; childbirth, delivery.

Patellar (pa-TELL-ar) **reflex** Extension of the leg by contraction of the quadriceps femoris muscle in response to tapping the patellar ligament. Also called the **knee jerk reflex.**

Patent ductus arteriosus Congenital anatomical heart defect in which the fetal connection between the aorta and pulmonary trunk remains open instead of closing completely after birth.

Pathogen (PATH-ō-jen) A disease-producing microbe.

Pathological (path′-ō-LOJ-i-kal) **anatomy** The study of structural changes caused by disease.

Pectoral (PEK-tō-ral) Pertaining to the chest or breast.

Pediatrician (pē′-dē-a-TRISH-un) A physician who specializes in the care and treatment of children.

Pedicel (PED-i-sel) Footlike structure, as on podocytes of a glomerulus.

Pelvic (PEL-vik) **cavity** Inferior portion of the abdominopelvic cavity that contains the urinary bladder, sigmoid colon, rectum, and internal female and male reproductive structures.

Pelvic splanchnic (PEL-vik SPLANGK-nik) **nerves** Consist of preganglionic parasympathetic axons from the levels of S2, S3, and S4 that supply the urinary bladder, reproductive organs, and the descending and sigmoid colon and rectum.

Pelvis The basinlike structure formed by the two hip bones, the sacrum, and the coccyx. The expanded, proximal portion of the ureter, lying within the kidney and into which the major calyces open.

Penis (PĒ-nis) The organ of urination and copulation in males; used to deposit semen into the female vagina.

Pepsin Protein-digesting enzyme secreted by chief cells of the stomach in the inactive form pepsinogen, which is converted to active pepsin by hydrochloric acid.

Peptic ulcer An ulcer that develops in areas of the gastrointestinal tract exposed to hydrochloric acid; classified as a gastric ulcer if in the lesser curvature of the stomach and as a duodenal ulcer if in the first part of the duodenum.

Percussion (per-KUSH-un) The act of striking (percussing) an underlying part of the body with short, sharp blows as an aid in diagnosing the part by the quality of the sound produced.

Perforating canal A minute passageway by means of which blood vessels and nerves from the periosteum penetrate into compact bone. Also called **Volkmann's** (FŌLK-manz) **canal.**

Pericardial (per′-i-KAR-dē-al) **cavity** Small potential space between the visceral and parietal layers of the serous pericardium that contains pericardial fluid.

Pericardium (per′-i-KAR-dē-um) A loose-fitting membrane that encloses the heart, consisting of a superficial fibrous layer and a deep serous layer.

Perichondrium (per′-i-KON-drē-um) The membrane that covers cartilage.

Perilymph (PER-i-limf) The fluid contained between the bony and membranous labyrinths of the inner ear.

Perimetrium (per′-i-MĒ-trē-um) The serosa of the uterus.

Perimysium (per′-i-MĪZ-ē-um) Invagination of the epimysium that divides muscles into bundles.

Perineum (per′-i-NĒ-um) The pelvic floor; the space between the anus and the scrotum in the male and between the anus and the vulva in the female.

Perineurium (per′-i-NOO-rē-um) Connective tissue wrapping around fascicles in a nerve.

Periodontal (per-ē-ō-DON-tal) **disease** A collective term for conditions characterized by degeneration of gingivae, alveolar bone, periodontal ligament, and cementum.

Periodontal ligament The periosteum lining the alveoli (sockets) for the teeth in the alveolar processes of the mandible and maxillae.

Periosteum (per′-ē-OS-tē-um) The membrane that covers bone and consists of connective tissue, osteogenic cells, and osteoblasts; is essential for bone growth, repair, and nutrition.

Peripheral (pe-RIF-er-al) Located on the outer part or a surface of the body.

Peripheral nervous system (PNS) The part of the nervous system that lies outside the central nervous system, consisting of nerves and ganglia.

Peristalsis (per'-i-STAL-sis) Successive muscular contractions along the wall of a hollow muscular structure.

Peritoneum (per'-i-tō-NĒ-um) The largest serous membrane of the body that lines the abdominal cavity and covers the viscera.

Peritonitis (per'-i-tō-NĪ-tis) Inflammation of the peritoneum.

Permissive (per-MIS-sive) **effect** A hormonal interaction in which the effect of one hormone on a target cell requires previous or simultaneous exposure to another hormone(s) to enhance the response of a target cell or increase the activity of another hormone.

Peroxisome (per-OK-si-sōm) Organelle similar in structure to a lysosome that contains enzymes that use molecular oxygen to oxidize various organic compounds; such reactions produce hydrogen peroxide; abundant in liver cells.

Perspiration Sweat; produced by sudoriferous (sweat) glands and containing water, salts, urea, uric acid, amino acids, ammonia, sugar, lactic acid, and ascorbic acid. Helps maintain body temperature and eliminate wastes.

pH A measure of the concentration of hydrogen ions (H^+) in a solution. The pH scale extends from 0 to 14, with a value of 7 expressing neutrality, values lower than 7 expressing increasing acidity, and values higher than 7 expressing increasing alkalinity.

Phagocytosis (fag'-ō-sī-TŌ-sis) The process by which phagocytes ingest particulate matter; the ingestion and destruction of microbes, cell debris, and other foreign matter.

Phalanx (FĀ-lanks) The bone of a finger or toe. *Plural is* **phalanges** (fa-LAN-jēz).

Pharmacology (far'-ma-KOL-ō-jē) The science of the effects and uses of drugs in the treatment of disease.

Pharynx (FAR-inks) The throat; a tube that starts at the internal nares and runs partway down the neck, where it opens into the esophagus posteriorly and the larynx anteriorly.

Phenotype (FĒ-nō-tīp) The observable expression of genotype; physical characteristics of an organism determined by genetic makeup and influenced by interaction between genes and internal and external environmental factors.

Phlebitis (fle-BĪ-tis) Inflammation of a vein, usually in a lower limb.

Phosphorylation (fos'-for-i-LĀ-shun) The addition of a phosphate group to a chemical compound; types include substrate-level phosphorylation, oxidative phosphorylation, and photo-phosphorylation.

Photopigment A substance that can absorb light and undergo structural changes that can lead to the development of a receptor potential. An example is rhodopsin. In the eye, also called **visual pigment.**

Photoreceptor Receptor that detects light shining on the retina of the eye.

Physiology (fiz'-ē-OL-ō-jē) Science that deals with the functions of an organism or its parts.

Pia mater (PĪ-a-MĀ-ter *or* PĒ-a MA-ter) The innermost of the three meninges (coverings) of the brain and spinal cord.

Pineal (PĪN-ē-al) **gland** A cone-shaped gland located in the roof of the third ventricle that secretes melatonin. Also called the **epiphysis cerebri** (ē-PIF-i-sis se-RĒ-brē).

Pinealocyte (pin-ē-AL-ō-sīt) Secretory cell of the pineal gland that releases melatonin.

Pinna (PIN-na) The projecting part of the external ear composed of elastic cartilage and covered by skin and shaped like the flared end of a trumpet. Also called the **auricle** (OR-i-kul).

Pinocytosis (pi'-nō-sī-TŌ-sis) A process by which most body cells can ingest membrane-surrounded droplets of interstitial fluid.

Pituicyte (pi-TOO-i-sīt) Supporting cell of the posterior pituitary.

Pituitary (pi-TOO-i-tār-ē) **gland** A small endocrine gland occupying the hypophyseal fossa above the sella turcica of the sphenoid bone and attached to the hypothalamus by the infundibulum. Also called the **hypophysis** (hī-POF-i-sis).

Pivot joint A synovial joint in which a rounded, pointed, or conical surface of one bone articulates with a ring formed partly by another bone and partly by a ligament, as in the joint between the atlas and axis and between the proximal ends of the radius and ulna. Also called a **trochoid** (TRŌ-koyd) **joint.**

Placenta (pla-SEN-ta) The special structure through which the exchange of materials between fetal and maternal circulations occurs. Also called the **afterbirth.**

Plantar flexion (PLAN-tar FLEK-shun) Bending the foot in the direction of the plantar surface (sole).

Plaque (PLAK) A layer of dense proteins on the inside of a plasma membrane in adherens junctions and desmosomes. A mass of bacterial cells, dextran (polysaccharide), and other debris that adheres to teeth (dental plaque). See also atherosclerotic plaque.

Plasma (PLAZ-ma) The extracellular fluid found in blood vessels; blood minus the formed elements.

Plasma cell Cell that develops from a B cell (lymphocyte) and produces antibodies.

Plasma (cell) membrane Outer, limiting membrane that separates the cell's internal parts from extracellular fluid or the external environment.

Platelet (PLĀT-let) A fragment of cytoplasm enclosed in a cell membrane and lacking a nucleus; found in the circulating blood; plays a role in hemostasis. Also called a **thrombocyte** (THROM-bō-sīt).

Platelet plug Aggregation of platelets (thrombocytes) at a site where a blood vessel is damaged that helps stop or slow blood loss.

Pleura (PLOOR-a) The serous membrane that covers the lungs and lines the walls of the chest and the diaphragm.

Pleural cavity Small potential space between the visceral and parietal pleurae.

Plexus (PLEK-sus) A network of nerves, veins, or lymphatic vessels.

Pluripotent stem cell Immature stem cell in red bone marrow that gives rise to precursors of all the different mature blood cells.

Pneumotaxic (noo-mō-TAK-sik) **area** A part of the respiratory center in the pons that continually sends inhibitory nerve impulses to the inspiratory area, limiting inhalation and facilitating exhalation.

Podiatry (pō-DĪ-a-trē) The diagnosis and treatment of foot disorders.

Polar body The smaller cell resulting from the unequal division of primary and secondary oocytes during meiosis. The polar body has no function and degenerates.

Polycythemia (pol'-ē-sī-THĒ-mē-a) Disorder characterized by an above-normal hematocrit (above 55%) in which hypertension, thrombosis, and hemorrhage can occur.

Polysaccharide (pol'-ē-SAK-a-rīd) A carbohydrate in which three or more monosaccharides are joined chemically.

Polyunsaturated fat A fatty acid that contains more than one double covalent bond between its carbon atoms; abundant in triglycerides of corn oil, safflower oil, and cottonseed oil.

Polyuria (pol'-ē-Ū-rē-a) An excessive production of urine.

Pons (PONZ) The part of the brain stem that forms a "bridge" between the medulla oblongata and the midbrain, anterior to the cerebellum.

Positive feedback A feedback mechanism in which the response enhances the original stimulus.

Postabsorptive (fasting) state Metabolic state after absorption of food is complete and energy needs of the body must be satisfied using molecules stored previously.

Postcentral gyrus Gyrus of cerebral cortex located immediately posterior to the central sulcus; contains the primary somatosensory area.

Posterior (pos-TĒR-ē-or) Nearer to or at the back of the body. Equivalent to **dorsal** in bipeds.

Posterior column–medial lemniscus pathways Sensory pathways that carry information related to proprioception, fine touch, two-point discrimination, pressure, and vibration. First-order neurons project from the spinal cord to the ipsilateral medulla in the posterior columns (gracile fasciculus and cuneate fasciculus). Second-order neurons project from the medulla to the contralateral thalamus in the medial lemniscus. Third-order neurons project from the thalamus to the somatosensory cortex (postcentral gyrus) on the same side.

Posterior pituitary Posterior lobe of the pituitary gland. Also called the **neurohypophysis** (noo-rō-hī-POF-i-sis).

Posterior root The structure composed of sensory axons lying between a spinal nerve and the dorsolateral aspect of the spinal cord. Also called the **dorsal (sensory) root.**

Posterior root ganglion (GANG-glē-on) A group of cell bodies of sensory neurons and their supporting cells located along the posterior root of a spinal nerve. Also called a **dorsal (sensory) root ganglion.**

Postganglionic neuron (pōst′-gang-lē-ON-ik NOO-ron) The second autonomic motor neuron in an autonomic pathway, having its cell body and dendrites located in an autonomic ganglion and its unmyelinated axon ending at cardiac muscle, smooth muscle, or a gland.

Postsynaptic (pōst-sin-AP-tik) **neuron** The nerve cell that is activated by the release of a neurotransmitter from another neuron and carries nerve impulses away from the synapse.

Precapillary sphincter (SFINGK-ter) A ring of smooth muscle fibers (cells) at the site of origin of true capillaries that regulate blood flow into true capillaries.

Precentral gyrus Gyrus of cerebral cortex located immediately anterior to the central sulcus; contains the primary motor area.

Preganglionic (prē′-gang-lē-ON-ik) **neuron** The first autonomic motor neuron in an autonomic pathway, with its cell body and dendrites in the brain or spinal cord and its myelinated axon ending at an autonomic ganglion, where it synapses with a postganglionic neuron.

Pregnancy Sequence of events that normally includes fertilization, implantation, embryonic growth, and fetal growth and terminates in birth.

Premenstrual syndrome (PMS) Severe physical and emotional stress occurring late in the postovulatory phase of the menstrual cycle and sometimes overlapping with menstruation.

Prepuce (PRĒ-poos) The loose-fitting skin covering the glans of the penis and clitoris. Also called the **foreskin.**

Presbyopia (prez-bē-Ō-pē-a) A loss of elasticity of the lens of the eye due to advancing age with resulting inability to focus clearly on near objects.

Presynaptic (prē-sin-AP-tik) **inhibition** Type of inhibition in which neurotransmitter released by an inhibitory neuron depresses the release of neurotransmitter by a presynaptic neuron.

Presynaptic (prē-sin-AP-tik) **neuron** A neuron that propagates nerve impulses toward a synapse.

Prevertebral ganglion (prē-VER-te-bral GANG-lē-on) A cluster of cell bodies of postganglionic sympathetic neurons anterior to the spinal column and close to large abdominal arteries. Also called a **collateral ganglion.**

Primary germ layer One of three layers of embryonic tissue, called ectoderm, mesoderm, and endoderm, that give rise to all tissues and organs of the body.

Primary motor area A region of the cerebral cortex in the precentral gyrus of the frontal lobe of the cerebrum that controls specific muscles or groups of muscles.

Primary somatosensory area A region of the cerebral cortex posterior to the central sulcus in the postcentral gyrus of the parietal lobe of the cerebrum that localizes exactly the points of the body where somatic sensations originate.

Prime mover The muscle directly responsible for producing a desired motion. Also called an **agonist** (AG-ō-nist).

Primitive gut Embryonic structure formed from the dorsal part of the yolk sac that gives rise to most of the gastrointestinal tract.

Principal cell Cell type in the distal convoluted tubules and collecting ducts of the kidneys that is stimulated by aldosterone and antidiuretic hormone.

Proctology (prok-TOL-ō-jē) The branch of medicine concerned with the rectum and its disorders.

Progeny (PROJ-e-nē) Offspring or descendants.

Progesterone (prō-JES-te-rōn) A female sex hormone produced by the ovaries that helps prepare the endometrium of the uterus for implantation of a fertilized ovum and the mammary glands for milk secretion.

Prognosis (prog-NŌ-sis) A forecast of the probable results of a disorder; the outlook for recovery.

Prolactin (prō-LAK-tin) **(PRL)** A hormone secreted by the anterior pituitary that initiates and maintains milk secretion by the mammary glands.

Prolapse (PRŌ-laps) A dropping or falling down of an organ, especially the uterus or rectum.

Proliferation (prō-lif′-er-Ā-shun) Rapid and repeated reproduction of new parts, especially cells.

Pronation (prō-NĀ-shun) A movement of the forearm in which the palm is turned posteriorly or inferiorly.

Prophase (PRŌ-fāz) The first stage of mitosis during which chromatid pairs are formed and aggregate around the metaphase plate of the cell.

Proprioception (prō-prē-ō-SEP-shun) The perception of the position of body parts, especially the limbs, independent of vision; this sense is possible due to nerve impulses generated by proprioceptors.

Proprioceptor (prō′-prē-ō-SEP-tor) A receptor located in muscles, tendons, joints, or the internal ear (muscle spindles, tendon organs, joint kinesthetic receptors, and hair cells of the vestibular apparatus) that provides information about body position and movements.

Prostaglandin (pros′-ta-GLAN-din) **(PG)** A membrane-associated lipid; released in small quantities and acts as a local hormone.

Prostate (PROS-tāt) A doughnut-shaped gland inferior to the urinary bladder that surrounds the superior portion of the male urethra and secretes a slightly acidic solution that contributes to sperm motility and viability.

Protein An organic compound consisting of carbon, hydrogen, oxygen, nitrogen, and sometimes sulfur and phosphorus; synthesized on ribosomes and made up of amino acids linked by peptide bonds.

Proteasome Tiny cellular organelle in cytosol and nucleus containing proteases that destroy unneeded, damaged, or faulty proteins.

Prothrombin (prō-THROM-bin) An inactive blood-clotting factor synthesized by the liver, released into the blood, and converted to active thrombin in the process of blood clotting by the activated enzyme prothrombinase.

Protraction (prō-TRAK-shun) The movement of the mandible or shoulder girdle forward on a plane parallel with the ground.

Proximal (PROK-si-mal) Nearer the attachment of a limb to the trunk; nearer to the point of origin or attachment.

Pseudopods (SOO-dō-pods) Temporary protrusions of the leading edge of a migrating cell; cellular projections that surround a particle undergoing phagocytosis.

Pterygopalatine ganglion (ter′-i-gō-PAL-a-tīn GANG-glē-on) A cluster of cell bodies of parasympathetic postganglionic neurons ending at the lacrimal and nasal glands.

Ptosis (TŌ-sis) Drooping, as of the eyelid or the kidney.

Puberty (PŪ-ber-tē) The time of life during which the secondary sex characteristics begin to appear and the capability for sexual reproduction is possible; usually occurs between the ages of 10 and 17.

Pubic symphysis A slightly movable cartilaginous joint between the anterior surfaces of the hip bones.

Puerperium (pū′-er-PER-ē-um) The period immediately after childbirth, usually 4–6 weeks.

Pulmonary (PUL-mo-ner′-ē) Concerning or affected by the lungs.

Pulmonary circulation The flow of deoxygenated blood from the right ventricle to the lungs and the return of oxygenated blood from the lungs to the left atrium.

Pulmonary edema (e-DĒ-ma) An abnormal accumulation of interstitial fluid in the tissue spaces and alveoli of the lungs due to increased pulmonary capillary permeability or increased pulmonary capillary pressure.

Pulmonary embolism (EM-bō-lizm) **(PE)** The presence of a blood clot or a foreign substance in a pulmonary arterial blood vessel that obstructs circulation to lung tissue.

Pulmonary ventilation The inflow (inhalation) and outflow (exhalation) of air between the atmosphere and the lungs. Also called **breathing**.

Pulp cavity A cavity within the crown and neck of a tooth, which is filled with pulp, a connective tissue containing blood vessels, nerves, and lymphatic vessels.

Pulse (PULS) The rhythmic expansion and elastic recoil of a systemic artery after each contraction of the left ventricle.

Pulse pressure The difference between the maximum (systolic) and minimum (diastolic) pressures; normally about 40 mmHg.

Pupil The hole in the center of the iris, the area through which light enters the posterior cavity of the eyeball.

Purkinje (pur-KIN-jē) **fiber** Muscle fiber (cell) in the ventricular tissue of the heart specialized for conducting an action potential to the myocardium; part of the conduction system of the heart.

Pus The liquid product of inflammation containing leukocytes or their remains and debris of dead cells.

Pyloric (pī-LOR-ik) **sphincter** A thickened ring of smooth muscle through which the pylorus of the stomach communicates with the duodenum. Also called the **pyloric valve.**

Pyogenesis (pi′-ō-JEN-e-sis) Formation of pus.

Pyramid (PIR-a-mid) A pointed or cone-shaped structure. One of two roughly triangular structures on the anterior aspect of the medulla oblongata composed of the largest motor tracts that run from the cerebral cortex to the spinal cord. A triangular structure in the renal medulla.

Pyramidal (pi-RAM-i-dal) **tracts (pathways)** *See* **Direct motor pathways**.

Q

QRS wave The deflection waves of an electrocardiogram that represent onset of ventricular depolarization.

Quadriplegia (kwod′-ri-PLĒ-jē-a) Paralysis of four limbs: two upper and two lower.

R

Radiographic (rā′-dē-ō-GRAF-ic) **anatomy** Diagnostic branch of anatomy that includes the use of x rays.

Rami communicantes (RĀ-mē kō-mū-ni-KAN-tēz) Branches of a spinal nerve. *Singular is* **ramus communicans** (RĀ-mus kō-MŪ-ni-kans).

Rapid eye movement (REM) sleep Stage of sleep in which dreaming occurs, lasting for 5 to 10 minutes several times during a sleep cycle; characterized by rapid movements of the eyes beneath the eyelids.

Reactivity (rē-ak-TI-vi-tē) Ability of an antigen to react specifically with the antibody whose formation it induced.

Receptor A specialized cell or a distal portion of a neuron that responds to a specific sensory modality, such as touch, pressure, cold, light, or sound, and converts it to an electrical signal (generator or receptor potential). A specific molecule or cluster of molecules that recognizes and binds a particular ligand.

Receptor-mediated endocytosis A highly selective process whereby cells take up specific ligands, which usually are large molecules or particles, by enveloping them within a sac of plasma membrane. Ligands are eventually broken down by enzymes in lysosomes.

Receptor potential Depolarization or hyperpolarization of the plasma membrane of a receptor that alters release of neurotransmitter from the cell; if the neuron that synapses with the receptor cell becomes depolarized to threshold, a nerve impulse is triggered.

Recessive allele An allele whose presence is masked in the presence of a dominant allele on the homologous chromosome.

Reciprocal innervation (re-SIP-rō-kal in-ner-VĀ-shun) The phenomenon by which action potentials stimulate contraction of one muscle and simultaneously inhibit contraction of antagonistic muscles.

Recombinant DNA Synthetic DNA, formed by joining a fragment of DNA from one source to a portion of DNA from another.

Recovery oxygen consumption Elevated oxygen use after exercise ends due to metabolic changes that start during exercise and continue after exercise. Previously called **oxygen debt.**

Recruitment (rē-KROOT-ment) The process of increasing the number of active motor units. Also called **motor unit summation.**

Rectouterine pouch A pocket formed by the parietal peritoneum as it moves posteriorly from the surface of the uterus and is reflected onto the rectum; the most inferior point in the pelvic cavity. Also called the **pouch** or **cul de sac of Douglas.**

Rectum (REK-tum) The last 20 cm (8 in.) of the gastrointestinal tract, from the sigmoid colon to the anus.

Red nucleus A cluster of cell bodies in the midbrain, occupying a large part of the tectum from which axons extend into the rubroreticular and rubrospinal tracts.

Red pulp That portion of the spleen that consists of venous sinuses filled with blood and thin plates of splenic tissue called splenic (Billroth's) cords.

Reduction The addition of electrons to a molecule or, less commonly, the removal of oxygen from a molecule that results in an increase in the energy content of the molecule.

Referred pain Pain that is felt at a site remote from the place of origin.

Reflex Fast response to a change (stimulus) in the internal or external environment that attempts to restore homeostasis.

Reflex arc The most basic conduction pathway through the nervous system, connecting a receptor and an effector and consisting of a receptor, a sensory neuron, an integrating center in the central nervous system, a motor neuron, and an effector.

Refraction (rē-FRAK-shun) The bending of light as it passes from one medium to another.

Refractory (re-FRAK-to-rē) **period** A time period during which an excitable cell (neuron or muscle fiber) cannot respond to a stimulus that is usually adequate to evoke an action potential.

Regional anatomy The division of anatomy dealing with a specific region of the body, such as the head, neck, chest, or abdomen.

Regurgitation (rē-gur′-ji-TĀ-shun) Return of solids or fluids to the mouth from the stomach; backward flow of blood through incompletely closed heart valves.

Relaxin (RLX) A female hormone produced by the ovaries and placenta that increases flexibility of the pubic symphysis and helps dilate the uterine cervix to ease delivery of a baby.

Releasing hormone Hormone secreted by the hypothalamus that can stimulate secretion of hormones of the anterior pituitary.

Remodeling Replacement of old bone by new bone tissue.

Renal (RĒ-nal) Pertaining to the kidneys.

Renal corpuscle (KOR-pus-l) A glomerular (Bowman's) capsule and its enclosed glomerulus.

Renal pelvis A cavity in the center of the kidney formed by the expanded, proximal portion of the ureter, lying within the kidney, and into which the major calyces open.

Renal pyramid A triangular structure in the renal medulla containing the straight segments of renal tubules and the vasa recta.

Renin (RĒ-nin) An enzyme released by the kidney into the plasma, where it converts angiotensinogen into angiotensin I.

Renin–angiotensin–aldosterone (RAA) pathway A mechanism for the control of blood pressure, initiated by the secretion of renin by juxtaglomerular cells of the kidney in response to low blood pressure; renin catalyzes formation of angiotensin I, which is converted to angiotensin II by angiotensin-converting enzyme (ACE), and angiotensin II stimulates secretion of aldosterone.

Repolarization (rē-pō-lar-i-ZĀ-shun) Restoration of a resting membrane potential after depolarization.

Reproduction (rē-prō-DUK-shun) The formation of new cells for growth, repair, or replacement; the production of a new individual.

Reproductive cell division Type of cell division in which gametes (sperm and oocytes) are produced; consists of meiosis and cytokinesis.

Residual (re-ZID-ū-al) **volume** The volume of air still contained in the lungs after a maximal exhalation; about 1200 mL.

Resistance (re-ZIS-tans) Hindrance (impedance) to blood flow as a result of higher viscosity, longer total blood vessel length, and smaller blood vessel radius. Ability to ward off disease. The hindrance encountered by electrical charges as they move from one point to another. The hindrance encountered by air as it moves through the respiratory passageways.

Respiration (res-pi-RĀ-shun) Overall exchange of gases between the atmosphere, blood, and body cells consisting of pulmonary ventilation, external respiration, and internal respiration.

Respiratory center Neurons in the pons and medulla oblongata of the brain stem that regulate the rate and depth of pulmonary ventilation.

Respiratory membrane Structure in the lungs consisting of the alveolar wall and its basement membrane and a capillary endothelium and its basement membrane through which the diffusion of respiratory gases occurs.

Resting membrane potential The voltage difference between the inside and outside of a cell membrane when the cell is not responding to a stimulus; in many neurons and muscle fibers it is −70 to −90 mV, with the inside of the cell negative relative to the outside.

Retention (rē-TEN-shun) A failure to void urine due to obstruction, nervous contraction of the urethra, or absence of sensation of desire to urinate.

Rete (RĒ-tē) **testis** The network of ducts in the testes.

Reticular (re-TIK-ū-lar) **activating system (RAS)** A portion of the reticular formation that has many ascending connections with the cerebral cortex; when this area of the brain stem is active, nerve impulses pass to the thalamus and widespread areas of the cerebral cortex, resulting in generalized alertness or arousal from sleep.

Reticular formation A network of small groups of neuronal cell bodies scattered among bundles of axons (mixed gray and white matter) beginning in the medulla oblongata and extending superiorly through the central part of the brain stem.

Reticulocyte (re-TIK-ū-lō-sīt) An immature red blood cell.

Reticulum (re-TIK-ū-lum) A network.

Retina (RET-i-na) The deep coat of the posterior portion of the eyeball consisting of nervous tissue (where the process of vision begins) and a pigmented layer of epithelial cells that contact the choroid.

Retinal (RE-ti-nal) A derivative of vitamin A that functions as the light-absorbing portion of the photopigment rhodopsin.

Retraction (rē-TRAK-shun) The movement of a protracted part of the body posteriorly on a plane parallel to the ground, as in pulling the lower jaw back in line with the upper jaw.

Retroflexion (re-trō-FLEK-shun) A malposition of the uterus in which it is tilted posteriorly.

Retrograde degeneration (RE-trō-grād dē-jen-er-Ā-shun) Changes that occur in the proximal portion of a damaged axon only as far as the first node of Ranvier; similar to changes that occur during Wallerian degeneration.

Retroperitoneal (re′-trō-per-i-tō-NĒ-al) External to the peritoneal lining of the abdominal cavity.

Rh factor An inherited antigen on the surface of red blood cells in Rh⁺ individuals; not present in Rh⁻ individuals.

Rhinology (rī-NOL-ō-jē) The study of the nose and its disorders.

Rhodopsin (rō-DOP-sin) The photopigment in rods of the retina, consisting of a glycoprotein called opsin and a derivative of vitamin A called retinal.

Ribonucleic (rī-bō-noo-KLĒ-ik) **acid (RNA)** A single-stranded nucleic acid made up of nucleotides, each consisting of a nitrogenous base (adenine, cytosine, guanine, or uracil), ribose, and a phosphate group; three types are messenger RNA (mRNA), transfer RNA (tRNA), and ribosomal RNA (rRNA), each of which has a specific role during protein synthesis.

Ribosome (RĪ-bō-sōm) An organelle in the cytoplasm of cells, composed of a small subunit and a large subunit that contain ribosomal RNA and ribosomal proteins; the site of protein synthesis.

Right heart (atrial) reflex A reflex concerned with maintaining normal venous blood pressure.

Right lymphatic (lim-FAT-ik) **duct** A vessel of the lymphatic system that drains lymph from the upper right side of the body and empties it into the right subclavian vein.

Rigidity (ri-JID-i-tē) Hypertonia characterized by increased muscle tone, but reflexes are not affected.

Rigor mortis State of partial contraction of muscles after death due to lack of ATP; myosin heads (cross bridges) remain attached to actin, thus preventing relaxation.

Rod One of two types of photoreceptor in the retina of the eye; specialized for vision in dim light.

Root canal A narrow extension of the pulp cavity lying within the root of a tooth.

Root of penis Attached portion of penis that consists of the bulb and crura.

Rotation (rō-TĀ-shun) Moving a bone around its own axis, with no other movement.

Round ligament (LIG-a-ment) A band of fibrous connective tissue enclosed between the folds of the broad ligament of the uterus, emerging from the uterus just inferior to the uterine tube, extending laterally along the pelvic wall and through the deep inguinal ring to end in the labia majora.

Round window A small opening between the middle and internal ear, directly inferior to the oval window, covered by the secondary tympanic membrane.

Rugae (ROO-gē) Large folds in the mucosa of an empty hollow organ, such as the stomach and vagina.

S

Saccule (SAK-ūl) The inferior and smaller of the two chambers in the membranous labyrinth inside the vestibule of the internal ear containing a receptor organ for static equilibrium.

Sacral plexus (SĀ-kral PLEK-sus) A network formed by the ventral branches of spinal nerves L4 through S3.

Sacral promontory (PROM-on-tor′-ē) The superior surface of the body of the first sacral vertebra that projects anteriorly into the pelvic cavity; a line from the sacral promontory to the superior border of the pubic symphysis divides the abdominal and pelvic cavities.

Saddle joint A synovial joint in which the articular surface of one bone is saddle shaped and the articular surface of the other bone is shaped like the legs of the rider sitting in the saddle, as in the joint between the trapezium and the metacarpal of the thumb.

Sagittal (SAJ-i-tal) **plane** A plane that divides the body or organs into left and right portions. Such a plane may be **midsagittal (median),** in which the divisions are equal, or **parasagittal,** in which the divisions are unequal.

Saliva (sa-LĪ-va) A clear, alkaline, somewhat viscous secretion produced mostly by the three pairs of salivary glands; contains various salts, mucin, lysozyme, salivary amylase, and lingual lipase (produced by glands in the tongue).

Salivary amylase (SAL-i-ver-ē AM-i-lās) An enzyme in saliva that initiates the chemical breakdown of starch.

Salivary gland One of three pairs of glands that lie external to the mouth and pour their secretory product (saliva) into ducts that empty into the oral cavity; the parotid, submandibular, and sublingual glands.

Salt A substance that, when dissolved in water, ionizes into cations and anions, neither of which are hydrogen ions (H^+) or hydroxide ions (OH^-).

Saltatory (sal-ta-TŌ-rē) **conduction** The propagation of an action potential (nerve impulse) along the exposed parts of a myelinated axon. The action potential appears at successive nodes of Ranvier and therefore seems to leap from node to node.

Sarcolemma (sar′-kō-LEM-ma) The cell membrane of a muscle fiber (cell), especially of a skeletal muscle fiber.

Sarcomere (SAR-kō-mēr) A contractile unit in a striated muscle fiber (cell) extending from one Z disc to the next Z disc.

Sarcoplasm (SAR-kō-plazm) The cytoplasm of a muscle fiber (cell).

Sarcoplasmic reticulum (sar′-kō-PLAZ-mik re-TIK-ū-lum) A network of saccules and tubes surrounding myofibrils of a muscle fiber (cell), comparable to endoplasmic reticulum; functions to reabsorb calcium ions during relaxation and to release them to cause contraction.

Saturated fat A fatty acid that contains only single bonds (no double bonds) between its carbon atoms; all carbon atoms are bonded to the maximum number of hydrogen atoms; prevalent in triglycerides of animal products such as meat, milk, milk products, and eggs.

Scala tympani (SKA-la TIM-pan-ē) The inferior spiral-shaped channel of the bony cochlea, filled with perilymph.

Scala vestibuli (ves-TIB-ū-lē) The superior spiral-shaped channel of the bony cochlea, filled with perilymph.

Schwann (SCHVON) **cell** A neuroglial cell of the peripheral nervous system that forms the myelin sheath and neurolemma around a nerve axon by wrapping around the axon in a jelly-roll fashion.

Sciatica (sī-AT-i-ka) Inflammation and pain along the sciatic nerve; felt along the posterior aspect of the thigh extending down the inside of the leg.

Sclera (SKLE-ra) The white coat of fibrous tissue that forms the superficial protective covering over the eyeball except in the most anterior portion; the posterior portion of the fibrous tunic.

Scleral venous sinus A circular venous sinus located at the junction of the sclera and the cornea through which aqueous humor drains from the anterior chamber of the eyeball into the blood. Also called the **canal of Schlemm** (SHLEM).

Sclerosis (skle-RŌ-sis) A hardening with loss of elasticity of tissues.

Scoliosis (skō′-lē-Ō-sis) An abnormal lateral curvature from the normal vertical line of the backbone.

Scrotum (SKRŌ-tum) A skin-covered pouch that contains the testes and their accessory structures.

Sebaceous (se-BĀ-shus) **gland** An exocrine gland in the dermis of the skin, almost always associated with a hair follicle, that secretes sebum. Also called an **oil gland.**

Sebum (SĒ-bum) Secretion of sebaceous (oil) glands.

Secondary response Accelerated, more intense cell-mediated or antibody-mediated immune response upon a subsequent exposure to an antigen after the primary response.

Secondary sex characteristic A characteristic of the male or female body that develops at puberty under the influence of sex hormones but is not directly involved in sexual reproduction; examples are distribution of body hair, voice pitch, body shape, and muscle development.

Second messenger An intracellular mediator molecule that is produced in response to a first messenger (hormone or neurotransmitter) binding to its receptor in the plasma membrane of a target cell. Initiates a cascade of chemical reactions that produce characteristic effects for that particular target cell.

Secretion (se-KRĒ-shun) Production and release from a cell or a gland of a physiologically active substance.

Selective permeability (per′-mē-a-BIL-i-tē) The property of a membrane by which it permits the passage of certain substances but restricts the passage of others.

Semen (SĒ-men) A fluid discharged at ejaculation by a male that consists of a mixture of sperm and the secretions of the seminiferous tubules, seminal vesicles, prostate, and bulbourethral (Cowper's) glands.

Semicircular canals Three bony channels (anterior, posterior, lateral), filled with perilymph, in which lie the membranous semicircular canals filled with endolymph. They contain receptors for equilibrium.

Semicircular ducts The membranous semicircular canals filled with endolymph and floating in the perilymph of the bony semicircular canals; they contain cristae that are concerned with dynamic equilibrium.

Semilunar (sem′-ē-LOO-nar) **valve** A valve between the aorta or the pulmonary trunk and a ventricle of the heart.

Seminal vesicle (SEM-i-nal VES-i-kul) One of a pair of convoluted, pouchlike structures, lying posterior and inferior to the urinary bladder and anterior to the rectum, that secrete a component of semen into the ejaculatory ducts. Also termed **seminal gland.**

Seminiferous tubule (sem′-i-NI-fer-us TOO-būl) A tightly coiled duct, located in the testis, where sperm are produced.

Senescence (se-NES-ens) The process of growing old.

Sensation A state of awareness of external or internal conditions of the body.

Sensory neurons (NOO-ronz) Neurons that carry sensory information from cranial and spinal nerves into the brain and spinal cord or from a lower to a higher level in the spinal cord and brain. Also called **afferent neurons.**

Septal defect An opening in the septum (interatrial or interventricular) between the left and right sides of the heart.

Septum (SEP-tum) A wall dividing two cavities.

Serous (SIR-us) **membrane** A membrane that lines a body cavity that does not open to the exterior. The external layer of an organ formed by a serous membrane. The membrane that lines the pleural, pericardial, and peritoneal cavities. Also called a **serosa** (se-RŌ-sa).

Sertoli (ser-TŌ-lē) **cell** A supporting cell in the seminiferous tubules that secretes fluid for supplying nutrients to sperm and the hormone inhibin, removes excess cytoplasm from spermatogenic cells, and mediates the effects of FSH and testosterone on spermatogenesis. Also called a **sustentacular** (sus′-ten-TAK-ū-lar) **cell.**

Serum Blood plasma minus its clotting proteins.

Sesamoid (SES-a-moyd) **bones** Small bones usually found in tendons.

Sex chromosomes The twenty-third pair of chromosomes, designated X and Y, which determine the genetic sex of an individual; in males, the pair is XY; in females, XX.

Sexual intercourse The insertion of the erect penis of a male into the vagina of a female. Also called **coitus** (KŌ-i-tus).

Shivering Involuntary contraction of skeletal muscles that generates heat. Also called **involuntary thermogenesis.**

Shock Failure of the cardiovascular system to deliver adequate amounts of oxygen and nutrients to meet the metabolic needs of the body due to inadequate cardiac output. It is characterized by hypotension; clammy, cool, and pale skin; sweating; reduced urine formation; altered mental state; acidosis; tachycardia; weak, rapid pulse; and thirst. Types include hypovolemic, cardiogenic, vascular, and obstructive.

Shoulder joint A synovial joint where the humerus articulates with the scapula.

Sigmoid colon (SIG-moyd KŌ-lon) The S-shaped part of the large intestine that begins at the level of the left iliac crest, projects medially, and terminates at the rectum at about the level of the third sacral vertebra.

Sign Any objective evidence of disease that can be observed or measured such as a lesion, swelling, or fever.

Sinoatrial (si-nō-Ā-trē-al) **(SA) node** A small mass of cardiac muscle fibers (cells) located in the right atrium inferior to the opening of the superior vena cava that spontaneously depolarize and generate a cardiac action potential about 100 times per minute. Also called the **pacemaker.**

Sinus (SĪ-nus) A hollow in a bone (paranasal sinus) or other tissue; a channel for blood (vascular sinus); any cavity having a narrow opening.

Sinusoid (SĪ-nū-soyd) A large, thin-walled, and leaky type of capillary, having large intercellular clefts that may allow proteins and blood cells to pass from a tissue into the bloodstream; present in the liver, spleen, anterior pituitary, parathyroid glands, and red bone marrow.

Skeletal muscle An organ specialized for contraction, composed of striated muscle fibers (cells), supported by connective tissue, attached to a bone by a tendon or an aponeurosis, and stimulated by somatic motor neurons.

Skin The external covering of the body that consists of a superficial, thinner epidermis (epithelial tissue) and a deep, thicker dermis (connective tissue) that is anchored to the subcutaneous layer.

Skull The skeleton of the head consisting of the cranial and facial bones.

Sleep A state of partial unconsciousness from which a person can be aroused; associated with a low level of activity in the reticular activating system.

Sliding-filament mechanism The explanation of how thick and thin filaments slide relative to one another during striated muscle contraction to decrease sarcomere length.

Small intestine A long tube of the gastrointestinal tract that begins at the pyloric sphincter of the stomach, coils through the central and inferior part of the abdominal cavity, and ends at the large intestine; divided into three segments: duodenum, jejunum, and ileum.

Smooth muscle A tissue specialized for contraction, composed of smooth muscle fibers (cells), located in the walls of hollow internal organs, and innervated by autonomic motor neurons.

Sodium-potassium ATPase An active transport pump located in the plasma membrane that transports sodium ions out of the cell and potassium ions into the cell at the expense of cellular ATP. It functions to keep the ionic concentrations of these ions at physiological levels. Also called the **sodium-potassium pump.**

Soft palate (PAL-at) The posterior portion of the roof of the mouth, extending from the palatine bones to the uvula. It is a muscular partition lined with mucous membrane.

Solution A homogeneous molecular or ionic dispersion of one or more substances (solutes) in a dissolving medium (solvent) that is usually liquid.

Somatic (sō-MAT-ik) **cell division** Type of cell division in which a single starting parent cell duplicates itself to produce two identical cells; consists of mitosis and cytokinesis.

Somatic nervous system (SNS) The portion of the peripheral nervous system consisting of somatic sensory (afferent) neurons and somatic motor (efferent) neurons.

Somite (SŌ-mīt) Block of mesodermal cells in a developing embryo that is distinguished into a myotome (which forms most of the skeletal muscles), dermatome (which forms connective tissues), and sclerotome (which forms the vertebrae).

Spasm (SPAZM) A sudden, involuntary contraction of large groups of muscles.

Spasticity (spas-TIS-i-tē) Hypertonia characterized by increased muscle tone, increased tendon reflexes, and pathological reflexes (Babinski sign).

Spermatic (sper-MAT-ik) **cord** A supporting structure of the male reproductive system, extending from a testis to the deep inguinal ring, that includes the ductus (vas) deferens, arteries, veins, lymphatic vessels, nerves, cremaster muscle, and connective tissue.

Spermatogenesis (sper'-ma-tō-JEN-e-sis) The formation and development of sperm in the seminiferous tubules of the testes.

Sperm cell A mature male gamete. Also termed **spermatozoon** (sper'-ma-tō-ZŌ-on).

Spermiogenesis (sper'-mē-ō-JEN-e-sis) The maturation of spermatids into sperm.

Sphincter (SFINGK-ter) A circular muscle that constricts an opening.

Sphincter of the hepatopancreatic ampulla A circular muscle at the opening of the common bile and main pancreatic ducts in the duodenum. Also called the **sphincter of Oddi** (OD-ē).

Sphygmomanometer (sfig'-mō-ma-NOM-e-ter) An instrument for measuring arterial blood pressure.

Spinal (SPĪ-nal) **cord** A mass of nerve tissue located in the vertebral canal from which 31 pairs of spinal nerves originate.

Spinal nerve One of the 31 pairs of nerves that originate on the spinal cord from posterior and anterior roots.

Spinal shock A period from several days to several weeks following transection of the spinal cord and characterized by the abolition of all reflex activity.

Spinothalamic (spī-nō-tha-LAM-ik) **tracts** Sensory (ascending) tracts that convey information up the spinal cord to the thalamus for sensations of pain, temperature, crude touch, and deep pressure.

Spinous (SPĪ-nus) **process** A sharp or thornlike process or projection. Also called a **spine.** A sharp ridge running diagonally across the posterior surface of the scapula.

Spiral organ The organ of hearing, consisting of supporting cells and hair cells that rest on the basilar membrane and extend into the endolymph of the cochlear duct. Also called the **organ of Corti** (KOR-tē).

Spirometer (spī-ROM-e-ter) An apparatus used to measure lung volumes and capacities.

Splanchnic (SPLANK-nik) Pertaining to the viscera.

Spleen (SPLĒN) Large mass of lymphatic tissue between the fundus of the stomach and the diaphragm that functions in formation of blood cells during early fetal development, phagocytosis of ruptured blood cells, and proliferation of B cells during immune responses.

Spongy (cancellous) bone tissue Bone tissue that consists of an irregular latticework of thin plates of bone called trabeculae; spaces between trabeculae of some bones are filled with red bone marrow; found inside short, flat, and irregular bones and in the epiphyses (ends) of long bones.

Sprain Forcible wrenching or twisting of a joint with partial rupture or other injury to its attachments without dislocation.

Squamous (SKWĀ-mus) Flat or scalelike.

Starling's law of the capillaries The movement of fluid between plasma and interstitial fluid is in a state of near equilibrium at the arterial and venous ends of a capillary; that is, filtered fluid and absorbed fluid plus that returned to the lymphatic system are nearly equal.

Starling's law of the heart The force of muscular contraction is determined by the length of the cardiac muscle fibers just before they contract; within limits, the greater the length of stretched fibers, the stronger the contraction.

Starvation (star-VĀ-shun) The loss of energy stores in the form of glycogen, triglycerides, and proteins due to inadequate intake of nutrients or inability to digest, absorb, or metabolize ingested nutrients.

Stasis (STĀ-sis) Stagnation or halt of normal flow of fluids, as blood or urine, or of the intestinal contents.

Static equilibrium (ē-kwi-LIB-rē-um) The maintenance of posture in response to changes in the orientation of the body, mainly the head, relative to the ground.

Stellate reticuloendothelial (STEL-āt re-tik'-ū-lō-en'-dō-THĒ-lē-al) **cell** Phagocytic cell bordering a sinusoid of the liver. Also called a **Kupffer** (KOOP-fer) **cell.**

Stenosis (sten-Ō-sis) An abnormal narrowing or constriction of a duct or opening.

Stereocilia (ste'-rē-ō-SIL-ē-a) Groups of extremely long, slender, nonmotile microvilli projecting from epithelial cells lining the epididymis.

Stereognosis (ste'-rē-og-NŌ-sis) The ability to recognize the size, shape, and texture of an object by touching it.

Sterile (STE-ril) Free from any living microorganisms. Unable to conceive or produce offspring.

Sterilization (ster'-i-li-ZĀ-shun) Elimination of all living microorganisms. Any procedure that renders an individual incapable of reproduction (for example, castration, vasectomy, hysterectomy, or oophorectomy).

Stimulus Any stress that changes a controlled condition; any change in the internal or external environment that excites a sensory receptor, a neuron, or a muscle fiber.

Stomach The J-shaped enlargement of the gastrointestinal tract directly inferior to the diaphragm in the epigastric, umbilical, and left hypochondriac regions of the abdomen, between the esophagus and small intestine.

Straight tubule (TOO-būl) A duct in a testis leading from a convoluted seminiferous tubule to the rete testis.

Stratum (STRĀ-tum) A layer.

Stratum basalis (ba-SAL-is) The layer of the endometrium next to the myometrium that is maintained during menstruation and gestation and produces a new stratum functionalis following menstruation or parturition.

Stratum functionalis (funk'-shun-AL-is) The layer of the endometrium next to the uterine cavity that is shed during menstruation and that forms the maternal portion of the placenta during gestation.

Stressor A stress that is extreme, unusual, or long-lasting and triggers the stress response.

Stress response Wide-ranging set of bodily changes, triggered by a stressor, that gears the body to meet an emergency. Also known as **general adaptation syndrome (GAS).**

Stretch receptor Receptor in the walls of blood vessels, airways, or organs that monitors the amount of stretching. Also termed **baroreceptor.**

Stretch reflex A monosynaptic reflex triggered by sudden stretching of muscle spindles within a muscle that elicits contraction of that same muscle. Also called a **tendon jerk.**

Stroke volume The volume of blood ejected by either ventricle in one systole; about 70 mL in an adult at rest.

Stroma (STRŌ-ma) The tissue that forms the ground substance, foundation, or framework of an organ, as opposed to its functional parts (parenchyma).

Subarachnoid (sub′-a-RAK-noyd) **space** A space between the arachnoid mater and the pia mater that surrounds the brain and spinal cord and through which cerebrospinal fluid circulates.

Subcutaneous (sub′-kū-TĀ-nē-us) Beneath the skin. Also called **hypodermic** (hi-pō-DER-mik).

Subcutaneous layer A continuous sheet of areolar connective tissue and adipose tissue between the dermis of the skin and the deep fascia of the muscles. Also called the **superficial fascia** (FASH-ē-a).

Subdural (sub-DOO-ral) **space** A space between the dura mater and the arachnoid mater of the brain and spinal cord that contains a small amount of fluid.

Sublingual (sub-LING-gwal) **gland** One of a pair of salivary glands situated in the floor of the mouth deep to the mucous membrane and to the side of the lingual frenulum, with a duct (Rivinus') that opens into the floor of the mouth.

Submandibular (sub′-man-DIB-ū-lar) **gland** One of a pair of salivary glands found inferior to the base of the tongue deep to the mucous membrane in the posterior part of the floor of the mouth, posterior to the sublingual glands, with a duct (Wharton's) situated to the side of the lingual frenulum. Also called the **submaxillary** (sub′-MAK-si-ler-ē) **gland.**

Submucosa (sub-mū-KŌ-sa) A layer of connective tissue located deep to a mucous membrane, as in the gastrointestinal tract or the urinary bladder; the submucosa connects the mucosa to the muscularis layer.

Submucosal plexus A network of autonomic nerve fibers located in the superficial part of the submucous layer of the small intestine. Also called the **plexus of Meissner** (MĪZ-ner).

Substrate A molecule upon which an enzyme acts.

Subthreshold stimulus A stimulus of such weak intensity that it cannot initiate an action potential (nerve impulse).

Sudoriferous (soo′-dor-IF-er-us) **gland** An apocrine or eccrine exocrine gland in the dermis or subcutaneous layer that produces perspiration. Also called a **sweat gland.**

Sulcus (SUL-kus) A groove or depression between parts, especially between the convolutions of the brain. *Plural is* **sulci** (SUL-sī).

Summation (sum-MĀ-shun) The addition of the excitatory and inhibitory effects of many stimuli applied to a neuron. The increased strength of muscle contraction that results when stimuli follow one another in rapid succession.

Superficial (soo′-per-FISH-al) Located on or near the surface of the body or an organ.

Superficial fascia (FASH-ē-a) A continuous sheet of fibrous connective tissue between the dermis of the skin and the deep fascia of the muscles. Also called **subcutaneous** (sub′-kū-TĀ-nē-us) **layer.**

Superficial inguinal (IN-gwi-nal) **ring** A triangular opening in the aponeurosis of the external oblique muscle that represents the termination of the inguinal canal.

Superior (soo-PĒR-ē-or) Toward the head or upper part of a structure. Also called **cephalad** (SEF-a-lad) or **craniad.**

Superior vena cava (VĒ-na CĀ-va) **(SVC)** Large vein that collects blood from parts of the body superior to the heart and returns it to the right atrium.

Supination (soo-pi-NĀ-shun) A movement of the forearm in which the palm is turned anteriorly or superiorly.

Surface anatomy The study of the structures that can be identified from the outside of the body.

Surfactant (sur-FAK-tant) Complex mixture of phospholipids and lipoproteins, produced by type II alveolar (septal) cells in the lungs, that decreases surface tension.

Susceptibility (sus-sep′-ti-BIL-i-tē) Lack of resistance to the damaging effects of an agent such as a pathogen.

Suspensory ligament (sus-PEN-so-rē LIG-a-ment) A fold of peritoneum extending laterally from the surface of the ovary to the pelvic wall.

Sutural (SOO-cher-al) **bone** A small bone located within a suture between certain cranial bones. Also called **Wormian** (WER-mē-an) **bone.**

Suture (SOO-cher) An immovable fibrous joint that joins skull bones.

Sympathetic (sim′-pa-THET-ik) **division** One of the two subdivisions of the autonomic nervous system, having cell bodies of preganglionic neurons in the lateral gray columns of the thoracic segment and the first two or three lumbar segments of the spinal cord; primarily concerned with process involving the expenditure of energy.

Sympathetic trunk ganglion (GANG-glē-on) A cluster of cell bodies of sympathetic postganglionic neurons lateral to the vertebral column, close to the body of a vertebra. These ganglia extend inferiorly through the neck, thorax, and abdomen to the coccyx on both sides of the vertebral column and are connected to one another to form a chain on each side of the vertebral column. Also called **sympathetic chain** or **vertebral chain ganglia.**

Sympathomimetic (sim′-pa-thō-mi-MET-ik) Producing effects that mimic those brought about by the sympathetic division of the autonomic nervous system.

Symphysis (SIM-fi-sis) A line of union. A slightly movable cartilaginous joint such as the pubic symphysis.

Symporter A transmembrane transporter protein that moves two substances, often Na^+ and another substance, in the same direction across a plasma membrane. Also called a **cotransporter.**

Symptom (SIMP-tum) A subjective change in body function not apparent to an observer, such as pain or nausea, that indicates the presence of a disease or disorder of the body.

Synapse (SYN-aps) The functional junction between two neurons or between a neuron and an effector, such as a muscle or gland; may be electrical or chemical.

Synapsis (sin-AP-sis) The pairing of homologous chromosomes during prophase I of meiosis.

Synaptic (sin-AP-tik) **cleft** The narrow gap at a chemical synapse that separates the axon terminal of one neuron from another neuron or muscle fiber (cell) and across which a neurotransmitter diffuses to affect the postsynaptic cell.

Synaptic delay The length of time between the arrival of the action potential at a presynaptic axon terminal and the membrane potential

(IPSP or EPSP) change on the postsynaptic membrane; usually about 0.5 msec.

Synaptic end bulb Expanded distal end of an axon terminal that contains synaptic vesicles. Also called a **synaptic knob.**

Synaptic vesicle Membrane-enclosed sac in a synaptic end bulb that stores neurotransmitters.

Synarthrosis (sin′-ar-THRŌ-sis) An immovable joint such as a suture, gomphosis, and synchondrosis.

Synchondrosis (sin′-kon-DRŌ-sis) A cartilaginous joint in which the connecting material is hyaline cartilage.

Syndesmosis (sin′-dez-MŌ-sis) A slightly movable joint in which articulating bones are united by fibrous connective tissue.

Syndrome (SIN-drōm) A group of signs and symptoms that occur together in a pattern that is characteristic of a particular disease or abnormal condition.

Synergist (SIN-er-jist) A muscle that assists the prime mover by reducing undesired action or unnecessary movement.

Synergistic (syn-er-JIS-tik) **effect** A hormonal interaction in which the effects of two or more hormones acting together is greater or more extensive than the sum of each hormone acting alone.

Synostosis (sin′-os-TŌ-sis) A joint in which the dense fibrous connective tissue that unites bones at a suture has been replaced by bone, resulting in a complete fusion across the suture line.

Synovial (si-NŌ-vē-al) **cavity** The space between the articulating bones of a synovial joint, filled with synovial fluid. Also called a **joint cavity.**

Synovial fluid Secretion of synovial membranes that lubricates joints and nourishes articular cartilage.

Synovial joint A fully movable or diarthrotic joint in which a synovial (joint) cavity is present between the two articulating bones.

Synovial membrane The deeper of the two layers of the articular capsule of a synovial joint, composed of areolar connective tissue that secretes synovial fluid into the synovial (joint) cavity.

System An association of organs that have a common function.

Systemic (sis-TEM-ik) Affecting the whole body; generalized.

Systemic anatomy The anatomic study of particular systems of the body, such as the skeletal, muscular, nervous, cardiovascular, or urinary systems.

Systemic circulation The routes through which oxygenated blood flows from the left ventricle through the aorta to all the organs of the body and deoxygenated blood returns to the right atrium.

Systemic vascular resistance (SVR) All the vascular resistance offered by systemic blood vessels. Also called **total peripheral resistance.**

Systole (SIS-tō-lē) In the cardiac cycle, the phase of contraction of the heart muscle, especially of the ventricles.

Systolic (sis-TOL-ik) **blood pressure** The force exerted by blood on arterial walls during ventricular contraction; the highest pressure measured in the large arteries, about 120 mmHg under normal conditions for a young adult.

T

T cell A lymphocyte that becomes immunocompetent in the thymus and can differentiate into a helper T cell or a cytotoxic T cell, both of which function in cell-mediated immunity.

T wave The deflection wave of an electrocardiogram that represents ventricular repolarization.

Tachycardia (tak′-i-KAR-dē-a) An abnormally rapid resting heartbeat or pulse rate (over 100 beats per minute).

Tactile (TAK-tīl) Pertaining to the sense of touch.

Tactile disc Modified epidermal cell in the stratum basale of hairless skin that functions as a cutaneous receptor for discriminative touch. Also called a **Merkel** (MER-kel) **disc.**

Teniae coli (TĒ-nē-ē KŌ-lī) The three flat bands of thickened, longitudinal smooth muscle running the length of the large intestine, except in the rectum. *Singular is* **tenia coli.**

Target cell A cell whose activity is affected by a particular hormone.

Tarsal bones The seven bones of the ankle. Also called **tarsals.**

Tarsal gland Sebaceous (oil) gland that opens on the edge of each eyelid. Also called a **Meibomian** (mī-BŌ-mē-an) **gland.**

Tarsal plate A thin, elongated sheet of connective tissue, one in each eyelid, giving the eyelid form and support. The aponeurosis of the levator palpebrae superioris is attached to the tarsal plate of the superior eyelid.

Tarsus (TAR-sus) A collective term for the seven bones of the ankle.

Tectorial (tek-TŌ-rē-al) **membrane** A gelatinous membrane projecting over and in contact with the hair cells of the spiral organ (organ of Corti) in the cochlear duct.

Teeth (TĒTH) Accessory structures of digestion, composed of calcified connective tissue and embedded in bony sockets of the mandible and maxilla, that cut, shred, crush, and grind food. Also called **dentes** (DEN-tēz).

Telophase (TEL-ō-fāz) The final stage of mitosis in which the daughter nuclei become established.

Tendon (TEN-don) A white fibrous cord of dense regular connective tissue that attaches muscle to bone.

Tendon organ A proprioceptive receptor, sensitive to changes in muscle tension and force of contraction, found chiefly near the junctions of tendons and muscles. Also called a **Golgi** (GOL-jē) **tendon organ.**

Tendon reflex A polysynaptic, ipsilateral reflex that protects tendons and their associated muscles from damage that might be brought about by excessive tension. The receptors involved are called tendon organs (Golgi tendon organs).

Tentorium cerebelli (ten-TŌ-rē-um ser′-e-BEL-ē) A transverse shelf of dura mater that forms a partition between the occipital lobe of the cerebral hemispheres and the cerebellum and that covers the cerebellum.

Teratogen (TER-a-tō-jen) Any agent or factor that causes physical defects in a developing embryo.

Terminal ganglion (TER-min-al GANG-glē-on) A cluster of cell bodies of parasympathetic postganglionic neurons either lying very close to the visceral effectors or located within the walls of the visceral effectors supplied by the postganglionic neurons.

Testis (TES-tis) Male gonad that produces sperm and the hormones testosterone and inhibin. Also called a **testicle.**

Testosterone (tes-TOS-te-rōn) A male sex hormone (androgen) secreted by interstitial endocrinocytes (Leydig cells) of a mature testis; needed for development of sperm; together with a second androgen termed **dihydrotestosterone (DHT),** controls the growth and development of male reproductive organs, secondary sex characteristics, and body growth.

Tetany (TET-a-nē) Hyperexcitability of neurons and muscle fibers (cells) caused by hypocalcemia and characterized by intermittent or continuous tonic muscular contractions; may be due to hypoparathyroidism.

Thalamus (THAL-a-mus) A large, oval structure located bilaterally on either side of the third ventricle, consisting of two masses of gray matter organized into nuclei; main relay center for sensory impulses ascending to the cerebral cortex.

Thermoreceptor (THER-mō-rē-sep-tor) Sensory receptor that detects changes in temperature.

Thigh The portion of the lower limb between the hip and the knee.

Third ventricle (VEN-tri-kul) A slitlike cavity between the right and left halves of the thalamus and between the lateral ventricles of the brain.

Thirst center A cluster of neurons in the hypothalamus that is sensitive to the osmotic pressure of extracellular fluid and brings about the sensation of thirst.

Thoracic (thō-RAS-ik) **cavity** Superior portion of the ventral body cavity that contains two pleural cavities, the mediastinum, and the pericardial cavity.

Thoracic duct A lymphatic vessel that begins as a dilation called the cisterna chyli, receives lymph from the left side of the head, neck, and chest, the left arm, and the entire body below the ribs, and empties into the left subclavian vein. Also called the **left lymphatic** (lim-FAT-ik) **duct.**

Thoracolumbar (thō´-ra-kō-LUM-bar) **outflow** The axons of sympathetic preganglionic neurons, which have their cell bodies in the lateral gray columns of the thoracic segments and first two or three lumbar segments of the spinal cord.

Thorax (THŌ-raks) The chest.

Threshold potential The membrane voltage that must be reached to trigger an action potential.

Threshold stimulus Any stimulus strong enough to initiate an action potential or activate a sensory receptor.

Thrombin (THROM-bin) The active enzyme formed from prothrombin that converts fibrinogen to fibrin during formation of a blood clot.

Thrombolytic (throm-bō-LIT-ik) **agent** Chemical substance injected into the body to dissolve blood clots and restore circulation; mechanism of action is direct or indirect activation of plasminogen; examples include tissue plasminogen activator (t-PA), streptokinase, and urokinase.

Thrombosis (throm-BŌ-sis) The formation of a clot in an unbroken blood vessel, usually a vein.

Thrombus A stationary clot formed in an unbroken blood vessel, usually a vein.

Thymus (THĪ-mus) A bilobed organ, located in the superior mediastinum posterior to the sternum and between the lungs, in which T cells develop immunocompetence.

Thyroglobulin (thī-rō-GLŌ-bū-lin) **(TGB)** A large glycoprotein molecule produced by follicular cells of the thyroid gland in which some tyrosines are iodinated and coupled to form thyroid hormones.

Thyroid cartilage (THĪ-royd KAR-ti-lij) The largest single cartilage of the larynx, consisting of two fused plates that form the anterior wall of the larynx.

Thyroid follicle (FOL-i-kul) Spherical sac that forms the parenchyma of the thyroid gland and consists of follicular cells that produce thyroxine (T_4) and triiodothyronine (T_3).

Thyroid gland An endocrine gland with right and left lateral lobes on either side of the trachea connected by an isthmus; located anterior to the trachea just inferior to the cricoid cartilage; secretes thyroxine (T_4), triiodothyronine (T_3), and calcitonin.

Thyroid-stimulating hormone (TSH) A hormone secreted by the anterior pituitary that stimulates the synthesis and secretion of thyroxine (T_4) and triiodothyronine (T_3).

Thyroxine (thī-ROK-sēn) **(T_4)** A hormone secreted by the thyroid gland that regulates metabolism, growth and development, and the activity of the nervous system.

Tic Spasmodic, involuntary twitching of muscles that are normally under voluntary control.

Tidal volume The volume of air breathed in and out in any one breath; about 500 mL in quiet, resting conditions.

Tissue A group of similar cells and their intercellular substance joined together to perform a specific function.

Tissue factor (TF) A factor, or collection of factors, whose appearance initiates the blood clotting process. Also called **thromboplastin** (throm-bō-PLAS-tin).

Tissue plasminogen activator (t-PA) An enzyme that dissolves small blood clots by initiating a process that converts plasminogen to plasmin, which degrades the fibrin of a clot.

Tongue A large skeletal muscle covered by a mucous membrane located on the floor of the oral cavity.

Tonicity (tō-NIS-i-tē) A measure of the concentration of impermeable solute particles in a solution relative to cytosol. When cells are bathed in an **isotonic solution,** they neither shrink nor swell.

Tonsil (TON-sil) An aggregation of large lymphatic nodules embedded in the mucous membrane of the throat.

Torn cartilage A tearing of an articular disc (meniscus) in the knee.

Total lung capacity The sum of tidal volume, inspiratory reserve volume, expiratory reserve volume, and residual volume; about 6000 mL in an average adult.

Trabecula (tra-BEK-ū-la) Irregular latticework of thin plates of spongy bone. Fibrous cord of connective tissue serving as supporting fiber by forming a septum extending into an organ from its wall or capsule. *Plural is* **trabeculae** (tra-BEK-ū-lē).

Trabeculae carneae (KAR-nē-ē) Ridges and folds of the myocardium in the ventricles.

Trachea (TRĀ-kē-a) Tubular air passageway extending from the larynx to the fifth thoracic vertebra. Also called the **windpipe.**

Tract A bundle of nerve axons in the central nervous system.

Transcription (trans-KRIP-shun) The first step in the expression of genetic information in which a single strand of DNA serves as a template for the formation of an RNA molecule.

Translation (trans-LĀ-shun) The synthesis of a new protein on a ribosome as dictated by the sequence of codons in messenger RNA.

Transverse colon (trans-VERS KŌ-lon) The portion of the large intestine extending across the abdomen from the right colic (hepatic) flexure to the left colic (splenic) flexure.

Transverse fissure (FISH-er) The deep cleft that separates the cerebrum from the cerebellum.

Transverse plane A plane that divides the body or organs into superior and inferior portions. Also called a **horizontal plane.**

Transverse tubules (TOO-būls) **(T tubules)** Small, cylindrical invaginations of the sarcolemma of striated muscle fibers (cells) that conduct muscle action potentials toward the center of the muscle fiber.

Trauma (TRAW-ma) An injury, either a physical wound or psychic disorder, caused by an external agent or force, such as a physical blow or emotional shock; the agent or force that causes the injury.

Tremor (TREM-or) Rhythmic, involuntary, purposeless contraction of opposing muscle groups.

Triad (TRĪ-ad) A complex of three units in a muscle fiber composed of a transverse tubule and the sarcoplasmic reticulum terminal cisterns on both sides of it.

Tricuspid (trī-KUS-pid) **valve** Atrioventricular (AV) valve on the right side of the heart.

Triglyceride (trī-GLI-cer-īd) A lipid formed from one molecule of glycerol and three molecules of fatty acids that may be either solid (fats) or liquid (oils) at room temperature; the body's most highly concentrated source of chemical potential energy. Found mainly within adipocytes. Also called a **neutral fat** or a **triacylglycerol.**

Trigone (TRĪ-gon) A triangular region at the base of the urinary bladder.

Triiodothyronine (trī-ī-ō-dō-THĪ-rō-nēn) **(T₃)** A hormone produced by the thyroid gland that regulates metabolism, growth and development, and the activity of the nervous system.

Trophoblast (TRŌF-ō-blast) The superficial covering of cells of the blastocyst.

Tropic (TRŌ-pik) **hormone** A hormone whose target is another endocrine gland.

Trunk The part of the body to which the upper and lower limbs are attached.

Tubal ligation (lī-GĀ-shun) A sterilization procedure in which the uterine (Fallopian) tubes are tied and cut.

Tubular reabsorption The movement of filtrate from renal tubules back into blood in response to the body's specific needs.

Tubular secretion The movement of substances in blood into renal tubular fluid in response to the body's specific needs.

Tumor suppressor gene A gene coding for a protein that normally inhibits cell division; loss or alteration of a tumor suppressor gene called *p53* is the most common genetic change in a wide variety of cancer cells.

Tunica albuginea (TOO-ni-ka al′-bū-JIN-ē-a) A dense white fibrous capsule covering a testis or deep to the surface of an ovary.

Tunica externa (eks-TER-na) The superficial coat of an artery or vein, composed mostly of elastic and collagen fibers. Also called the **adventitia.**

Tunica interna (in-TER-na) The deep coat of an artery or vein, consisting of a lining of endothelium, basement membrane, and internal elastic lamina. Also called the **tunica intima** (IN-ti-ma).

Tunica media (MĒ-dē-a) The intermediate coat of an artery or vein, composed of smooth muscle and elastic fibers.

Twitch contraction Brief contraction of all muscle fibers (cells) in a motor unit triggered by a single action potential in its motor neuron.

Tympanic antrum (tim-PAN-ik AN-trum) An air space in the middle ear that leads into the mastoid air cells or sinus.

Type II cutaneous mechanoreceptor A sensory receptor embedded deeply in the dermis and deeper tissues that detects stretching of skin. Also called a **Ruffini corpuscle.**

U

Umbilical cord The long, ropelike structure containing the umbilical arteries and vein that connect the fetus to the placenta.

Umbilicus (um-BIL-i-kus *or* um-bil-Ī-kus) A small scar on the abdomen that marks the former attachment of the umbilical cord to the fetus. Also called the **navel.**

Upper limb The appendage attached at the shoulder girdle, consisting of the arm, forearm, wrist, hand, and fingers. Also called **upper extremity.**

Up-regulation Phenomenon in which there is an increase in the number of receptors in response to a deficiency of a hormone or neurotransmitter.

Uremia (ū-RĒ-mē-a) Accumulation of toxic levels of urea and other nitrogenous waste products in the blood, usually resulting from severe kidney malfunction.

Ureter (Ū-rē-ter) One of two tubes that connect the kidney with the urinary bladder.

Urethra (ū-RĒ-thra) The duct from the urinary bladder to the exterior of the body that conveys urine in females and urine and semen in males.

Urinary (Ū-ri-ner-ē) **bladder** A hollow, muscular organ situated in the pelvic cavity posterior to the pubic symphysis; receives urine via two ureters and stores urine until it is excreted through the urethra.

Urine The fluid produced by the kidneys that contains wastes and excess materials; excreted from the body through the urethra.

Urogenital (ū′-rō-JEN-i-tal) **triangle** The region of the pelvic floor inferior to the pubic symphysis, bounded by the pubic symphysis and the ischial tuberosities, and containing the external genitalia.

Urology (ū-ROL-ō-jē) The specialized branch of medicine that deals with the structure, function, and diseases of the male and female urinary systems and the male reproductive system.

Uterine (Ū-ter-in) **tube** Duct that transports ova from the ovary to the uterus. Also called the **Fallopian** (fal-LŌ-pē-an) **tube** or **oviduct.**

Uterosacral ligament (ū′-ter-ō-SĀ-kral LIG-a-ment) A fibrous band of tissue extending from the cervix of the uterus laterally to the sacrum.

Uterus (Ū-te-rus) The hollow, muscular organ in females that is the site of menstruation, implantation, development of the fetus, and labor. Also called the **womb.**

Utricle (Ū-tri-kul) The larger of the two divisions of the membranous labyrinth located inside the vestibule of the inner ear, containing a receptor organ for static equilibrium.

Uvea (Ū-vē-a) The three structures that together make up the vascular tunic of the eye.

Uvula (Ū-vū-la) A soft, fleshy mass, especially the V-shaped pendant part, descending from the soft palate.

V

Vagina (va-JĪ-na) A muscular, tubular organ that leads from the uterus to the vestibule, situated between the urinary bladder and the rectum of the female.

Vallate papilla (VAL-āt pa-PIL-a) One of the circular projections that is arranged in an inverted V-shaped row at the back of the tongue; the largest of the elevations on the upper surface of the tongue containing taste buds. Also called **circumvallate papilla.**

Valence (VĀ-lens) The combining capacity of an atom; the number of deficit or extra electrons in the outermost electron shell of an atom.

Varicocele (VAR-i-kō-sēl) A twisted vein; especially, the accumulation of blood in the veins of the spermatic cord.

Varicose (VAR-i-kōs) Pertaining to an unnatural swelling, as in the case of a varicose vein.

Vasa recta (VĀ-sa REK-ta) Extensions of the efferent arteriole of a juxtamedullary nephron that run alongside the loop of the nephron (Henle) in the medullary region of the kidney.

Vasa vasorum (va-SŌ-rum) Blood vessels that supply nutrients to the larger arteries and veins.

Vascular (VAS-kū-lar) Pertaining to or containing many blood vessels.

Vascular spasm Contraction of the smooth muscle in the wall of a damaged blood vessel to prevent blood loss.

Vascular (venous) sinus A vein with a thin endothelial wall that lacks a tunica media and externa and is supported by surrounding tissue.

Vascular tunic (TOO-nik) The middle layer of the eyeball, composed of the choroid, ciliary body, and iris. Also called the **uvea** (Ū-ve-a).

Vasectomy (va-SEK-tō-mē) A means of sterilization of males in which a portion of each ductus (vas) deferens is removed.

Vasoconstriction (vāz-ō-kon-STRIK-shun) A decrease in the size of the lumen of a blood vessel caused by contraction of the smooth muscle in the wall of the vessel.

Vasodilation (vāz′-ō-DĪ-lā-shun) An increase in the size of the lumen of a blood vessel caused by relaxation of the smooth muscle in the wall of the vessel.

Vasomotion (vāz-ō-MŌ-shun) Intermittent contraction and relaxation of the smooth muscle of the metarterioles and precapillary sphincters that result in an intermittent blood flow.

Vein A blood vessel that conveys blood from tissues back to the heart.

Vena cava (VĒ-na KĀ-va) One of two large veins that open into the right atrium, returning to the heart all of the deoxygenated blood from the systemic circulation except from the coronary circulation.

Ventral (VEN-tral) Pertaining to the anterior or front side of the body; opposite of dorsal.

Ventral body cavity Cavity near the ventral aspect of the body that contains viscera and consists of a superior thoracic cavity and an inferior abdominopelvic cavity.

Ventral ramus (RĀ-mus) The anterior branch of a spinal nerve, containing sensory and motor fibers to the muscles and skin of the anterior surface of the head, neck, trunk, and the limbs.

Ventricle (VEN-tri-kul) A cavity in the brain filled with cerebrospinal fluid. An inferior chamber of the heart.

Ventricular fibrillation (ven-TRIK-ū-lar fib-ri-LĀ-shun) Asynchronous ventricular contractions; unless reversed by defibrillation, results in heart failure.

Venule (VEN-ūl) A small vein that collects blood from capillaries and delivers it to a vein.

Vermiform appendix (VER-mi-form a-PEN-diks) A twisted, coiled tube attached to the cecum.

Vermilion (ver-MIL-yon) The area of the mouth where the skin on the outside meets the mucous membrane on the inside.

Vermis (VER-mis) The central constricted area of the cerebellum that separates the two cerebellar hemispheres.

Vertebral (VER-te-bral) **canal** A cavity within the vertebral column formed by the vertebral foramina of all the vertebrae and containing the spinal cord. Also called the **spinal canal.**

Vertebral column The 26 vertebrae of an adult and 33 vertebrae of a child; encloses and protects the spinal cord and serves as a point of attachment for the ribs and back muscles. Also called the **backbone, spine,** or **spinal column.**

Vesicle (VES-i-kul) A small bladder or sac containing liquid.

Vesicouterine (ves′-ik-ō-Ū-ter-in) **pouch** A shallow pouch formed by the reflection of the peritoneum from the anterior surface of the uterus, at the junction of the cervix and the body, to the posterior surface of the urinary bladder.

Vestibular (ves-TIB-ū-lar) **apparatus** Collective term for the organs of equilibrium, which includes the saccule, utricle, and semicircular ducts.

Vestibular membrane The membrane that separates the cochlear duct from the scala vestibuli.

Vestibule (VES-ti-būl) A small space or cavity at the beginning of a canal, especially the inner ear, larynx, mouth, nose, and vagina.

Villus (VIL-lus) A projection of the intestinal mucosal cells containing connective tissue, blood vessels, and a lymphatic vessel; functions in the absorption of the end products of digestion. *Plural is* **villi** (VIL-ī).

Viscera (VIS-er-a) The organs inside the ventral body cavity. *Singular is* **viscus** (VIS-kus).

Visceral (VIS-er-al) Pertaining to the organs or to the covering of an organ.

Visceral effectors (e-FEK-torz) Organs of the ventral body cavity that respond to neural stimulation, including cardiac muscle, smooth muscle, and glands.

Vital capacity The sum of inspiratory reserve volume, tidal volume, and expiratory reserve volume; about 4800 mL.

Vital signs Signs necessary to life that include temperature (T), pulse (P), respiratory rate (RR), and blood pressure (BP).

Vitamin An organic molecule necessary in trace amounts that acts as a catalyst in normal metabolic processes in the body.

Vitreous (VIT-rē-us) **body** A soft, jellylike substance that fills the vitreous chamber of the eyeball, lying between the lens and the retina.

Vocal folds Pair of mucous membrane folds below the ventricular folds that function in voice production. Also called **true vocal cords.**

Voltage-gated channel An ion channel in a plasma membrane composed of integral proteins that functions like a gate to permit or restrict the movement of ions across the membrane in response to changes in the voltage.

Vulva (VUL-va) Collective designation for the external genitalia of the female. Also called the **pudendum** (poo-DEN-dum).

W

Wallerian (wal-LE-rē-an) **degeneration** Degeneration of the portion of the axon and myelin sheath of a neuron distal to the site of injury.

Wandering macrophage (MAK-rō-fāj) Phagocytic cell that develops from a monocyte, leaves the blood, and migrates to infected tissues.

Wave summation (sum-MĀ-shun) The increased strength of muscle contraction that results when muscle action potentials occur one after another in rapid succession.

White matter Aggregations or bundles of myelinated and unmyelinated axons located in the brain and spinal cord.

White pulp The regions of the spleen composed of lymphatic tissue, mostly B lymphocytes.

White ramus communicans (RĀ-mus kō-MŪ-ni-kans) The portion of a preganglionic sympathetic axon that branches from the anterior ramus of a spinal nerve to enter the nearest sympathetic trunk ganglion.

X

X-chromosome inactivation The random and permanent inactivation of one X chromosome in each cell of a developing female embryo. Also called **lyonization.**

Xiphoid (ZĪ-foyd) Sword-shaped. The inferior portion of the sternum is the **xiphoid process.**

Y

Yolk sac An extraembryonic membrane composed of the exocoelomic membrane and hypoblast. It transfers nutrients to the embryo, is a source of blood cells, contains primordial germ cells that migrate into the gonads to form primitive germ cells, forms part of the gut, and helps prevent desiccation of the embryo.

Z

Zona fasciculata (ZŌ-na fa-sik′-ū-LA-ta) The middle zone of the adrenal cortex consisting of cells arranged in long, straight cords that secrete glucocorticoid hormones, mainly cortisol.

Zona glomerulosa (glo-mer′-ū-LŌ-sa) The outer zone of the adrenal cortex, directly under the connective tissue covering, consisting of cells arranged in arched loops or round balls that secrete mineralo-corticoid hormones, mainly aldosterone.

Zona pellucida (pe-LOO-si-da) Clear glycoprotein layer between a secondary oocyte and the surrounding granulosa cells of the corona radiata.

Zona reticularis (ret-ik′-ū-LAR-is) The inner zone of the adrenal cortex, consisting of cords of branching cells that secrete sex hormones, chiefly androgens.

Zygote (ZĪ-g̱ot) The single cell resulting from the union of male and female gametes; the fertilized ovum.

Credits

Illustration Credits

Chapter 1 CO art by Keith Kasnot. Table 1.2: Keith Kasnot. 1.1: Tomo Narashima. 1.2–1.4: Jared Schneidman Design. 1.5, 1.6: Kevin Somerville. 1.7: Lynn O'Kelley. 1.8–1.13: Kevin Somerville.

Chapter 2 CO art by Keith Kasnot. 2.1: Imagineering. 2.2, 2.3: Jared Schneidman Design. 2.4: Imagineering. 2.5: Jared Schneidman Design. 2.6–2.11: Imagineering. 2.12–2.15: Jared Schneidman Design. 2.16: Imagineering. 2.17: Jared Schneidman Design. 2.18: Imagineering. 2.19–2.21: Jared Schneidman Design. 2.22: Imagineering. 2.23: Jared Schneidman Design. 2.24: Imagineering. 2.25: Jared Schneidman Design.

Chapter 3 CO art by Keith Kasnot. 3.1, 3.2: Tomo Narashima. 3.3–3.7: Imagineering. 3.8: Jared Schneidman Design. 3.5: Adapted from Bruce Alberts et al., Essential Cell Biology, F12.5, p375 and F12.12, p380 (New York: Garland Publishing Inc., 1998). ©1998 Garland Publishing Inc. 3.9–3.16: Imagineering. 3.17–3.21: Tomo Narashima. 3.22: Imagineering. 3.23–3.25: Tomo Narashima. 3.26–3.33: Imagineering. 3.34: Hilda Muinos.

Chapter 4 CO art by Keith Kasnot. Table 4.1, Table 4.2: Nadine Sokol, Kevin Somerville. Table 4.3: Nadine Sokol, Kevin Somerville, Imagineering. Table 4.4, Table 4.5: Nadine Sokol. 4.1: Kevin Somerville. 4.2: Imagineering. 4.3: Kevin Somerville. 4.4, 4.5: Imagineering.

Chapter 5 CO art by Keith Kasnot. 5.1–5.7: Kevin Somerville. 5.9: Imagineering.

Chapter 6 CO art by Keith Kasnot. 6.1: Leonard Dank. 6.2: Lauren Keswick. 6.3–6.9: Kevin Somerville. 6.10: Leonard Dank. 6.11: Kevin Somerville. 6.12: Jared Schneidman Design. 6.13: Kevin Somerville.

Chapter 7 CO art by Keith Kasnot. Table 7.1: Nadine Sokol. 7.1–7.24: Leonard Dank.

Chapter 8 CO art by Keith Kasnot. Table 8.1, 8.1–8.17: Leonard Dank.

Chapter 9 CO art by Keith Kasnot. 9.1–9.4, 9.11–9.14: Leonard Dank.

Chapter 10 CO art by Keith Kasnot. 10.1, 10.3: Kevin Somerville. 10.4, 10.6: Imagineering. 10.7: Hilda Muinos. 10.8–10.10: Imagineering. 10.11: Kevin Somerville. 10.12: Imagineering. 10.13–10.15: Jared Schneidman Design. 10.16, 10.18: Imagineering. 10.19: Beth Willert. 10.20: Kevin Somerville.

Chapter 11 CO art by Keith Kasnot. Table 11.1: Kevin Somerville. 11.1–11.17: Leonard Dank. 11.18: Leonard Dank, Kevin Somerville. 11.19–11.23: Leonard Dank.

Chapter 12 CO art by Keith Kasnot. Table 12.1: Kevin Somerville. Table 12.3: Jared Schneidman Design. 12.1: Kevin Somerville. 12.2: Jared Schneidman Design. 12.3: Kevin Somerville. 12.4, 12.5: Imagineering. 12.6: Kevin Somerville. 12.7: Sharon Ellis. 12.8–12.12: Imagineering. 12.13: Jared Schneidman Design. 12.14: Imagineering. 12.15: Jared Schneidman Design. 12.16: Nadine Sokol. 12.17: Imagineering.

Chapter 13 CO art by Keith Kasnot. 13.1–13.5: Sharon Ellis. 13.6–13.9: Leonard Dank. 13.10: Kevin Somerville. 13.11: Sharon Ellis. 13.12, 13.13: Steve Oh, Myriam Kirkman-Oh. 13.14: Imagineering. 13.15, 13.16: Steve Oh, Myriam Kirkman-Oh. 13.17: Imagineering.

Chapter 14 CO art by Keith Kasnot. Table 14.1, Table 14.3: Hilda Muinos. 14.1, 14.2: Kevin Somerville. 14.3: Sharon Ellis. 14.4: Kevin Somerville, Imagineering. 14.5–14.8: Sharon Ellis. 14.9: Imagineering. 14.10, 14.11, 14.13–14.15: Sharon Ellis. 14.17: Hilda Muinos. 14.18–14.24: Sharon Ellis. 14.25, 14.26: Kevin Somerville.

Chapter 15 CO art by Keith Kasnot. Table 15.3: Kevin Somerville. 15.1: Imagineering. 15.2, 15.3: Kevin Somerville. 15.4: Leonard Dank. 15.5: Imagineering. 15.6: Kevin Somerville. 15.7: Jared Schneidman Design. 15.8: Kevin Somerville. 15.9, 15.10: Sharon Ellis. 15.11: Imagineering. Adapted from Purves et al., Neuroscience 2e, F26.1 and F26.2, p498 (Sunderland, MA: Sinauer Associates, 1997). ©1997 Sinauer Associates.

Chapter 16 CO art by Keith Kasnot. Table 16.1: Kevin Somerville. Table 16.2: Imagineering. 16.1: Tomo Narashima. 16.2: Molly Borman. 16.3: Imagineering. Adapted from "Making Sense of Taste" by David V. Smith and Robert F. Margolskee, Scientific American, March 2001, page 38. 16.5: Sharon Ellis. 16.6: Tomo Narashima . 16.9: Lynn O'Kelley. 16.10: Tomo Narashima. 16.11, 16.12: Jared Schneidman Design. 16.13: Lynn O'Kelley. 16.14: Jared Schneidman Design. 16.15: Lynn O'Kelley. 16.16: Imagineering. Adapted from Seeley et al., Anatomy and Physiology 4e, F15.22, p480 (New York: WCB McGraw-Hill, 1998) ©1998 The McGraw-Hill Companies. 16.17–16.21: Tomo Narashima. 16.22, 16.23: Tomo Narashima, Sharon Ellis. 16.24, 16.25: Kevin Somerville.

Chapter 17 CO art by Keith Kasnot. 17.1: Jared Schneidman Design. 17.2, 17.3: Imagineering. 17.4: Kevin Somerville. 17.5: Sharon Ellis. 17.6: Imagineering.

Chapter 18 CO art by Keith Kasnot. Table 18.4: Nadine Sokol, Imagineering. Table 18.5–18.8: Nadine Sokol. Table 18.9: Imagineering. Table 18.10: Nadine Sokol. 18.1: Steve Oh. 18.2: Jared Schneidman Design. 18.3, 18.4: Imagineering. 18.5: Lynn O'Kelley. 18.6: Imagineering. 18.7: Jared Schneidman Design. 18.8: Lynn O'Kelley. 18.9: Jared Schneidman Design. 18.10: Molly Borman. 18.11, 18.12: Jared Schneidman Design. 18.13: Molly Borman. 18.14: Jared Schneidman Design. 18.15: Molly Borman. 18.16, 18.17: Jared Schneidman Design. 18.18: Molly Borman. 18.19: Jared Schneidman Design. 18.20: Nadine Sokol. 18.21: Kevin Somerville.

Chapter 19   CO art by Keith Kasnot. Table 19.3: Jared Schneidman Design. 19.1: Hilda Muinos, Nadine Sokol. 19.3, 19.4: Nadine Sokol. 19.5, 19.6, 19.8: Jared Schneidman Design. 19.9: Nadine Sokol. 19.11: Imagineering. 19.12: Jean Jackson. 19.13: Nadine Sokol.

Chapter 20   CO art by Keith Kasnot. 20.1, 20.2: Kevin Somerville, Imagineering. 20.3–20.6: Kevin Somerville. 20.7: Nadine Sokol, Imagineering. 20.8, 20.9: Kevin Somerville. 20.10: Kevin Somerville, Imagineering. 20.11, 20.12: Burmar Technical Corp. 20.13, 20.14: Imagineering. 20.15: Imagineering, Kevin Somerville. 20.16: Hilda Muinos. 20.18: Kevin Somerville. 20.20: Hilda Muinos.

Chapter 21   CO art by Keith Kasnot. Exhibits 21.2–21.12: Keith Ciociola. Table 21.2: Imagineering. 21.1: Kevin Somerville. 21.2: Hilda Muinos. 21.3: Nadine Sokol, Imagineering. 21.4: Kevin Somerville. 21.6, 21.7: Jared Schneidman Design. 21.8: Imagineering. 21.9: Kevin Somerville. 21.10–21.12: Jared Schneidman Design. 21.13: Kevin Somerville. 21.14–21.16: Jared Schneidman Design. 21.17–21.30: Kevin Somerville. 21.31: Kevin Somerville, Keith Ciociola. 21.32: Kevin Somerville.

Chapter 22   CO art by Keith Kasnot. Table 22.3: Jean Jackson. 22.1: Molly Borman. 22.2: Sharon Ellis. 22.3: Molly Borman. 22.4: Nadine Sokol. 22.5: Steve Oh. 22.6: Molly Borman. 22.7: Steve Oh. 22.8: Kevin Somerville. 22.9, 22.10: Molly Borman. 22.11–21.20: Jared Schneidman Design. 22.21: Nadine Sokol, Imagineering.

Chapter 23   CO art by Keith Kasnot. 23.1: Molly Borman, Kevin Somerville. 23.2, 23.4, 23.5: Molly Borman. 23.6: Steve Oh. 23.8, 23.10: Molly Borman. 23.11, 23.12: Kevin Somerville. 23.13: Jared Schneidman Design. 23.14: Kevin Somerville. 23.15: Jared Schneidman Design, Imagineering. 23.16–21.24: Jared Schneidman Design. 23.25: Imagineering. 23.26: Jared Schneidman Design. 23.27: Kevin Somerville. 23.28: Jared Schneidman Design. 23.29: Kevin Somerville.

Chapter 24   CO art by Keith Kasnot. 24.1–24.3: Steve Oh. 24.4: Nadine Sokol. 24.5: Molly Borman. 24.6: Steve Oh. 24.7, 24.8, 24.10: Nadine Sokol. 24.11: Steve Oh. 24.12: Kevin Somerville. 24.13: Imagineering. 24.14–24.16: Jared Schneidman Design. 24.17: Steve Oh, Jared Schneidman Design. 24.18: Jared Schneidman Design. 24.19: Kevin Somerville. 24.20, 24.21: Jared Schneidman Design. 24.23: Kevin Somerville. 24.25, 24.26: Jared Schneidman Design. 24.27: Molly Borman. 24.28: Kevin Somerville.

Chapter 25   CO art by Keith Kasnot. 25.1–25.20: Imagineering.

Chapter 26   CO art by Keith Kasnot. Table 26.1: Nadine Sokol. 26.1, 26.2: Kevin Somerville. 26.3, 26.4: Steve Oh. 26.5: Imagineering. 26.6: Kevin Somerville. 26.7: Nadine Sokol. 26.8: Kevin Somerville. 26.9: Imagineering. 26.10–26.20: Jared Schneidman Design. 26.21: Steve Oh. 26.22: Kevin Somerville.

Chapter 27   CO art by Keith Kasnot. 27.1–27.8: Jared Schneidman Design.

Chapter 28   CO art by Keith Kasnot. Table 28.1, 28.1, 28.2: Imagineering. 28.3–28.6: Kevin Somerville. 28.7: Jared Schneidman Design. 28.8: Kevin Somerville. 28.9, 28.10–28.15, 28.17: Jared Schneidman Design. 28.18, 28.21–28.24: Kevin Somerville. 28.25–28.28: Jared Schneidman Design. 28.29, 28.30: Kevin Somerville.

Chapter 29   CO art by Keith Kasnot. Table 29.2, 29.1–29.15: Kevin Somerville. 29.16: Jared Schneidman Design. 29.17, 29.18: Kevin Somerville. 29.19–29.26: Jared Schneidman Design.

Focus on Homeostasis icons   Imagineering.

Photo Credits

Chapter 1   Figure 1.1: John Wilson White. Figure 1.8a: From Stephen A. Kieffer and E. Robert Heitzman, *An Atlas of Cross–Sectional Anatomy.* Harper & Row, Publishers, New York, 1979. Figure 1.8b: Lester V. Bergman/Project Masters, Inc. Figure 1.8c: Martin Rotker. Figure 1.11: Mark Nielsen. Page 21 (top left): Biophoto Associates/Photo Researchers. Page 21 (top right): Scott Camazine/Photo Researchers. Page 21 (bottom left): Simon Fraser/Photo Researchers. Page 21 (bottom right): Courtesy Andrew Joseph Tortora and Damaris Soler. Page 22: Howard Sochurek/Medical Images, Inc.

Chapter 3   Figure 3.6: Andy Washnik. Figure 3.17c: Courtesy Kent McDonald, UC Berkeley Electron Microscope Laboratory. Figure 3.20b: D. W. Fawcett/Photo Researchers. Figure 3.21b: Biophoto Associates/Photo Researchers. Figure 3.23b: Courtesy Daniel S. Friend, Harvard Medical School. Figure 3.24b: D.W. Fawcett/Photo Researchers. Figure 3.25c: CNRI/Photo Researchers. Figure 3.33: Courtesy Michael Ross, University of Florida.

Chapter 4   Page 109 (top): Biophoto Associates/Photo Researchers. Pages 109 (bottom), 110, 111 (bottom), 112, 113 and 114 (bottom), 121, 122, 123 (top), 124 (bottom), 125, 126 (top), 127 and 130: Courtesy Michael Ross, University of Florida. Page 111 (top): Biophoto Associates/Photo Researchers. Page 114 (top): Lester V. Bergman/The Bergman Collection. Page 123 (bottom): Courtesy Andrew J. Kuntzman. Page 124 (top): Ed Reschke. Page 126 (bottom): John Burbidge/Photo Researchers. Page 131: Biophoto Associates/Photo Researchers. Page 132: © Ed Reschke.

Chapter 5   Figure 5.3b: © L.V. Bergman/Bergman Collection. Figure 5.4b: Science Photo Library/Photo Researchers. Figure 5.8a: Alain Dex/Photo Researchers. Figure 5.8b: Biophoto Associates/Photo Researchers.

Chapter 6   Figure 6.1b: Mark Nielsen. Figure 6.8a: The Bergman Collection. Figure 6.8b: Courtesy Michael Ross, University of Florida. Figure 6.14: P. Motta, Dept. of Anatomy, University La Sapienza, Rome/Photo Researchers.

Chapter 9   Figures 9.5, 9.6, 9.8, 9.7, 9.9 and 9.10: John Wilson White.

Chapter 10   Figure 10.2: Courtesy Fujita. Figure 10.5: Courtesy Denah Appelt and Clara Franzini-Armstrong. Figure 10.17: © John Wiley & Sons. Page 296: Biophoto Associates/Photo Researchers.

Chapter 12   Figure 12.3c: Science VU/Visuals Unlimited. Figure 12.6c: Dennis Kunkel/Phototake. Figure 12.6d: Martin Rotker/Phototake.

Chapter 13   Figure 13.1b: Mark Nielsen. Figure 13.3b: Jean Claude Revy/Phototake. Figure 13.10b: Copyright by Richard Kessel and Randy Kardon, *Tissues and Organs: A Text-Atlas of Scanning Electron Microscopy,* W. H. Freeman and Company, 1979. All rights reserved. Reprinted by permission.

Chapter 14   Figure 14.1b: Mark Nielsen. Figure 14.9e: From Stephen A. Kieffer and E. Robert Heitzman, *An Atlas of Cross-Sectional Anatomy,* Harper and Row, Publishers, 1979. Reproduced with permission. Figure 14.12: From N. Gluhbegovic and T.H. Williams, *The Human Brain: A Photographic Guide,* Harper and Row, Publishers, 1980. Reproduced with permission. Figure 14.16: From *Nature,* November 26, 1992, Vol. 360, page 340. Reproduced with permission from *Nature* and Robert Zatorre, Department of Neuropsychology, McGill University.

Chapter 16 Figure 16.4: John Moore. Figure 16.8: Courtesy Michael Ross, University of Florida. Figure 16.16a: From N. Gluhbegovic and T. H. Williams, *The Human Brain: A Photographic Guide,* Harper and Row, Publishers, 1980.

Chapter 18 Figure 18.5: Mark Nielsen. Figures 18.10b, 18.13b, 18.15b and 18.18c: Courtesy Michael Ross, University of Florida. Figure 18.22a: From *New England Journal of Medicine,* February 18, 1999, vol. 340, No. 7, Page 524. Photo provided courtesy of Robert Gagel, Department of Internal Medicine, University of Texas M.D. Anderson Cancer Center, Houston, Texas. Figures 18.22b,c and d: © The Bergman Collection/Project Masters, Inc. Figure 18.22e: Biophoto Associates/Photo Researchers.

Chapter 19 Figure 19.2: From Lennart Nilsson, *Our Body Victorious,* Boehringer Ingelheim International GmbH. Reproduced with permission. Figure 19.7: John Cunningham/Visuals Unlimited. Figure 19.10: From Lennart Nilsson, *The Incredible Machine,* Boehringer Ingelheim International GmbH. Reproduced with permission. Figure 19.14: Lewin/Royal Free Hospital/Photo Researchers.

Chapter 20 Figures 20.3b, 20.4b, 20.6e and 20.8c: Mark Nielsen. Figure 20.17: Gregg Adams/Stone. Figure 20.19a: © Vu/Cabisco/Visuals Unlimited. Figure 20.19b: W. Ober/Visuals Unlimited.

Chapter 21 Figure 21.1d: Dennis Strete. Figure 21.1e: Courtesy Michael Ross, University of Florida. Figure 21.5: Mark Nielsen.

Chapter 22 Figures 22.5, 22.6 and 22.7: Courtesy Michael Ross, University of Florida. Figure 22.9b: National Cancer Institute/Photo Researchers.

Chapter 23 Figures 23.1b and 23.9: Mark Nielsen. Figure 23.3: Courtesy Lynne Marie Barghesi. Figure 23.7: John Cunningham/Visuals Unlimited. Figure 23.11: Biophoto Associates/Photo Researchers.

Chapter 24 Figures 24.5b, 24.9, 24.19c, 24.24c–d and Figure 24.28c,d: Courtesy Michael Ross, University of Florida. Figures 24.11b and 24.22b: Mark Nielsen. Figure 24.12c: Ed Reschke. Figure 24.24a: Fred E. Hossler/Visuals Unlimited. Figure 24.24b: Willis/Biological Photo Service.

Chapter 26 Figure 26.3b: Mark Nielsen. Figure 26.6b: Dennis Strete. Figure 26.8b: Courtesy Michael Ross, University of Florida.

Chapter 28 Figures 28.3, 28.5b–c and 28.13b: Mark Nielsen. Figure 28.6a: Ed Reschke. Figure 28.16: Biophoto Associates/Photo Researchers. Figure 28.19: P. Motta/Photo Researchers. Figure 28.20: Courtesy Andrew J. Kuntzman, Wright State University.

Chapter 29 Figure 29.1b: David Phillips/Photo Researchers. Figure 29.1c: Myriam Wharman/Phototake. Figure 29.11b: Siu, Biomedical Comm./Custom Medical Stock Photo. Figures 29.14a, g and h: Photo provided courtesy of Kohei Shiota, Congenital Anomaly Research Center, Kyoto University, Graduate School of Medicine. Figures 29.14b–e: Courtesy National Museum of Health and Medicine, Armed Forces Institute of Pathology. Figure 29.14f: Photo by Lennart Nilsson/Albert Bonniers Förlag AB, *A Child is Born,* Dell Publishing Company. Reproduced with permission.

Index

Note   Page numbers in **boldface** type indicate a major discussion. A *t* following a page number indicates a table, an *f* following a page number indicates a figure, and an *e* following a page number indicates an exhibit.

EPONYMS USED IN THIS TEXT

In the life sciences, an eponym is the name of a structure, drug, or disease that is based on the name of a person. For example, you may be more familiar with the Achilles tendon than you are with its more anatomically descriptive term, the calcaneal tendon. Because eponyms remain in frequent use, this listing correlates common eponyms with their anatomical terms.

EPONYM	ANATOMICAL TERM
Achilles tendon	calcaneal tendon
Adam's apple	thyroid cartilage
ampulla of Vater (VA-ter)	hepatopancreatic ampulla
Bartholin's (BAR-tō-linz) gland	greater vestibular gland
Billroth's (BIL-rōtz) cord	splenic cord
Bowman's (BŌ-manz) capsule	glomerular capsule
Bowman's (BŌ-manz) gland	olfactory gland
Broca's (BRŌ-kaz) area	motor speech area
Brunner's (BRUN-erz) gland	duodenal gland
bundle of His (HISS)	artrioventricular (AV) bundle
canal of Schlemm (SHLEM)	scleral venous sinus
circle of Willis (WIL-is)	cerebral arterial circle
Cooper's (KOO-perz) ligament	suspensory ligament of the breast
Cowper's (KOW-perz) gland	bulbourethral gland
crypt of Lieberkühn (LE-ber-kyūn)	intestinal gland
duct of Santorini (san'-tō-RĒ-ne)	accessory duct
duct of Wirsung (VĒR-sung)	pancreatic duct
Eustachian (yoo-STĀ-kē-an)	auditory tube
Fallopian (fal-LŌ-pē-an) tube	uterine tube
gland of Littré (LĒ-tra)	urethral gland
Golgi (GOL-jē) tendon organ	tendon organ
Graafian (GRAF-ē-an) follicle	mature ovarian follicle
Hassall's (HAS-alz) corpuscle	thymic corpuscle
Haversian (ha-VĒR-shun) canal	central canal
Haversian (ha-VĒR-shun) system	osteon
Heimlich (HĪM-lik) maneuver	abdomial thrust maneuver
islet of Langerhans (LANG-er-hanz)	pancreatic islet

EPONYM	ANATOMICAL TERM
Kupffer (KOOP-fer) cell	stellate reticuloendothelial cell
Leydig (LĪ-dig) cell	interstitial endocrinocyte
loop of Henle (HEN-lē)	loop of the nephron
Luschka's (LUSH-kaz) aperture	lateral aperture
Magendie's (ma-JEN-dēz) aperture	median aperture
Meibomian (mi-BŌ-mē-an) gland	tarsal gland
Meissner (MĪS-ner) corpuscle	corpuscle of touch
Merkel (MER-kel) disc	tactile disc
Müllerian (mil-E rē-an) duct	paramesonephric duct
organ of Corti (KOR-tē)	spiral organ
Pacinian (pa-SIN-ē-an) corpuscle	lamellated corpuscle
Peyer's (PĪ-erz) patch	aggregated lymphatic follicle
plexus of Auerbach (OW-er-bak)	myenteric plexus
plexus of Meissner (MĪS-ner)	submucosal plexus
pouch of Douglas	rectouterine pouch
Purkinje (pur-KIN-jē) fiber	conduction myofiber
Rathke's (rath-KĒZ) pouch	hypophyseal pouch
Ruffini (roo-FĒ-ne) corpuscle	type II cutaneous mechanoreceptor
Sertoli (ser-TŌ-lē) cell	sustentacular cell
Skene's (SKĒNZ) gland	paraurethral gland
sphincter of Oddi (OD-dē)	sphincter of the hepatopancreatic ampulla
Volkmann's (FŌLK-manz) canal	perforating canal
Wernicke's (VER-ni-kēz) area	auditory association area
Wharton's (HWAR-tunz) jelly	mucous connective tissue
Wolffian duct	mesonephric duct
Wormian (WER-mē-an) bone	sutural bone

COMBINING FORMS, WORD ROOTS, PREFIXES, AND SUFFIXES

Many of the terms used in anatomy and phsiology are compound words; that is, they are made up of word roots and one or more prefixes or suffixes. For example, *leukocyte* is formed from the word roots *leuk-* meaning "white", a connecting vowel (o), and *cyte* meaning "cell." Thus, a leukocyte is a white blood cell. The following list includes some of the most commonly used combining forms, word roots, prefixes, ad suffixes used in the study of anatomy and physiology. Each entry includes a usage example. Learning the meanings of these fundamental word parts will help you remember terms that, at first glance, may seem long or complicated.

COMBINING FORMS AND WORD ROOTS

Acous-, Acu- hearing Acoustics.
Acr- extremity Acromegaly.
Aden- gland Adenoma.
Alg-, Algia- pain Neuralgia.
Angi- vessel Angiocardiography.
Anthr- joint Arthropathy.
Aut-, Auto- self Autolysis.
Audit- hearing Auditory canal.

Bio- life, living Biopsy.
Blast- germ, bud Blastula.
Blephar- eyelid Blepharitis.
Brachi- arm Brachial plexus.
Bronch- trachea, windpipe Bronchoscopy.
Bucc- cheek Buccal.

Capit- head Decapitate.
Carcin- cancer Carcinogenic.
Cardi-, Cardia-, Cardio- heart Cardiogram.
Cephal- head Hydrocephalus.
Cerebro- brain Cerebrospinal fluid.
Chole- bile, gall Cholecystogram.
Chondr-, cartilage Chondrocyte.
Cor-, Coron- heart Coronary.
Cost- rib Costal.
Crani- skull Craniotomy.
Cut- skin Subcutaneous.
Cyst- sac, bladder Cystoscope.

Derma-, Dermato- skin Dermatosis.
Dura- hard Dura mater.

Enter- intestine Enteritis.
Erythr- red Erythrocyte.

Gastr- stomach Gastrointestinal.
Gloss- tongue Hypoglossal.
Glyco- sugar Glycogen.
Gyn-, Gynec- female, woman Gynecology.

Hem-, Hemat- blood Hematoma.
Hepar-, Hepat- liver Hepatitis.
Hist-, Histio- tissue Histology.
Hydr- water Dehydration.
Hyster- uterus Hysterectomy.

Ischi- hip, hip joint Ischium.

Kines- motion Kinesiology.

Labi- lip Labial.
Lacri- tears Lacrimal glands.
Laparo- loin, flank, abdomen Laparoscopy.
Leuko- white Leukocyte.
Lingu- tongue Sublingual glands.
Lip- fat Lipid.
Lumb- lower back, loin Lumbar.

Macul- spot, blotch Macula.
Malign- bad, harmful Malignant.
Mamm-, Mast- breast Mammography, Mastitis.
Meningo- membrane Meningitis.
Myel- marrow, spinal cord Myeloblast.
My-, Myo- muscle Myocardium.

Necro- corpse, dead Necrosis.
Nephro- kidney Nephron.
Neuro- nerve Neurotransmitter.

Ocul- eye Binocular.
Odont- tooth Orthodontic.
Onco- mass, tumor Oncology.
Oo- egg Oocyte.
Opthalm- eye Ophthalmology.
Or- mouth Oral.
Osm- odor, sense of small Anosmia.
Os-, Osseo-, Osteo- bone Osteocyte.
Ot- ear Otitus media.

Palpebr- eyelid Palpebra.
Patho- disease Pathogen.
Pelv- basin Renal pelvis.
Phag- to eat Phagocytosis.
Phleb- vein Phlebitis.
Phren- diaphragm Phrenic.
Pilo- hair Depilatory.
Pneumo- lung, air Pneumothorax.
Pod- foot Podocyte.
Procto- anus, rectum Proctology.
Pulmon- lung Pulmonary.

Ren- kidneys Renal artery.
Rhin- nose Rhinitis.

Scler-, Sclero- hard Atherosclerosis.
Sep-, Spetic- toxic condition due to micoorganisms Septicemia.
Soma-, Somato- body Somatotropin.
Sten- narrow Stenosis.
Stasis-, Stat- stand still Homeostasis.

Tegument- skin, covering Integumentary.
Therm- heat Thermogenesis.
Thromb- clot, lump Thrombus.

Vas- vessel, duct Vasoconstriction.

Zyg- joined Zygote.

PREFIXES

A-, An- without, lack of, deficient Anesthesia.
Ab- away from, from Abnormal.
Ad-, Af- to, toward Adduction, Afferent neuron.
Alb- white Albino.
Alveol- cavity, socket Alveolus.
Andro- male, masculine Androgen.
Ante- before Antebrachial vein.
Anti- against Anticoagulant.

Bas- base, foundation Basal ganglia.
Bi- two, double Biceps.
Brady- slow Bradycardia.

Cata- down, lower, under Catabolism.
Circum- around Circumduction.
Cirrh- yellow Cirrhosis of the liver.
Co-, Con-, Com with, together Congenital.
Contra- against, opposite Contraception.
Crypt- hidden, concealed Cryptorchidism.
Cyano- blue Cyanosis.

De- down, from Deciduous.
Demi-, hemi- half Hemiplegia.
Di-, Diplo- two Diploid.
Dis- separation, apart, away from Dissection.
Dys- painful, difficult Dyspnea.

E-, Ec-, Ef- out from, out of Efferent neuron.
Ecto-, Exo- outside Ectopic pregnancy.
Em-, En- in, on Emmetropia.
End-, Endo- within, inside Endocardium.
Epi- upon, on, above Epidermis.
Eu- good, easy, normal Eupnea.
Ex-, Exo- outside, beyond Exocrine gland.
Extra- outside, beyond, in addition to Extracellular fluid.

Fore- before, in front of Forehead.

Gen- originate, produce, form Genitalia.
Gingiv- gum Gingivitis.

Hemi- half Hemiplegia.
Heter-, Hetero- other, different Heterozygous.
Homeo-, Homo- unchanging, the same, steady Homeostasis.
Hyper- over, above, excessive Hyperglycemia.
Hypo- under, beneath, deficient Hypothalamus.

In-, Im- in, inside, not Incontinent.
Infra- beneath Infraorbital.
Inter- among, between Intercostal.
Intra- within, inside Intracellular fluid.
Ipsi- same Ipsilateral.
Iso- equal, like Isotonic.

Juxta- near to Juxtaglomerular apparatus.

Later- side Lateral.

Macro- large, great Macrophage.
Mal- bad, abnormal Malnutrition.
Medi-, Meso- middle Medial.
Mega-, Megalo- great, large Magakaryocyte.
Melan- black Melanin.
Meta- after, beyond Metacarpus.
Micro- small Microfilament.
Mono- one Monounsaturated fat.

Neo- new Neonatal.

Oligo- small, few Oliguria.
Ortho- straight, normal Orthopedics.
Para- near, beyond, beside Paranasal sinus.
Peri- around Pericardium.
Poly- much, many, too much Polycythemia.
Post- after, beyond Postnatal.
Pre-, Pro- before, in front of Presynaptic.
Pseudo- false Pseudostratified.
Retro- backward, behind Retroperitoneal.
Semi- half Semicircular canals.
Sub- under, beneath, below Submucosa.
Super- above, beyond Superficial.
Supra- above, over Suprarenal.
Sym-, Syn- with, together Symphysis.
Tachy- rapid Tachycardia.
Trans- across, through, beyond Transudation.
Tri- three Trigone.

SUFFIXES

-able capable of, having ability to Viable.
-ac, -al pertaining to Cardiac.
-algia painful condition Myalgia.
-an, -ian pertaining to Circadian.
-ant having the characteristic of Malignant.
-ary connected with Ciliary.
-asis, -asia, -esis, -osis condition or state of Hemostasis.

-asthenia weakness Myasthenia.
-ation process, action, condition Inhalation.
-centesis puncture, usually for drainage Amniocentesis.
-cid, -cide, -cis, cut, kill destroy Spermicide.
-ectomize, -ectomy excision of, removal of Thyroidectomy.
-emia condition of blood Anemia.
-esthesia sensation, feeling Anesthesia.
-fer carry Efferent arteriole.
-gen agent that produces or originates Pathogen.
-genic producing Pyogenic.
-gram record Electrocardiogram.
-graph instrument for recording Electroencephalograph.
-ia state, condition Hypermetropia.
-ician person associated with Pediatrician.
-ics art of, science of Optics.
-ism condition, state Rheumatism.
-itis inflammation Neuritis.
-logy the study or science of Physiology.
-lysis dissolution, loosening, destruction Hemolysis.
-malacia softening Osteomalacia.
-megaly enlarged Cardiomegaly.
-mers, -meres parts Polymers.

-oma tumor Fibroma.
-osis condition, disease Necrosis.
-ostomy create an opening Colostomy.
-otomy surgical incision Tracheotomy.
-pathy disease Myopathy.
-penia deficiency Thrombocytopenia
-philic to like, have an affinity for Hydrophilic.
-phobe, -phobia fear of, aversion to Photophobia.
-plasia, -plasty forming, molding Rhinoplasty.
-pnea breath Apnea.
-poiesis making Hemopoiesis.
-ptosis falling, sagging Blepharoptosis.
-rrhage bursting forth, abnormal discharge Hemorrhage.
-rrhea flow, discharge Diarrhea.
-scope instrument for viewing Bronchoscope.
-stomy creation of a mouth or artificial opening Tracheostomy.
-tomy cutting into, incision into Laparotomy.
-tripsy crushing Lithotripsy.
-trophy relating to nutrition or growth Atrophy.
-uria urine Polyuria.